Structural Optimization

Volume 1
Optimality Criteria

MATHEMATICAL CONCEPTS AND METHODS IN SCIENCE AND ENGINEERING

Series Editor: Angelo Miele
Mechanical Engineering and Mathematical Sciences
Rice University

Recent volumes in this series:

22 **APPLICATIONS OF FUNCTIONAL ANALYSIS IN ENGINEERING** • *J. L. Nowinski*

23 **APPLIED PROBABILITY** • *Frank A. Haight*

24 **THE CALCULUS OF VARIATIONS AND OPTIMAL CONTROL: An Introduction** • *George Leitmann*

25 **CONTROL, IDENTIFICATION, AND INPUT OPTIMIZATION** • *Robert Kalaba and Karl Spingarn*

26 **PROBLEMS AND METHODS OF OPTIMAL STRUCTURAL DESIGN** • *N. V. Banichuk*

27 **REAL AND FUNCTIONAL ANALYSIS, Second Edition Part A: Real Analysis** • *A. Mukherjea and K. Pothoven*

28 **REAL AND FUNCTIONAL ANALYSIS, Second Edition Part B: Functional Analysis** • *A. Mukherjea and K. Pothoven*

29 **AN INTRODUCTION TO PROBABILITY THEORY WITH STATISTICAL APPLICATIONS** • *Michael A. Golberg*

30 **MULTIPLE-CRITERIA DECISION MAKING: Concepts, Techniques, and Extensions** • *Po-Lung Yu*

31 **NUMERICAL DERIVATIVES AND NONLINEAR ANALYSIS** • *Harriet Kagiwada, Robert Kalaba, Nima Rasakhoo, and Karl Spingarn*

32 **PRINCIPLES OF ENGINEERING MECHANICS Volume 1: Kinematics—The Geometry of Motion** • *Millard F. Beatty, Jr.*

33 **PRINCIPLES OF ENGINEERING MECHANICS Volume 2: Dynamics—The Analysis of Motion** • *Millard F. Beatty, Jr.*

34 **STRUCTURAL OPTIMIZATION Volume 1: Optimality Criteria** • *Edited by M. Save and W. Prager*

Structural Optimization

Volume 1
Optimality Criteria

Edited by
M. Save
Polytechnic Institute of Mons
Mons, Belgium
and
W. Prager

Technical Editor
W. H. Warner
University of Minnesota
Minneapolis, Minnesota

With a contribution by G. Sacchi

PLENUM PRESS • NEW YORK AND LONDON

Library of Congress Cataloging in Publication Data

Main entry under title:

Structural optimization.

(Mathematical concepts and methods in science and engineering; 34)
Bibliography: v. 1, p.
Includes index.
Contents: v. 1. Optimality criteria.
1. Structural design—Mathematical models. I. Save, M. II. Prager, William, 1903-1980. III. Sacchi, G. (Gianantonnio) IV. Series.
TA658.2.S757 1985 624.1′771 85-22590
ISBN 0-306-41861-4

A Division of Plenum Publishing Corporation
233 Spring Street, New York, N.Y. 10013

Printed in the United States of America

To the memory of my wife,
who died in the course of this work,
I dedicate my personal contibution

M.S.

Que serais-je sans toi
Qui vins à ma rencontre
Que serais-je sans toi
Qu'un coeur au bois dormant. . . .

Louis Aragon

Preface

After the IUTAM Symposium on Optimization in Structural Design held in Warsaw in 1973, it was clear to me that the time had come for organizing into a consistent body of thought the enormous quantity of results obtained in this domain, studied from so many different points of view, with so many different methods, and at so many levels of practical applicability. My colleague and friend Gianantonnio Sacchi from Milan and I met with Professor Prager in Savognin in July 1974, where I submitted to them my first ideas for a treatise on structural optimization: It should cover the whole domain from basic theory to practical applications, and deal with various materials, various types of structures, various functions required of the structures, and various types of cost. Obviously, this was to be a team effort, to total three or four volumes, to be written in a balanced manner as textbooks and handbooks. Nothing similar existed at that time, and, indeed, nothing has been published to date. Professor Prager was immediately in favor of such a project. He agreed to write a first part on optimality criteria with me and to help me in the general organization of the series. Since Professor Sacchi was willing to write the text on variational methods, it remained to find authors for parts on the mathematical programming approach to structural optimization (and, more generally, on numerical methods) and on practical optimal design procedures in metal and concrete. Last, but not least, a publisher was needed. Professor Prager found an optimized solution to this problem through his contacts with Professor A. Miele, editor of the series *Mathematical Concepts and Methods in Science and Engineering* published by Plenum Press.

The setting up of a complete international team of specialists was a difficult task that took four years of contacts, with acceptances, withdrawals, and replacements. Professor Prager's role in this context was also decisive.

In the meantime we wrote a large part of our volume on optimality criteria. A formal general contract was signed with the publisher at the end of 1979. Most unfortunately, Professor Prager died, rather suddenly, on March 16, 1980. Two days earlier he had sent me his last letter with examples of optimization for multiple loadings. The present treatise is dedicated to his memory.

Volume 1 is aimed at being primarily a textbook that stresses the mechanical aspects of optimization problems. The theory is believed to be presented in a rigorous, though engineering-oriented manner, progressing from simple to more complicated problems. In order to emphasize unification of the approach to various behavioral constraints, the division of the book is based on the topology of structures. Only simple problems are presented and solved, analytically as a rule, even if often computer-aided. They should help in understanding the concepts and solution methods, and, we hope, may be inspiring as reference solutions to more practical designs. Many examples are recent or even unpublished. Exercises are presented at the end of most sections. Their solutions are given in the last chapter. All references are assembled, in alphabetical and chronological order, in a single list at the end of the book.

The technical editor for this treatise is Professor W. H. Warner of the University of Minnesota. The editors are indebted to him for a job well done. Professor Sacchi is pleased to acknowledge the help of Professor C. Cinquini and of Dr. G. Sacchi in writing Chapter 7.

Volume 2 will deal with numerical methods of structural optimization, including the use of optimality criteria, and Volume 3 will consider practical applications to metal and reinforced concrete structures, taking the requirements of Codes of Practice into account.

M. SAVE

Contents

4. Optimal Design of Trusses *W. Prager*

5. Optimal Design of Grillages *W. Prager*

6. Optimal Design of Plates, Shells, and Disks *M. Save*

7. Lagrange Multiplier Methods for Optimization with Constraints *G. Sacchi*

8. Review of Some Extensions of Design Procedures *M. Save*

1

Review of Some Basic Concepts and Theorems of Structural Analysis

1.1. Static and Kinematic Variables

In this section some basic concepts of structural theory will be reviewed. Throughout the general discussion the example of a propped cantilever beam will be used to illustrate the concepts. This horizontal beam is built-in at its left end $x = 0$ and has a fixed simple support at its right end $x = l$.

Whereas the members of a structure are three-dimensional bodies, structural theory, in general, treats them as one- or two-dimensional continua. Rods, beams, and arches belong to the first class; disks, plates, and shells, to the second. A point of a one- or two-dimensional continuum may be specified by a single parameter or by a pair of parameters. In this chapter the letter x will be used to denote these parameters. For an arch, for instance, x may be the arc length measured from a reference point on the arch, and dx will then be used to denote the line element of the arch. For a plate, x may stand for the pair of rectangular coordinates specifying a point, and dx will denote the area element of the plate. Adopting a unified terminology, we shall call dx the *volume* of the considered element and use the term *specific* in the sense of *per unit volume*.

Let $P_\alpha(x)$ ($\alpha = 1, 2, \ldots$) denote the specific intensities of the *distributed generalized loads* acting at the point x. Assuming, for instance, that our example beam is subject to distributed rightward horizontal and downward vertical forces and a distributed counterclockwise couple, we may denote their specific intensities by $P_1(x)$, $P_2(x)$, and $lP_3(x)$. The generalized loads P_1, P_2, and P_3 then have the same dimension. For the sake of brevity, concentrated forces and couples will not be considered in this chapter.

To the generalized loads $P_\alpha(x)$ there correspond *generalized displacements* $p_\alpha(x)$ in the sense that the work of the former on the latter is given by the integral of

$$w^{(e)}(x) = \sum_{\alpha} P_\alpha(x) p_\alpha(x) \tag{1.1}$$

over the structure. The quantity $w^{(e)}(x)$ will be called the *specific external work* at the point x. For our example beam the generalized displacements are the rightward horizontal and downward vertical displacements and l times the counterclockwise rotation of the cross section. Note that the definition that led to dimensionally homogeneous generalized loads resulted in dimensionally homogeneous generalized displacements.

The generalized displacements, which are supposed to be small in comparison to the dimensions of the structure, are subject to *kinematic conditions of continuity*. For the example beam we shall require the displacements to be continuous and to have piecewise continuous first derivatives with respect to x.

The generalized displacements are also subject to *kinematic constraints*, which may be *external* or *internal*. For our beam the external kinematic constraints are

$$p_1(0) = p_2(0) = p_3(0) = 0, \qquad p_1(l) = p_2(l) = 0. \tag{1.2}$$

To each constraint corresponds a *reaction*. For the example beam, the reaction corresponding to, say, the constraint $p_2(l) = 0$ is a vertical force at $x = l$. Because the work of the reactions on the displacements $p_\alpha(x)$ vanishes, the constraints are called *workless*. Internal kinematic constraints will be discussed later.

We shall define the state of stress at the point x by generalized stresses $Q_j(x)$ $(j = 1, 2, \ldots)$, in whose choice we have some freedom. For our beam we may, for instance, take as Q_1 the axial force, as Q_2 the shear force, and as Q_3 the bending moment divided by the span l in order to have dimensionally homogeneous generalized stresses. The usual sign conventions for these generalized stresses are indicated in Fig. 1.1, which shows positive stresses.

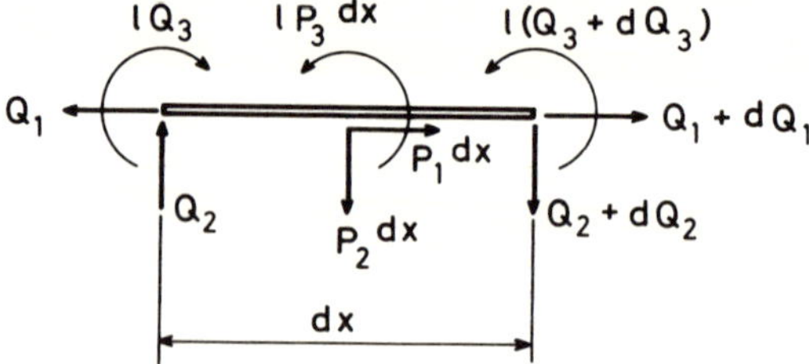

Figure 1.1. Generalized forces.

The generalized stresses and their derivatives are related to the generalized loads by the *equations of equilibrium.* For the example beam these equations are

$$Q_1' + P_1 = 0, \qquad Q_2' + P_2 = 0, \qquad lQ_3' - Q_2 + lP_3 = 0, \tag{1.3}$$

where the prime denotes differentiation with respect to x.

The generalized stresses are subject to *static conditions of continuity,* which follow from the equations of equilibrium. For our beam these conditions require the generalized stresses to be continuous and their derivatives to be piecewise continuous with jumps only where the generalized loads are discontinuous.

The generalized stresses, moreover, are subject to *static constraints.* For the example beam the only static constraint is $Q_3(l) = 0$. To the generalized stresses Q_j there "correspond" *generalized strains* q_j in the sense that the *specific internal work* of the former on the latter is

$$w^{(i)}(x) = \sum_j Q_j(x) q_j(x). \tag{1.4}$$

To relate the generalized strains to the generalized displacements and their derivatives, we stipulate the validity of the *principle of virtual work,*

$$\int \{w^{(i)}(x) - w^{(e)}(x)\}\, dx = 0, \tag{1.5}$$

where the integration is extended over the structure. We now use the equations of equilibrium to express the generalized loads in the definition of $w^{(e)}$ by the generalized stresses and their derivatives, remove the latter by integration by parts, and remember that the exterior kinematic constraints are workless. The desired formulas for the generalized strains then follow from the requirement that the coefficient of each generalized stress in the integrand must vanish because the variation of the generalized loads with x, and hence that of the generalized stresses, is arbitrary.

If we first assume the loads on the example beam to be continuous, the generalized stresses and their first derivatives are continuous. We then have

$$\begin{aligned}\int_0^l w^{(e)}\, dx &= -\int_0^l [Q_1' p_1 + Q_2' p_2 + (Q_3' - Q_2/l) p_3]\, dx \\ &= -(Q_1 p_1 + Q_2 p_2 + Q_3 p_3)_0^l \\ &\quad + \int_0^l [Q_1 p_1' + Q_2(p_2' + p_3/l) + Q_3 p_3']\, dx.\end{aligned} \tag{1.6}$$

On account of the external kinematic constraints and the static constraint, the term in the second line of Eq. (1.6) vanishes. Substitution of the integral in the third line of Eq. (1.6) into Eq. (1.5) then furnishes

$$\int_0^l [Q_1(q_1 - p_1') + Q_2(q_2 - p_2' - p_3/l) + Q_3(q_3 - p_3')]\, dx = 0 \qquad (1.7)$$

and, hence, the following definitions of the generalized strains:

$$q_1 = p_1', \qquad q_2 = p_2' + p_3/l, \qquad q_3 = p_3'. \qquad (1.8)$$

If, say, the load P_1 and, hence, Q_1' have discontinuities at $x = a$, separate integrations by parts must be performed over the beam segments to the left and right of $x = a$. In the second line of Eq. (1.6) we must then add the term $[Q_1 p_1]_{a-0}^{a+0}$, which does, however, vanish because Q_1 and p_1 are continuous at $x = a$.

In treating a specific problem the analyst may decide to neglect certain generalized strains. For example, if the right-hand support of the example beam had horizontal mobility, we might decide to treat the material center line of the beam as inextensible. This amounts to setting $q_1 = 0$ and thus to introducing the *internal kinematic constraint* $p_1' = 0$, whose imposition changes the status of the axial force from that of a generalized stress to that of a *reaction* to this constraint. The considered constraint is *workless*, because the reaction to it (that is, the axial force) does no work on displacements satisfying the constraint.

If we assume mobility of the right-hand support, the axial force vanishes at this end. The variation of the axial force along the beam follows from this boundary condition and the first equation of equilibrium, Eq. (1.3). For our example beam, however, with its fixed right-hand support, we can no longer determine the variation of the axial force along the beam if we treat its material center line as inextensible.

The adoption of Bernoulli's hypothesis, according to which cross sections remain normal to the bent center line of the beam, is equivalent to the imposition of the *internal* kinematic constraint $q_2 \equiv p_2' + p_3/l = 0$. Indeed, at each point of the center line, the counterclockwise rotation p_3/l of the cross section then equals the counterclockwise rotation $-p_2'$ of the element of the center line. Because q_3 vanishes identically, the shear force ceases to be a generalized stress and becomes a reaction to this constraint.

A field $Q_j(x)$ of generalized stresses that satisfies the static conditions of continuity, the static constraints, and the equations of equilibrium for the given loads is said to be *statically admissible* for these loads. A field $p_\alpha(x)$ of generalized displacements is called *kinematically admissible* if it satisfies the kinematic conditions of continuity and the external and internal

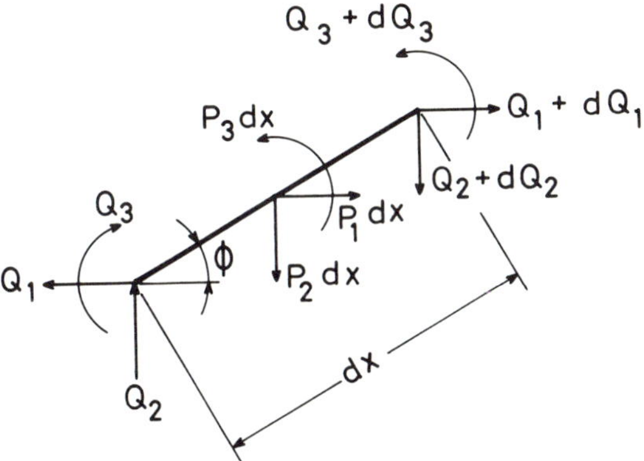

Figure 1.2. Generalized forces, skew bar element.

kinematic constraints. A field $q_j(x)$ of generalized strains is called kinematically admissible if it is derived from a kinematically admissible field of generalized displacements.

Generalized loads $P_\alpha(x)$ and generalized stresses $Q_j(x)$ that are statically admissible for these loads will be said to constitute a set of *associated static variables.* Similarly, kinematically admissible displacements $p_\alpha(x)$ and the generalized strains $q_j(x)$ derived from them will be said to constitute a set of *associated kinematic variables.* In view of the way in which the generalized strains are derived from the generalized displacements, the principle of virtual work is valid for *independent* sets of associated static and kinematic variables; there is no need for these two sets to obey the relation of cause and effect.

Exercise 1.1. For the inclined beam in Fig. 1.2, the specific intensities of the distributed loads P_1, P_2 and the distributed couple P_3 are given as functions of the distance x along the beam. Determine the expressions for the generalized strains q_1, q_2, q_3 corresponding to the generalized stresses Q_1, Q_2, Q_3 shown in the figure.

1.2. Linearly Elastic Structures

1.2.1. Basic Concepts and Relations

The mechanical behavior of an *elastic* structure is characterized by the fact that the work done by the loads in deforming the structure is stored in the structure as *elastic energy* that may be recovered when the structure is unloaded. For a *linearly* elastic structure the specific elastic energy may be written as a homogeneous quadratic function of the generalized strains

$$\phi(q) = \tfrac{1}{2} \sum_j \sum_k C_{jk} q_j q_k, \tag{1.9}$$

called *specific strain energy*, or as a homogeneous quadratic function of the generalized stresses

$$\psi(Q) = \tfrac{1}{2} \sum_j \sum_k \tilde{C}_{jk} Q_j Q_k, \tag{1.10}$$

called *specific complementary energy*. The *elastic coefficients* C_{jk} and $\tilde{C}_{jk}$ in these functions satisfy the symmetry relations

$$C_{jk} = C_{kj}, \qquad \tilde{C}_{jk} = \tilde{C}_{kj}. \tag{1.11}$$

By their mechanical meaning ϕ and ψ are *positive definite*; that is, they can not assume negative values, and they vanish only when all generalized strains and stresses vanish.

Infinitesimal changes dq_j of the generalized strains q_j change the specific strain energy by $d\phi = \Sigma_j\, (\partial\phi/\partial_j)\, dq_j$. On the other hand, $d\phi$ equals the work $\Sigma_j\, Q_j\, dq_j$ done by the generalized stresses Q_j on the strain changes dq_j. Accordingly,

$$Q_j = \frac{\partial\phi}{\partial q_j} = \sum_k C_{jk} q_k, \tag{1.12}$$

where the first symmetry relation (1.11), has been used in the transition to the last term. Similarly,

$$q_j = \frac{\partial\psi}{\partial Q_j} = \sum_k \tilde{C}_{jk} Q_k. \tag{1.13}$$

Since ϕ is positive definite, the square symmetric matrix C_{jk} has an inverse, and substitution of Eq. (1.12) into Eq. (1.13) shows that this inverse is the square symmetric matrix with elements $\tilde{C}_{jk}$.

The *strain energy* $\Phi(q)$ of a kinematically admissible strain field $q_j(x)$ is obtained by integrating the specific strain energy $\phi(x)$ of this field over the structure. Similarly, the *complementary energy* $\Psi(Q)$ of a statically admissible stress field $Q_j(x)$ is the integral of the specific complementary energy $\psi(x)$ of this field over the structure. Note that the use of Eqs. (1.9), (1.10), (1.12), and (1.13) in these definitions yields

$$2\Phi(q) = \int \sum_j Q_j q_j\, dx = 2\Psi(Q). \tag{1.14}$$

If $q_j(x)$ and $q_j^*(x)$ are *distinct* kinematically admissible strain fields, the strain energy computed from the kinematically admissible field $q_j^*(x) - q_j(x)$ is positive:

$$\frac{1}{2}\int \sum_j \sum_k [C_{jk}(q_j^* - q_j)(q_k^* - q_k)]\,dx > 0. \qquad (1.15)$$

Denoting the strain energies of the two fields by $\Phi(q)$ and $\Phi(q^*)$ and using the first symmetry relation, Eq. (1.11), we rewrite Eq. (1.15) in the form

$$\Phi(q^*) + \Phi(q) > \int \sum_j (Q_j q_j^*)\,dx, \qquad (1.16)$$

where Q_j are the stresses corresponding to the strains q_j in accordance with Eq. (1.12). Subtraction of the first equation in Eq. (1.14) from Eq. (1.16) finally furnishes the inequality

$$\Phi(q^*) - \Phi(q) > \int \sum_j \{Q_j(q_j^* - q_j)\}\,dx. \qquad (1.17)$$

We leave it to the reader to derive the analogous inequality

$$\Psi(Q^*) - \Psi(Q) > \int \sum_j \{(Q_j^* - Q_j)q_j\}\,dx, \qquad (1.18)$$

where $Q_j^*(x)$ and $Q_j(x)$ are *distinct* statically admissible stress fields for given loads $P_\alpha(x)$, $\Psi(Q^*)$ and $\Psi(Q)$ are their complementary energies, respectively, and $q_j(x)$ is the strain field derived from the stress field $Q_j(x)$ in accordance with Eq. (1.13).

The remainder of this section is devoted to general theorems concerning a linearly elastic structure whose elastic coefficients are given functions of position and whose supports preclude any rigid-body motion of the entire structure. Sections 1.2.1–1.2.6 concern the following problem: given the time-independent loads $P_\alpha(x)$ acting on the structure, determine the displacements $p_\alpha(x)$, strains $q_j(x)$, and stresses $Q_j(x)$ caused by these loads. This problem will be referred to as the *fundamental problem* of structural *elastostatics*. Sections 1.2.7–1.2.9 are concerned with free and forced *vibrations* of linearly elastic structures.

1.2.2. Uniqueness Theorem

To constitute a solution of the fundamental problem, the fields $p_\alpha(x)$, $Q_j(x)$, and $q_j(x)$ must satisfy the following conditions. The displacements

$p_\alpha(x)$ must be kinematically admissible, and the stresses $Q_j(x)$ must be statically admissible for the given loads $P_\alpha(x)$. Furthermore, the stresses $Q_j(x)$ and the strains $q_j(x)$, which are derived from the displacements $p_\alpha(x)$, must satisfy the stress–strain relation of Eq. (1.12).

It can be shown that there exist unique displacement, stress, and strain fields satisfying these conditions. Since the existence proof is outside the scope of this review, we restrict ourselves to proving uniqueness.

If there existed distinct solution fields $\bar{p}_\alpha$, $\bar{q}_j$, $\bar{Q}_j$ and p_α, q_j, Q_j, their differences would constitute a solution for vanishing loads, and the principle of virtual work would yield

$$\int \sum_j [(\bar{Q}_j - Q_j)(\bar{q}_j - q_j)]\, dx = 0. \tag{1.19}$$

According to Eq. (1.14) the left side of (1.19) is twice the strain energy of the field $\bar{q}_j - q_j$. Since this energy is positive definite, it follows from Eq. (1.19) that $\bar{q}_j - q_j$ vanishes identically. Applied to the field $\bar{q}_j - q_j$, the stress-strain relation of Eq. (1.12) then shows that $\bar{Q}_j - Q_j$ also vanishes identically. It finally follows from $\bar{q}_j - q_j \equiv 0$ that the field $\bar{p}_\alpha - p_\alpha$ can only represent a rigid-body motion of the structure. Since this is supposed to be excluded by the supports, the field $\bar{p}_\alpha - p_\alpha$ must also vanish identically.

1.2.3. Principle of Minimum Complementary Energy

Let the displacement, strain, and stress fields p_α, q_j, Q_j be the solution of the fundamental problem for the loads P_α, and denote by $Q_j^* \not\equiv Q_j$ an arbitrary statically admissible stress field for these loads. Application of the principle of virtual work to the kinematic variables p_α, q_j and the stresses $Q_j^* - Q_j$, which are statically admissible for vanishing loads, then furnishes

$$0 = \int \sum_j [Q_j^* - Q_j)q_j]\, dx < \Psi(Q^*) - \Psi(Q), \tag{1.20}$$

where Eq. (1.18) has been used in the transition to the last term. According to Eq. (1.20) the stress field of the solution of the fundamental problem for given loads has a smaller complementary energy than any other stress field that is statically admissible for these loads. This is the *principle of minimum complementary energy.*

1.2.4. Principle of Minimum Potential Energy

Let the displacement, strain, and stress fields p_α, q_j, Q_j be the solution fields of the fundamental problem for the loads P_α, and denote by p_α^* and

q_j^* any kinematically admissible displacement field that is distinct from p_α and the corresponding strain field, respectively. Application of the principle of virtual work to the static variables P_α, Q_j and the kinematic variables $p_\alpha^* - p_\alpha$, $q_j^* - q_j$ and subsequent use of the inequality in Eq. (1.17) then furnishes

$$\Pi(p^*) \equiv \Phi(q^*) - \int \sum_\alpha (P_\alpha p_\alpha^*)\, dx > \Phi(q) - \int \sum_\alpha (P_\alpha p_\alpha)\, dx. \qquad (1.21)$$

The expression $\Pi(p^*)$ is called the *potential energy* of the kinematically admissible displacement field $p^*(x)$. According to Eq. (1.21) the displacement field of the solution of the fundamental problem has a smaller potential energy than any other kinematically admissible displacement field. This is the *principle of minimum potential energy.*

1.2.5. Reciprocity Theorem

Let p_α, q_j, Q_j and $\bar{p}_\alpha$, $\bar{q}_j$, $\bar{Q}_j$ denote the solution fields of the fundamental problems for the loads P_α and $\bar{P}_\alpha$. Applying the principle of virtual work to the static variables P_α, Q_j and the kinematic variables $\bar{p}_\alpha$, $\bar{q}_j$ and using Eq. (1.12), we obtain

$$\int \sum_\alpha (P_\alpha \bar{p}_\alpha)\, dx = \int \sum_j (Q_j \bar{q}_j)\, dx = \int \sum_j \sum_k (C_{jk} \bar{q}_j q_k)\, dx. \qquad (1.22)$$

Similarly,

$$\int \sum_\alpha (\bar{P}_\alpha p_\alpha)\, dx = \int \sum_j (\bar{Q}_j q_j)\, dx = \int \sum_j \sum_k (C_{jk} q_j \bar{q}_k)\, dx. \qquad (1.23)$$

In view of the symmetry of C_{jk}, the rightmost terms in Eqs. (1.22) and (1.23) have the same value. The work of the first loading on the displacements caused by the second loading thus equals the work of the second loading on the displacements caused by the first loading. This is the *reciprocity theorem* of Betti (1872).

1.2.6. Principle of Stationary Mutual Potential Energy

Letting P_α, p_α, q_j, Q_j and $\bar{P}_\alpha$, $\bar{p}_\alpha$, $\bar{q}_j$, $\bar{Q}_j$ have the same meaning as in Section 1.2.5, we consider two arbitrary kinematically admissible displacement fields p_α^*, $\bar{p}_\alpha^*$ and the corresponding strain fields q_j^*, $\bar{q}_j^*$. Shield and Prager (1970) defined the mutual potential energy for these fields and the loads P_α and $\bar{P}_\alpha$ as

$$U(p^*, \bar{p}^*) = \tfrac{1}{2}\int \sum_j \sum_k (C_{jk} q_j^* \bar{q}_k^*)\, dx - \tfrac{1}{2}\int \sum_\alpha (P_\alpha \bar{p}_\alpha^* + \bar{P}_\alpha p_\alpha^*)\, dx. \tag{1.24}$$

In view of Eqs. (1.11) and (1.12), the mutual potential energy of the solution fields p_α, q_j and $\bar{p}_\alpha$, $\bar{q}_j$ is

$$U(p, \bar{p}) = \tfrac{1}{2}\int \sum_j (Q_j \bar{q}_j)\, dx - \tfrac{1}{2}\int \sum_\alpha (P_\alpha \bar{p}_\alpha + \bar{P}_\alpha p_\alpha)\, dx. \tag{1.25}$$

Using the principle of virtual work, we may rewrite Eq. (1.25) in any of the following forms:

$$\begin{aligned} U(p, \bar{p}) &= -\tfrac{1}{2}\int \sum_\alpha (P_\alpha \bar{p}_\alpha)\, dx = -\tfrac{1}{2}\int \sum_\alpha (\bar{P}_\alpha p_\alpha)\, dx \\ &= -\tfrac{1}{2}\int \sum_j \sum_k (C_{jk} q_j \bar{q}_k)\, dx. \end{aligned} \tag{1.26}$$

Note that

$$\begin{aligned} &\tfrac{1}{2}\int \sum_j \sum_k \{C_{jk}(q_j^* - q_j)(\bar{q}_k^* - \bar{q}_k)\}\, dx \\ &\quad = \tfrac{1}{2}\int \sum_j \sum_k \{C_{jk}(q_j^* \bar{q}_k^* - q_j \bar{q}_k)\}\, dx - \tfrac{1}{2}\int \sum_\alpha (P_\alpha \bar{p}_\alpha^* + \bar{P}_\alpha p_\alpha^*)\, dx \\ &\quad = U(p^*, \bar{p}^*) - U(p, \bar{p}), \end{aligned} \tag{1.27}$$

where Eqs. (1.11) and (1.12) and the principle of virtual work have been used in the transition to the second line, and Eq. (1.26) in the transition to the third.

We now restrict the starred fields to the neighborhood of the solution fields, replacing, say, p_α^* by the sum of p_α and a small variation δp_α, etc., and $U(p^*, \bar{p}^*)$ by $U(p, \bar{p}) + \delta U + \delta^2 U$, where $\delta^2 U$ is a homogeneous second-order expression in the considered variations. It then follows from (1.27) that

$$\delta U = 0, \tag{1.28}$$

$$\delta^2 U = \tfrac{1}{2}\int \sum_j \sum_k (C_{jk}\, \delta q_j\, \delta \bar{q}_k)\, dx. \tag{1.29}$$

Equation (1.28) shows that the mutual potential energy of Eq. (1.24) is stationary in the neighborhood of $p^*_\alpha = p_\alpha$, $\bar{p}^*_\alpha = \bar{p}_\alpha$. The integrand in Eq. (1.29), however, is generally not restricted in sign. Accordingly, it cannot be asserted that the mutual potential energy assumes its smallest or greatest value for $p^*_\alpha = p_\alpha$, $\bar{p}^*_\alpha = \bar{p}_\alpha$.

1.2.7. Natural Modes

For the kind of linearly elastic structure specified at the end of Section 1.2.1, consider a *natural vibration* during which all generalized displacements vary with time in proportion to cos ωt. Accordingly, the generalized displacements simultaneously assume their amplitude values p_α at times when the generalized velocities vanish, while the generalized velocities simultaneously assume their amplitude values ωp_α at instants when the generalized displacements vanish.

The amplitude of the kinetic energy of the structure will be written as

$$\frac{\omega^2}{2}\int \sum_\alpha \sum_\beta \{m_{\alpha\beta}(x)p_\alpha(x)p_\beta(x)\}\,dx,$$

where $m_{\alpha\beta}(x)$ is the typical element of the *mass matrix.* For the example beam of Section 1.1, for instance, m_{11} and m_{22} equal the specific mass of the beam, and m_{33} equals the specific moment of inertia divided by L^2, while all other elements of the mass matrix vanish. Note, however, that m_{12} vanishes here only because the displacements p_1 and p_2 are orthogonal. Note, furthermore, that the expression $m_{\alpha\beta}p_\alpha p_\beta$, which is $2/\omega^2$ times the specific kinetic energy, is *positive definite.*

Since there is no dissipation of energy in the considered elastic structure, the amplitudes of the kinetic and elastic energies have the same values, and we have

$$\int \sum_j \sum_k (C_{jk}q_j q_k)\,dx - \omega^2 \int \sum_\alpha \sum_\beta (m_{\alpha\beta}p_\alpha p_\beta)\,dx = 0, \tag{1.30}$$

where the strains q_j correspond to the displacements p_α. This equation shows that we may interpret the expressions $P_\alpha = \omega^2 \Sigma_\beta\, m_{\alpha\beta}p_\beta$ as *inertia loads* that are in equilibrium with the stresses $Q_j = \Sigma_k\, C_{jk}p_k$, and that the *natural mode* p_α may be treated as the static response of the structure to these inertia loads.

If $p_\alpha^{(g)}$ and $p_\alpha^{(h)}$ are natural modes with frequencies ω_g and ω_h, and if $P_\alpha^{(g)} = \omega_g^2\, \Sigma_\beta\, m_{\alpha\beta}p_\beta^{(g)}$ and $P_\alpha^{(h)} = \omega_h^2\, \Sigma_\beta\, m_{\alpha\beta}p_\beta^{(h)}$ are the corresponding inertia loads, we have

$$\int \sum_{\alpha} (P_{\alpha}^{(g)} p_{\alpha}^{(h)})\, dx = \int \sum_{\alpha} (P_{\alpha}^{(h)} p_{\alpha}^{(g)})\, dx,$$

by Betti's theorem. Substitution of the expressions for $P_{\alpha}^{(g)}$ and $P_{\alpha}^{(h)}$ then yields the equation

$$(\omega_g^2 - \omega_h^2) \int \sum_{\alpha} \sum_{\beta} (m_{\alpha\beta} p_{\alpha}^{(g)} p_{\beta}^{(h)})\, dx = 0, \tag{1.31}$$

which shows that natural modes with *distinct* frequencies are *orthogonal* in the sense that

$$\int \sum_{\alpha} \sum_{\beta} (m_{\alpha\beta} p_{\alpha}^{(g)} p_{\beta}^{(h)})\, dx = 0. \tag{1.32}$$

If $p_{\alpha}^{(g)}$ and p_{α}^{*} are linearly independent natural modes with the same frequency, any linear combination of them is also a natural mode with this frequency, and p_{α}^{*} may be replaced by a linear combination $p_{\alpha}^{(h)}$ that is orthogonal to $p_{\alpha}^{(g)}$. This procedure may obviously be extended to the case of several natural modes with the same frequency. Natural modes thus may always be *orthogonalized* to satisfy the condition of Eq. (1.32).

Let $\omega_1 \leq \omega_2 \leq \cdots$ be the ordered natural frequencies of a linearly elastic structure, and let $p_{\alpha}^{(1)}, p_{\alpha}^{(2)}, \ldots$ be the corresponding orthogonalized and normalized modes satisfying the conditions

$$\int \sum_{\alpha} \sum_{\beta} (m_{\alpha\beta} p_{\alpha}^{(g)} p_{\beta}^{(h)})\, dx = \begin{cases} 1 & \text{for } h = g, \\ 0 & \text{for } h \neq g. \end{cases} \tag{1.33}$$

Note that there corresponds to this relation for the displacements of natural modes the following relation for their strains:

$$\int \sum_{j} \sum_{k} (C_{jk} q_j^{(g)} q_k^{(h)})\, dx = \begin{cases} \omega_g^2 & \text{for } h = g, \\ 0 & \text{for } h \neq g, \end{cases} \tag{1.34}$$

where Eq. (1.30) has been used for the case $g = h$.

Any kinematically admissible displacement field $\bar{p}_{\alpha}(x)$ can be approximated by a weighted sum of natural modes:

$$\bar{p}_{\alpha} = c_1 p_{\alpha}^{(1)} + c_2 p_{\alpha}^{(2)} + \cdots + c_n p_{\alpha}^{(n)}. \tag{1.35}$$

We mention, without proof, that this approximation may be made arbitrarily good in the mean-square sense by choosing n sufficiently great. To determine the coefficient c_h, we multiply Eq. (1.33) by $\Sigma_\beta\, m_{\alpha\beta} p_\beta^{(h)}$, sum over α, and use Eq. (1.32). Thus,

$$c_h = \int \sum_\alpha \sum_\beta (m_{\alpha\beta} p_\alpha p_\beta^{(h)})\, dx. \tag{1.36}$$

1.2.8. Rayleigh's Principle

For an arbitrary kinematically admissible displacement field $\bar{p}_\alpha(x)$ and the corresponding strain field $\bar{q}_j(x)$, let us evaluate the functional

$$F(\bar{p}, \bar{q}) = \int \sum_j \sum_k (C_{ik} \bar{q}_j \bar{q}_k)\, dx - \omega_1^2 \int \sum_\alpha \sum_\beta (m_{\alpha\beta} \bar{p}_\alpha \bar{p}_\beta)\, dx. \tag{1.37}$$

Substituting the series $\bar{p}_\alpha = c_1 p_\alpha^{(1)} + c_2 p_\alpha^{(2)} + \cdots$ and $\bar{q}_j = c_1 q_j^{(1)} + c_2 q_j^{(2)} + \cdots$ into Eq. (1.37) and using Eqs. (1.33) and (1.34), we see that

$$F = \sum_g (\omega_g^2 - \omega_1^2) c_g^2 \tag{1.38}$$

is nonnegative because $\omega_g^2 - \omega_1^2 \geq 0$ and $c_g^2 \geq 0$. On the other hand, for $\bar{p}_\alpha = p_\alpha^{(1)}$ we have $F = 0$. The functional of Eq. (1.37) thus assumes its smallest value, zero, for $\bar{p}_\alpha = p_\alpha^{(1)}$. Another way of putting this result is to state that, of all kinematically admissible displacement fields $\bar{p}_\alpha$, the field $p_\alpha^{(1)}$ of the fundamental natural mode minimizes the Rayleigh quotient

$$R \equiv \frac{\int \Sigma_j \Sigma_k (C_{jk} \bar{q}_j \bar{q}_k)\, dx}{\int \Sigma_\alpha \Sigma_\beta\, (m_{\alpha\beta} \bar{p}_\alpha \bar{p}_\beta)\, dx}. \tag{1.39}$$

This minimum principle is due to Lord Rayleigh (1878).

1.2.9. Minimum Principle for Steady-State Forced Vibrations

Consider a linearly elastic structure subject to harmonically varying loads $P_\alpha(x) \cos \omega t$, where ω is a given frequency *below* the fundamental natural frequency ω_1 of the structure. In the so-called *steady-state response* of the structure to these loads, the displacements, strains, and stresses have the forms $p_\alpha(x) \cos \omega t$, $q_j(x) \cos \omega t$, and $Q_j(x) \cos \omega t$, respectively, where $p_\alpha(x)$, $q_j(x)$, and $Q_j(x)$ represent the *static* response of the structure to the combination of the loads $P_\alpha(x)$ and the inertia loads $\omega^2 \Sigma_\beta\, m_{\alpha\beta}(x) p_\beta(x)$.

Accordingly,

$$\int \sum_j \sum_k (C_{jk} q_j q_k)\, dx - \int \sum_\alpha (P_\alpha p_\alpha)\, dx - \omega^2 \int \sum_\alpha \sum_\beta (m_{\alpha\beta} p_\alpha p_\beta)\, dx = 0. \quad (1.40)$$

If $p_\alpha^*(x)$ is an arbitrary kinematically admissible displacement field, it follows from $\omega < \omega_1$ and Rayleigh's principle that

$$\int \sum_j \sum_k (C_{jk} q_j^* q_k^*)\, dx - \omega^2 \int \sum_\alpha \sum_\beta (m_{\alpha\beta} p_\alpha^* p_\beta^*)\, dx > 0. \quad (1.41)$$

When applied to the loads $P_\alpha + \omega^2 \Sigma_\beta\, m_{\alpha\beta} p_\beta$ and the corresponding stresses $Q_j = \Sigma_k\, C_{jk} q_k$, on the one hand, and to the kinematically admissible displacements p_α^* and the corresponding strains q_j^*, on the other hand, the principle of virtual work yields

$$\int \sum_j \sum_k (C_{jk} q_j^* q_k)\, dx = \int \sum_\alpha (P_\alpha p_\alpha^*)\, dx + \omega^2 \int \sum_\alpha \sum_\beta (m_{\alpha\beta} p_\alpha^* p_\beta)\, dx. \quad (1.42)$$

Since the displacement $p_\alpha - p_\alpha^*$ is kinematically admissible, we may substitute it for p_α^* in Eq. (1.41), provided we change the strong inequality to a weak one to allow for the case $p_\alpha^* = p_\alpha$. Accordingly,

$$\int \sum_j \sum_k [C_{jk}(q_j q_k - 2q_j^* q_k + q_j^* q_k^*)]\, dx$$
$$- \omega^2 \int \sum_\alpha \sum_\beta [m_{\alpha\beta}(p_\alpha p_\beta - 2p_\alpha^* p_\beta + p_\alpha^* p_\beta^*)]\, dx \geq 0. \quad (1.43)$$

Subtracting two times Eq. (1.38) from Eq. (1.43) and using Eq. (1.42), we obtain

$$\int \sum_j \sum_k (C_{jk} q_j^* q_k^*)\, dx - 2 \int \sum_\alpha (P_\alpha p_\alpha^*)\, dx - 2\omega^2 \int \sum_\alpha \sum_\beta (m_{\alpha\beta} p_\alpha^* p_\beta^*)\, dx$$
$$\geq \int \sum_j \sum_k (C_{jk} q_j q_k)\, dx - 2 \int \sum_\alpha (P_\alpha p_\alpha)\, dx - 2\omega^2 \int \sum_\alpha \sum_\beta (m_{\alpha\beta} p_\alpha p_\beta)\, dx. \quad (1.44)$$

The minimum principle expressed by Eq. (1.44) is identical to the principle of minimum potential energy applied to the loading consisting of the amplitudes of the given loads and the inertia loads, although the latter depend on the displacements. The principle is due to Icerman (1969).

Exercise 1.2. Prove the continued inequality

$$2\int \sum_{\alpha} (P_\alpha p_\alpha^*)\, dx - 2\Phi(q^*) \le \int \sum_{\alpha} (P_\alpha p_\alpha)\, dx \le 2\Psi(Q^{**}), \qquad (1.45)$$

where $p_\alpha(x)$ is the displacement field of the solution of the fundamental problem for the loads $P_\alpha(x)$, the stress field $Q_j^{**}(x)$ is statically admissible for these loads, and the displacement field $p_\alpha^*(x)$ with the strain field $q_j^*(x)$ is kinematically admissible.

Exercise 1.3. In the continued inequality of Eq. (1.45) replace p_α^* by cp_α^* and determine the constant c to maximize the lower bound for $W = \int \Sigma_\alpha (P_\alpha p_\alpha)\, dx$.

Exercise 1.4. In (1.45) replace Q_j^{**} by $Q_j^{**} + cQ_j^0$, where the stress field Q_j^0 is statically admissible for vanishing loads. Determine the constant c to minimize the upper bound for W (see Exercise 1.3).

Exercise 1.5. Show that of all kinematically admissible displacement fields $\bar{p}_\alpha$ that are orthogonal to the first natural mode $p_\alpha^{(1)}$, the second natural mode $p_\alpha^{(2)}$ minimizes the Rayleigh quotient of Eq. (1.39).

1.3. Rigid, Perfectly Plastic Structures

1.3.1. Basic Concepts and Relations

In Sections 1.3 and 1.4 the time rate of a variable will be indicated by a superimposed dot. For example, $\dot{p}_\alpha$ will denote a generalized velocity; $\dot{q}_j$, a generalized strain rate.

The discussion of the mechanical behavior of an element of a rigid, perfectly plastic structure is facilitated by the use of the following geometrical representation. In an n-dimensional Euclidean *stress space* the state of stress is represented by the *stress point* with rectangular Cartesian coordinates $Q_1, Q_2, \ldots, Q_n$. The radius vector of this point is called the stress vector and is denoted by $\mathbf{Q}$. Similarly, the rate of deformation is represented by the *strain rate vector* $\dot{\mathbf{q}}$, which has components $\dot{q}_1, \dot{q}_2, \ldots, \dot{q}_n$ with respect to the same rectangular axes. For the sake of brevity the expressions "state of stress $\mathbf{Q}$" (or briefly "stress $\mathbf{Q}$") and "strain rate $\dot{\mathbf{q}}$" will be used.

The *dissipation function* $D(\dot{\mathbf{q}})$ of a rigid, perfectly plastic structural element gives the rate at which mechanical energy is dissipated per unit volume in plastic flow with the strain rate $\dot{\mathbf{q}}$. Note that this function, which represents the specific power of dissipation, is nonnegative. Because the rigid, perfectly plastic structural element lacks viscosity, the dissipation function is homogeneous of order one; that is,

$$D(c\dot{\mathbf{q}}) = cD(\dot{\mathbf{q}}) \qquad \text{for any } c \geq 0. \tag{1.46}$$

To be *attainable* in the considered structural element, a state of stress $\mathbf{Q}$ must satisfy the *yield inequality*

$$\boldsymbol{Q} \cdot \dot{\boldsymbol{q}} \leq D(\dot{\mathbf{q}}) \tag{1.47}$$

for *all* strain rates $\dot{\mathbf{q}}$. (The center dot in Eq. (1.47) indicates scalar multiplication.) A state of stress $\mathbf{Q}$ that satisfies Eq. (1.47) as a strict inequality for *all* $\dot{\mathbf{q}}$ cannot cause plastic flow. On the other hand, if plastic flow with the strain rate $\dot{\mathbf{q}}$ can occur under the stress $\mathbf{Q}$, this strain rate and stress satisfy Eq. (1.47) with equality holding.

Stress points with position vectors $\mathbf{Q}$ that satisfy the yield inequality for a *given* strain rate vector $\dot{\mathbf{q}}$ fill a half-space that is bounded by a hyperplane normal to $\dot{\mathbf{q}}$. Because $D(\dot{\mathbf{q}})$ is homogeneous of order one in the components of $\dot{\mathbf{q}}$, the position of this hyperplane depends only on the direction, not the magnitude, of $\dot{\mathbf{q}}$. Note that the origin of the stress space, which represents the stress-free state, is a point of the considered half-space.

The intersection of the half-spaces corresponding to *all* strain rate directions is called the *yield domain.* As the intersection of convex domains, the yield domain is convex. The boundary of the yield domain is called the *yield locus.* Interior points of the yield domain represent states of stress under which the considered structural element remains rigid, while points of the yield locus represent states of stress capable of producing plastic flow.

It will be good to illustrate these concepts by an example. Consider combined extension and flexure of a beam with the sandwich section shown in Fig. 1.3, assuming that the thicknesses t_1 and t_2 of the cover plates are small in comparison to the height $2h$ of the core and the core does not transmit axial stresses. For the sake of dimensional homogeneity we use the axial force Q_1 and the quotient Q_2 of the bending moment by h as the

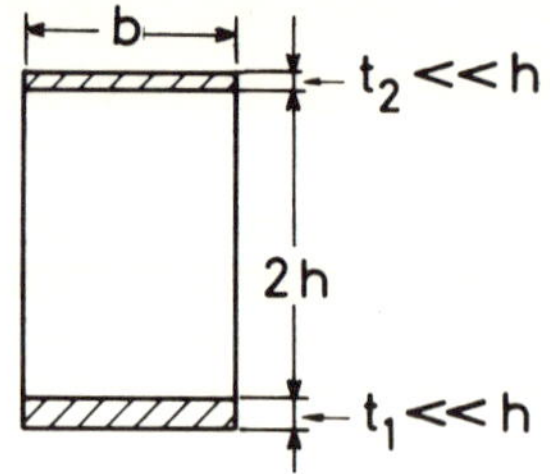

Figure 1.3. Sandwich section.

generalized stresses. The corresponding generalized strain rates are the rate of axial extension $\dot{q}_1$ and the product $\dot{q}_2$ of h by the rate of curvature. Because the rates of extension of the cover plates are $\dot{q}_1 + \dot{q}_2$ and $\dot{q}_1 - \dot{q}_2$, the yield inequality takes the form

$$D(\dot{\mathbf{q}}) \equiv Y_1|\dot{q}_1 + \dot{q}_2| + Y_2|\dot{q}_1 - \dot{q}_2| \geq Q_1\dot{q}_1 + Q_2\dot{q}_2, \tag{1.48}$$

where $Y_1 = \sigma_0 b t_1$ and $Y_2 = \sigma_0 b t_2$, while σ_0 is the common intensity of the yield stresses in uniaxial tension and compression. Note that the dissipation function is homogeneous of order one, but not linear in $\dot{q}_1$ and $\dot{q}_2$.

To obtain points of the yield locus we use the equality sign in Eq. (1.48), remove the absolute value signs by suitably restricting the signs and relative magnitudes of $\dot{q}_1$ and $\dot{q}_2$, and compare the coefficients of these strain rate components on the two sides of the resulting equation. For $\dot{q}_1 > |\dot{q}_2|$, for instance, we thus obtain the stress point with coordinates $Q_1 = Y_1 + Y_2$, $Q_2 = Y_1 - Y_2$. Similarly, for $\dot{q}_2 > |\dot{q}_1|$, we obtain the stress point $Q_1 = Y_1 - Y_2$, $Q_2 = Y_1 + Y_2$. For $\dot{q}_1 = \dot{q}_2 > 0$, however, the comparison of coefficients furnishes $Q_1 + Q_2 = 2Y_1$—that is, the equation of the line segment joining these stress points. Considering the other possibilities of removing the absolute value signs from Eq. (1.48), we obtain two further stress points, which follow from those obtained above by symmetry with respect to the origin of the stress plane. The yield locus thus is the rectangle shown in Fig. 1.4.

A point of the yield locus will be called regular or singular, depending on whether the yield locus has a unique normal at this point. The point E in Fig. 1.4 is a regular point; it represents a state of stress capable of producing plastic flow with a strain rate vector along the exterior normal of the yield locus at E. The vertex A in Fig. 1.4 is a singular point. As a

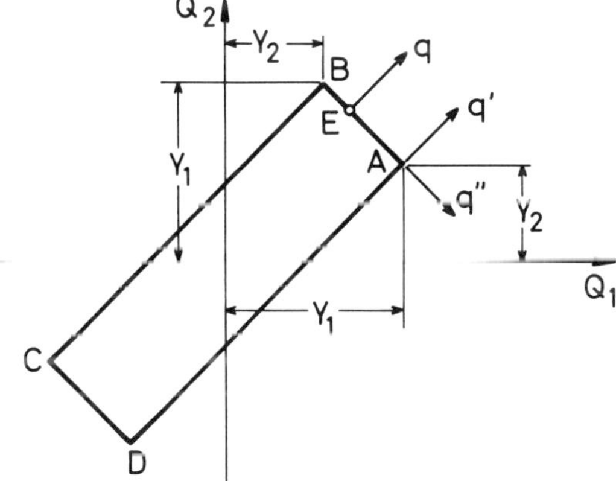

Figure 1.4. Yield polygon.

point of both sides AB and AD, it represents a state of stress capable of producing plastic flow with a strain rate vector $\dot{\mathbf{q}}'$ along the exterior normal of AB or a strain rate vector $\dot{\mathbf{q}}''$ along the exterior normal of AD. We have seen, moreover, that A is also associated with any strain rate vector satisfying $\dot{q}_1 > |\dot{q}_2|$. Accordingly, the vertex A is associated with any strain rate vector of the form $\alpha\dot{\mathbf{q}}' + \beta\dot{\mathbf{q}}''$, $\alpha \geq 0$, $\beta \geq 0$. Such a combination is hereafter called a nonnegative linear combination.

The following *normality rule* extends these statements to stress spaces of any number of dimensions. A regular point of the yield locus is associated with a strain rate vector along the exterior normal of the yield locus at this point. At a singular point two or more faces of the yield locus meet, and the strain rate vector may be any nonnegative linear combination of vectors along the exterior normals of these faces at the singular point.

Note that we have a one-to-one correspondence between the stress point and the direction of the strain rate vector only if the yield locus has neither singular points nor plane faces.

1.3.2. Fundamental Theorems of Limit Analysis

Limit analysis is concerned with the load-carrying capacities of rigid, perfectly plastic structures; its basic problem is as follows. If a rigid, perfectly plastic structure remains rigid under the given loading $P_\alpha(x)$, determine a factor $\Lambda > 1$ such that plastic flow is possible under the loads $\Lambda P_\alpha(x)$, but not under any loading $\Lambda^* P_\alpha(x)$, with $\Lambda^* < \Lambda$. The factor Λ is called the *load factor for plastic collapse* for the given loading.

In the following discussion a stress field $\mathbf{Q}(x)$ will be called *attainable* if, throughout the structure, the stress point with the radius vector $\mathbf{Q}(x)$ is a point of the yield domain for the point x of the structure. Furthermore, a velocity field $\dot{p}_\alpha(x)$ will be said to be normalized for the loading $P_\alpha(x)$ if

$$\sum_\alpha \int P_\alpha \dot{p}_\alpha \, dx = 1. \tag{1.49}$$

A strain rate field will be called normalized if it is derived from a normalized velocity field.

The *static theorem* of limit analysis states that the load factor for plastic collapse indicates the greatest multiple of the given loading for which there exists an attainable, statically admissible stress field.

If this theorem were false, an attainable stress field Q_j^* would be statically admissible for the loading $\Lambda^* P_\alpha$, while the load factor for plastic collapse Λ would satisfy the inequality

$$\Lambda < \Lambda^*. \tag{1.50}$$

If $\dot{p}_\alpha$ and $\dot{q}_j$ are normalized, kinematically admissible velocity and strain rate fields for plastic collapse under the loading ΛP_α, we have

$$\Lambda = \int D(\mathbf{q})\, dx. \tag{1.51}$$

On the other hand, the principle of virtual work applied to the static variables $\Lambda^* P_\alpha$, Q_j^* and the kinematic variables $\dot{p}_\alpha$, $\dot{q}_j$ and Eq. (1.47) furnish

$$\Lambda^* = \int \mathbf{Q}^* \cdot \dot{\mathbf{q}}\, dx \leq \int D(\dot{\mathbf{q}})\, dx. \tag{1.52}$$

The contradiction between Eq. (1.50) and the inequality following from Eqs. (1.51) and (1.52) establishes the static theorem.

The *kinematic theorem* of limit analysis states that the load factor for plastic collapse is given by the minimum value of the power of dissipation $\int D(\dot{\mathbf{q}})\, dx$ over all normalized, kinematically admissible strain rate fields.

If this theorem were false, a normalized, kinematically admissible strain rate field $\dot{q}_j^*$ would furnish the minimum value

$$\Lambda^* = \int D(\dot{\mathbf{q}}^*)\, dx \tag{1.53}$$

of the power of dissipation, while the load factor for plastic collapse would satisfy the inequality

$$\Lambda > \Lambda^*. \tag{1.54}$$

Denoting by $\dot{p}_\alpha^*$ the normalized, kinematically admissible velocity field furnishing the strain rate field $\dot{q}_j^*$, and by Q_j the attainable, statically admissible stress field during collapse under the loading ΛP_α, we now apply the principle of virtual work to the static variables ΛP_α, Q_j and the kinematic variables $\dot{p}_\alpha^*$, $\dot{q}_j^*$ and use Eq. (1.47). Thus,

$$\Lambda = \int \mathbf{Q} \cdot \dot{\mathbf{q}}^*\, dx \leq \int D(\dot{\mathbf{q}}^*)\, dx. \tag{1.55}$$

The contradiction between Eq. (1.54) and the inequality following from Eqs. (1.53) and (1.55) establishes the kinematic theorem.

The static and kinematic theorems have the following important corollary. If an attainable stress field Q_j^* is statically admissible for the loading $\Lambda^* P_\alpha$ and capable of producing the kinematically admissible strain rate field $\dot{q}_j^*$ that is normalized for the loads P_α, the load factor for plastic collapse Λ is equal to Λ^*.

Indeed, according to the kinematic and static theorems,

$$\Lambda^* \leq \Lambda \leq \int D(\dot{\mathbf{q}}^*)\, dx, \tag{1.56}$$

and application of the principle of virtual work to the static variables $\Lambda^* P_\alpha$, Q_j^* and the kinematic variables $\dot{q}_\alpha^*$, $\dot{q}_j^*$ yields

$$\Lambda^* = \int \mathbf{Q}^* \cdot \dot{\mathbf{q}}^*\, dx = \int D(\dot{\mathbf{q}}^*)\, dx, \tag{1.57}$$

where Eq. (1.57) is a consequence of the fact that the stresses $\mathbf{Q}^*$ produce the strain rates $\dot{\mathbf{q}}^*$. Substitution of Eq. (1.56) into Eq. (1.57) shows that $\Lambda = \Lambda^*$.

Some remarks on the history of limit analysis will be found in Section 1.4.

Exercise 1.6. If the stress vectors $\mathbf{Q}$ and $\mathbf{Q}^*$ represent attainable states of stress, and the stress $\mathbf{Q}$ is capable of producing plastic flow with the strain rate vector $\dot{\mathbf{q}}$, show that

$$(\mathbf{Q} - \mathbf{Q}^*) \cdot \dot{\mathbf{q}} \geq 0 \tag{1.58}$$

(principle of maximum local power of dissipation).

Exercise 1.7. Show that Eq. (1.58) becomes a strong inequality if $\mathbf{Q}^* \neq \mathbf{Q}$, $\dot{\mathbf{q}} \neq 0$, and if the yield locus is *strictly convex*; that is, if it does not have two points at which the exterior normals have the same direction.

Exercise 1.8. Consider a rigid, perfectly plastic structure at each point of which the yield locus is strictly convex. If the loading P_α can produce alternative modes of plastic collapse with the stress fields Q_j and Q_j^* and the normalized velocity and strain rate fields $\dot{p}_\alpha$, $\dot{q}_j$ and $\dot{p}_\alpha^*$, $\dot{q}_j^*$, prove that

$$\int (\mathbf{Q} - \mathbf{Q}^*) \cdot (\dot{\mathbf{q}} - \dot{\mathbf{q}}^*)\, dx = 0 \tag{1.59}$$

and use this equation to show that $\mathbf{Q} = \mathbf{Q}^*$ except, possibly, where $\dot{\mathbf{q}} = \dot{\mathbf{q}}^* = 0$.

Exercise 1.9. For the problem in Exercise 1.8 show that the velocity field of one solution is compatible with the stress field of the other solution.

1.4. Elastic, Perfectly Plastic Structures

1.4.1. Trusses

Consider a statically indeterminate, elastic, perfectly plastic truss that is supported in a manner that would exclude displacements of the joints if all bars were rigid. Denote by $\mathbf{P}_\alpha$ the vector of the load acting at joint α, and by Q_i and q_i the axial force and the elongation, respectively, of bar i of the truss. For the elastic, perfectly plastic bar, the force–elongation diagram has the form shown in Fig. 1.5. The *state point* with coordinates Q_i, q_i represents the instantaneous state of the bar. If Y_i and $-Y_i$ are the axial forces under which the bar yields in tension or compression, the state point is restricted to the *attainable strip* $-Y_i \leq Q_i \leq Y_i$. When the state point is an interior point of this strip, the bar responds in a purely elastic manner to a given time rate $\dot{Q}_i$ of the axial force; that is, the rate of elongation is $\dot{q}_i = \tilde{C}_i \dot{Q}_i$, with $\tilde{C}_i = l_i / EA_i$, where l_i and A_i are length and cross-sectional area of bar i, and E is Young's modulus. When, on the other hand, the state point is a boundary point of the attainable strip, the bar responds in a purely elastic or purely plastic manner to the given rate $\dot{Q}_i$, depending on whether $Q_i \dot{Q}_i$ is negative or zero. In the first case we again have $\dot{q}_i = \tilde{C}_i \dot{Q}_i$; in the second case the plastic rate of elongation has the sign of the axial force, but an arbitrary magnitude. Summing up, we have the following relations, in which the superscripts (e) and (p) indicate elastic and plastic responses:

$$\dot{q}_i^{(e)} = \tilde{C}_i \dot{Q}_i, \tag{1.60a}$$

$$\dot{q}_i^{(p)} = 0 \quad \text{if } |Q_i| \leq Y_i \quad \text{but } Q_i \dot{Q}_i < 0, \tag{1.60b}$$

$$\operatorname{sgn} \dot{q}_i^{(p)} = \operatorname{sgn} Q_i \quad \text{if } |Q_i| = Y_i \text{ and } \dot{Q}_i = 0. \tag{1.60c}$$

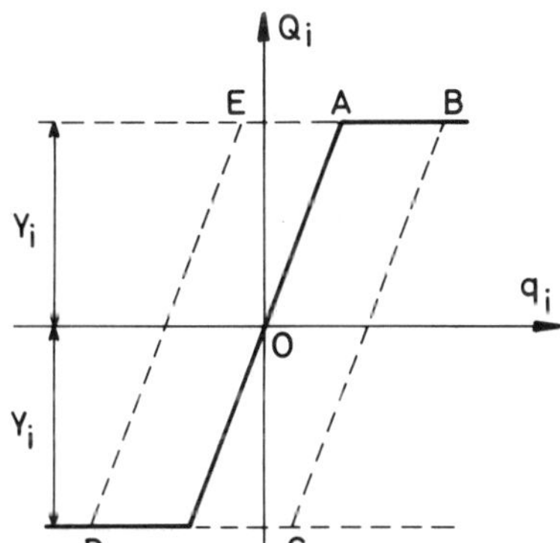

Figure 1.5. Perfectly plastic stress-strain law.

According to these relations, the state point can move only to the right along the upper boundary and only to the left along the lower boundary of the attainable strip. Between these boundaries the trajectory of the state point is straight with slope $\dot{Q}_i/\dot{q}_i = 1/\tilde{C}_i$. In Fig. 1.5, OABCDEAO is a typical closed trajectory of the state point that starts and ends with the stress-free state.

Starting with the virgin state, in which all bar forces and elongations vanish, let the truss be subjected to time-dependent loads $\mathbf{P}_\alpha(t)$ that vanish for $t \leq 0$ and vary so slowly with time that inertia forces need not be regarded. Denote by $Q_i'(t)$ the bar forces that would be generated by this loading if all bars behaved in a purely elastic manner regardless of the magnitudes of the axial forces. In the elastic, perfectly plastic truss the considered loading generates bar forces that may be written in the form

$$Q_i(t) = Q_i'(t) + Q_i''(t), \tag{1.61}$$

where the forces $Q_i''(t)$ are statically admissible for vanishing loads and, hence, constitute a state of *self-stress.* Note that the force $Q_i(t)$ in Eq. (1.61) is subject to the continued inequality $-Y_i \leq Q_i(t) \leq Y_i$, but the forces $Q_i'(t)$ and $Q_i''(t)$ are not restricted in this manner.

To the bar forces Q_i' and Q_i'' correspond the elastic elongations

$$q_i' = \tilde{C}_i Q_i', \qquad q_i'' = \tilde{C}_i Q_i'', \tag{1.62}$$

and the total elastic elongation of bar i is

$$q_i^{(e)} = q_i' + q_i''. \tag{1.63}$$

Note that the elongations q_i' and

$$q_i = q_i' + q_i'' + q_i^{(p)} \tag{1.64}$$

are kinematically admissible in the sense that they may be derived from joint displacements $\mathbf{p}_\alpha'$ and $\mathbf{p}_\alpha$ that satisfy the kinematic constraints at the supports of the truss. It follows that the elongations q_i'' or $q_i^{(p)}$, however, need not be kinematically admissible.

Instead of specifying the precise variations of the loads with time, let us specify a finite number n of *basic loadings* $\mathbf{P}_\alpha^{(1)}, \mathbf{P}_\alpha^{(2)}, \ldots, \mathbf{P}_\alpha^{(n)}$ at the typical joint α and restrict the discussion to loading programs of the form

$$\mathbf{P}_\alpha(t) = \lambda_1(t)\mathbf{P}_\alpha^{(1)} + \lambda_2(t)\mathbf{P}_\alpha^{(2)} + \cdots + \lambda_n(t)\mathbf{P}_\alpha^{(n)}, \tag{1.65}$$

where the sum of the *nonnegative* coefficients $\lambda_1(t), \lambda_2(t), \ldots, \lambda_n(t)$ does not exceed unity:

$$\lambda_j(t) \geq 0, \quad \sum_j \lambda_j(t) \leq 1 \qquad (j = 1, 2, \ldots, n). \tag{1.66}$$

Loading programs of this kind will be called *admissible* for the given basic loadings.

Depending on the basic loadings, the mechanical energy dissipated in plastic deformation during an indefinitely continued admissible program of loading may be finite or infinite. In the first case the truss *shakes down*; that is, it eventually reaches a state in which it responds in a purely elastic manner to any admissible continuation of the loading program. In the second case, however, successive plastic deformations of a bar may have constant or alternating signs and are thus likely to change the shape of the truss to an unacceptable extent or cause low-cycle fatigue. It is therefore important to establish a criterion for the given basic loadings that assures shakedown of the truss under any loading program that is admissible for these loadings. The following theorem contains a criterion of this kind.

Shakedown Theorem. The truss will shake down if there exists a *fixed* state of self-stress $\bar{Q}_i''$ such that superposition of this state and the purely elastic response $Q_i'(t)$ to any admissible program of loading will not, in any bar, produce a force $\bar{Q}_i(t)$ that violates the condition

$$-Y_i < \bar{Q}_i(t) = Q_i'(t) + \bar{Q}_i'' < Y_i \tag{1.67}$$

at any time.

To prove this theorem we compare the fixed state of self-stress $\bar{Q}_i''$ with the varying state of self-stress $Q_i''(t)$ in Eq. (1.61) and consider the fictitious complementary energy

$$\Psi(t) = \tfrac{1}{2} \sum_i \tilde{C}_i \{Q_i''(t) - \bar{Q}_i''\}^2. \tag{1.68}$$

Its time rate of change is

$$\dot{\Psi} = \sum_i \tilde{C}_i \{Q_i'' - \bar{Q}_i''\} \dot{Q}_i'', \tag{1.69}$$

where

$$\tilde{C}_i \dot{Q}_i'' = \dot{q}_i'' = \dot{q}_i - \dot{q}_i' - \dot{q}_i^{(p)} \tag{1.70}$$

by Eqs. (1.62) and (1.64). Now, $Q_i'' - \bar{Q}_i''$ is a state of self-stress, and the elongation rates $\dot{q}_i$ and $\dot{q}'_i$ are kinematically admissible. According to the principle of virtual work, we thus have

$$\sum_i (Q''_i - \bar{Q}''_i)(\dot{q}_i - \dot{q}'_i) = 0. \tag{1.71}$$

Moreover,

$$Q''_i - \bar{Q}''_i = Q'_i + Q''_i - (Q'_i + \bar{Q}''_i) = Q_i - \bar{Q}_i \tag{1.72}$$

by Eq. (1.61) and Eq. (1.67). Using Eqs. (1.71) and (1.72) in Eq. (1.69) furnishes

$$\dot{\Psi} = -\sum_i (Q_i - \bar{Q}_i)\dot{q}_i^{(p)}. \tag{1.73}$$

Whenever bar i deforms plastically, $Q_i\dot{q}_i^{(p)} = Y_i|\dot{q}_i^{(p)}|$. On the other hand, $\bar{Q}_i\dot{q}_i^{(p)} < Y_i|\dot{q}_i^{(p)}|$ on account of Eq. (1.67). Relation (1.73) thus shows that the fictitious complementary energy of Eq. (1.68) decreases whenever bars of the truss deform plastically. Since this energy is positive definite, however, it cannot decrease indefinitely; that is, plastic deformation must eventually cease. Since it may cease before Ψ has been reduced to zero, the state of self-stress Q''_i to which the truss shakes down may differ from the test case of self-stress Q''_i.

For the application of the shakedown theorem, the following remark is important. If to a test state of self-stress $\bar{Q}''_i$ there correspond bar forces $\bar{Q}_i^{(j)}$ $(j = 1, 2, \ldots, n)$ for each of the given basic loadings satisfying the continued inequality

$$-Y_i < \bar{Q}_i^{(j)} < Y_i, \tag{1.74}$$

the condition of Eq. (1.67) will be satisfied for any admissible program of loading. This follows from Eq. (1.66) and the equality

$$\bar{Q}_i = \sum_j \lambda_j \bar{Q}_i^{(j)}. \tag{1.75}$$

1.4.2. Other Structures

In a given state Q_i, q_i the typical bar of an elastic, perfectly plastic truss responds in either a purely elastic or a purely plastic manner to a given time rate $\dot{Q}_i$ of the axial force. This kind of behavior is obviously essential for the shakedown theorem established in Section 1.4.1. To show that a more general structural element consisting of an elastic, perfectly plastic material does not necessarily behave in this manner, consider a beam of

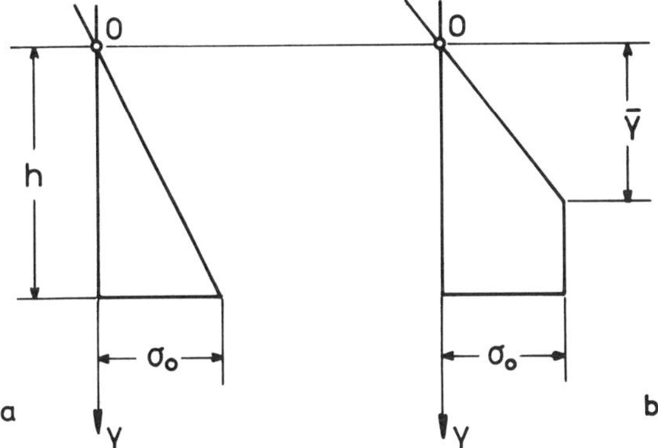

Figure 1.6. (a) Stress distribution at maximum elastic bending moment. (b) Elastic–plastic stress distribution.

rectangular cross section with height $2h$ and breadth b. Figure 1.6 shows typical stress distributions over the lower half of the height. As the bending moment M increases starting from zero, the stress is at first proportional to the distance y from the median layer of the beam. When the stress in the bottom layer equals the yield stress σ_0 (Fig. 1.6a), the bending moment M and the curvature κ have the values

$$M_0 = 2\sigma_0 bh^2/3, \qquad \kappa_0 = \sigma_0/Eh, \tag{1.76}$$

where E is Young's modulus. For greater values of the bending moment, the stress distribution has the form shown in Fig. 1.6b, and

$$M_0 = \tfrac{1}{3}\sigma_0 b(3h^2 - \bar{y}^2) = \tfrac{1}{3}\sigma_0 b[3h^2 - (\sigma_0/E\kappa)^2] \tag{1.77}$$

since $E\kappa\bar{y} = \sigma_0$. Figure 1.7 shows M/M_0 as function of κ/κ_0. The approach to the yield moment $Y = 3M_0/2$ is asymptotic. When the beam is in a state

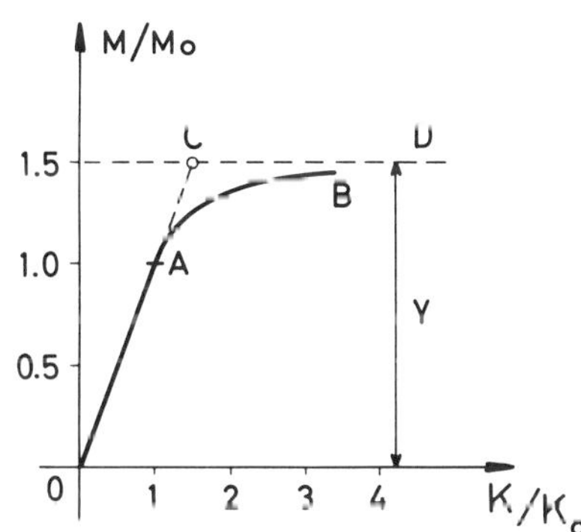

Figure 1.7. Moment versus curvature relation for perfectly plastic rectangular cross section.

represented by a point of the curve AB in Fig. 1.7, its response to an infinitesimal increase of the bending moment has an elastic, as well as a plastic, component, and the replacement of the actual diagram of M/M_0 versus κ/κ_0 by the line OCD in Fig. 1.7 would represent a rather strong idealization.

Consider next a rectangular sandwich section of total height $2h$ and breadth b. If the common thickness of the cover plates is denoted by t, the stress in the bottom layer equals the yield stress σ_0, when the bending moment M and the curvature κ have the values

$$M_0 = \frac{2b\sigma_0}{h}\int_{h-t}^{h} y^2\,dy = 2\sigma_0 bh^2\frac{t}{h} - \frac{t^2}{h^2} + \frac{t^3}{3h^3},$$
$$\kappa_0 = \frac{\sigma_0}{Eh}. \tag{1.78}$$

The yield moment

$$Y = 2b\sigma_0\int_{h-t}^{h} y\,dy = \frac{\sigma_0 bh^2(2 - t/h)t}{h} \tag{1.79a}$$

is reached when the stress in the top layer of the lower cover plate reaches the value σ_0; that is, when the curvature equals

$$\kappa_Y = \frac{\sigma_0}{E(h-t)} = \frac{\kappa_0}{1 - t/h}. \tag{1.79b}$$

For $t/h = 0.1$, for instance, the transition from fully elastic to fully plastic behavior takes place while the bending moment grows from M_0 to $Y = 1.06\ M_0$ and the curvature increases from κ_0 to $\kappa_Y = 1.11\ \kappa_0$. The replacement of the actual diagram of M versus κ by a straight segment through the origin and a parallel to the κ-axis thus appears legitimate for sufficiently small t/h. The actual elastic, perfectly plastic behavior of sandwich plates and shells can be idealized in an analogous manner.

For the sake of brevity the following discussion will be restricted to sandwich structures with cover plates or sheets that are very thin in comparison to the core. [For a more complete discussion the reader is referred to a paper by König (1966).] Except for the obvious changes of terms, the shakedown theorem of Section 1.4.1 applies to sandwich structures of this kind. To illustrate its application, consider a propped cantilever beam of the constant yield moment Y that is subject to the basic loadings shown in Fig. 1.8(a)–(c). Because extreme values of the bending moments can only

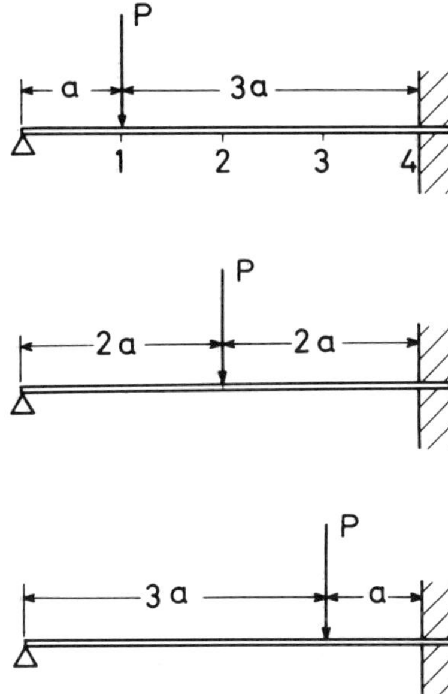

Figure 1.8. Propped cantilever: alternative repeated loadings.

occur at the cross sections marked 1 to 4, we need only consider the bending moments at these sections. In the following, we shall render the bending moments and the yield moment dimensionless by dividing them by *Pa*. Let $Q_i'^{(j)}$ be the dimensionless bending moment that the *j*th loading would produce at section *i* if the beam behaved in a purely elastic manner. The first three lines of Table 1.1 give these *elastic* bending moments. Since the load-free state must also be considered, the extreme nonnegative and nonpositive values $Q_i'^{+}$ and $Q_i'^{-}$ of the dimensionless elastic bending moments are given in the next two lines of Table 1.1, and the last line, in which *c* denotes an as yet unknown constant, gives the dimensionless bending moments $\bar{Q}_i''$ of the test state of self-stress stipulated by the shakedown

Table 1.1. Dimensionless Bending Moments

i	1	2	3	4
$Q_i'^{(1)}$	0.6328	0.2656	−0.1016	−0.4658
$Q_i'^{(2)}$	0.3125	0.6250	−0.0625	−0.7500
$Q_i'^{(3)}$	0.0895	0.1719	0.2578	−0.6563
$Q_i'^{+}$	0.6328	0.6250	0.2578	0
$Q_i'^{-}$	0	0	−0.1016	0.7500
$\bar{Q}_i''$	c	2c	3c	4c

theorem, which requires that the inequalities

$$Q_i'^{+} + \bar{Q}_i'' < Y/Pa \tag{1.80a}$$

and

$$Q_i'^{-} + \bar{Q}_i'' > -Y/Pa \tag{1.80b}$$

be satisfied for $i = 1, 2, 3, 4$. To determine a value P_0 such that the beam will shake down for $P < P_0$, we change these strong inequalities to weak ones and replace P by P_0. Since the inequalities contain only c and P_0 as unknowns, only two of them will, in general, be satisfied as equations, while the others must be satisfied as strong inequalities. It turns out that Eq. (1.80a) for $i = 2$ and Eq. (1.80b) for $i = 4$ are satisfied as equations. Accordingly,

$$0.6250 + 2c = Y/P_0a, \qquad 0.7500 - 4c = Y/P_0a, \tag{1.81}$$

and hence $P_0 = 3Y/a$ and $c = 1/48$. We leave it to the reader to check that, with these values of P_0 and c, the remaining inequalities are satisfied as strong inequalities. The beam will shake down under any program of loading that is admissible for the considered basic loading, provided $P < 3Y/a$.

1.4.3. Historical Remarks

The shakedown theorem was established by Bleich (1932) for simply indeterminate structures and by Melan (1936) for structures of any degree of indeterminacy and for the elastic, perfectly plastic continuum (1938). Whereas Bleich and Melan only considered failure by alternating plastic deformation, Horne (1950) pointed out the possibility of failure by progressive plastic deformation. Prager (1957) extended the shakedown theorem to combined cycles of loading and temperature. A kinematic counterpart of the static shakedown theorem of Melan was given by Koiter (1956).

When only nonnegative multiples $\lambda(t)P_\alpha$ of a single basic loading P_α are to be considered, the shakedown theorem is closely related to the static theorem of limit analysis. The former furnishes a load factor λ^* such that the mechanical energy dissipated in plastic deformation remains finite when $\lambda(t)$ oscillates between zero and $\lambda^* - \varepsilon$, where ε is an arbitrarily small positive number. The static theorem of limit analysis, on the other hand, furnishes a load factor λ^{**} such that the dissipated energy remains finite when $\lambda(t)$ increases monotonically from zero to $\lambda^{**} - \varepsilon$. If a subsequent reduction of the load factor to zero does not cause any plastic deformation, λ^* and λ^{**} have the same value. In this sense Melan (1936) may be said to have given the first proof of the static theorem of limit analysis.

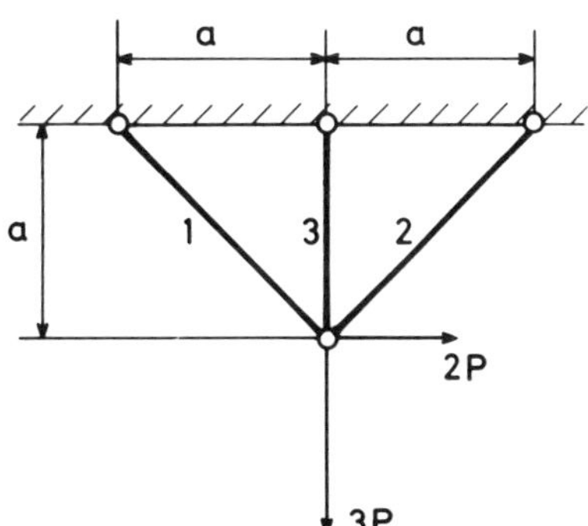

Figure 1.9. Three-bar truss: alternative repeated loadings.

Before the theory of elasticity was developed in the nineteenth century, methods, which may be regarded as precursors of limit analysis, were used to assess the safety of arches and retaining walls. For a discussion of these methods and the development leading to the fundamental theorems of limit analysis, see Prager (1974). The first direct proofs of these theorems were given by Gvozdev (1938) in a paper that appeared in the proceedings of a conference on plastic deformation. This Russian paper, which appeared on the eve of World War II, only became known to Western researchers after the fundamental theorems of limit analysis had been derived once more by Greenberg and Prager (1949) and Drucker *et al.* (1952).

Exercise 1.10. The bars of the truss in Fig. 1.9 have the same cross section. The basic loadings are (1) a vertical downward load $3P$, and (2) a horizontal rightward load $2P$. Determine the smallest value P_0 such that the truss will shake down if $P < P_0$.

2

Problem Formulation and Optimality Criteria

2.1. Terminology

The following concepts prove useful in the formulation of problems of optimal design.

2.1.1. Structural System

The same task may be fulfilled by structural systems of different kinds. For example, a ceiling may be constructed as a solid plate or as a thin plate carried by a grillage of beams. In principle, the optimal choice of a structural system is part of structural optimization. Frequently, however, this choice is dictated by other considerations; for example, by the fact that the designer has considerable experience with one kind of system and lacks comparable experience with other kinds. Moreover, the optimal choice of a structural system does not seem to be accessible to analytical treatment. Besides calling on his experience, the designer can do little more than optimize designs based on different systems and compare the results. In the vast majority of problems of structural optimization discussed in the literature, the structural system is therefore assumed to be given, and we shall make the same assumption throughout this book.

2.1.2. Layout

The term *layout of a structure* denotes the arrangement of the centerlines or median surfaces of its members, the types and locations of its supports, and the boundaries of segments or regions where certain cross-sectional dimensions or material properties will be constant or vary in a prescribed way—for instance, linearly. Note that dimensions, such as the cross-sectional areas of the bars of a truss or the thickness of a plate, are not specified by the layout of the structure.

2.1.3. Geometrical Constraints

Geometrical constraints concern the layout of the structure and may be more or less severe. At one end of the range we have the completely specified layout; at the other end, the layout that is left to the choice of the designer, except, perhaps, for symmetry constraints or for constraints on the space available for the structure. For example, for a continuous beam on four supports the positions of all supports may be prescribed, or the choice of the positions of the intermediate supports may be left to the designer, subject, perhaps, to the symmetry constraint that the two outer spans should be equal. Similarly, a plate may need to have a circular hole of a given diameter, the center of which may be specified or chosen anywhere within a given region.

2.1.4. Technological Constraints

Technological constraints include parameters such as the cross-sectional area of a rod or the bending stiffness of a beam. They aim at avoiding designs that are difficult to execute or that violate the basic assumptions of the theory used in optimization. Consider, for example, a rotationally symmetric circular plate with ringwise constant thickness. In the absence of technological constraints, the optimal thickness may turn out to be exceedingly small in some rings and far too great in comparison to the diameter of the plate in other rings. To avoid the resulting violations of the basic assumptions of plate theory, we may prescribe lower and upper bounds for the plate thickness.

2.1.5. Behavioral Constraints

Behavioral constraints specify the desired performance characteristics of the structure. Typical behavioral constraints are a *lower bound* on the buckling load of a rod or the load factor at plastic collapse of a beam, or an *upper bound* on the greatest deflection of a plate.

2.1.6. Purpose

A structure that is subject to *only one loading* and *only one behavioral constraint* is called a *single-purpose structure.* All other structures are classified as *multiple-purpose structures.* Structures of the latter kind are, of course, the rule, and single-purpose structures are the exception. In the development of the theory of structural optimization, however, single-purpose structures play an important role.

2.1.7. Design Variables and Functions

The complete specification of a structure involves parameters with values fixed in advance or chosen by the designer. The latter are called *design variables.* For example, the complete specification of a rod that is to consist of cylindrical segments involves the length of the rod and the positions and diameters of the interfaces of the segments. The length of the rod is determined by its function and is therefore not a design variable. The diameters, on the other hand, are obvious design variables, but the positions of the interfaces may or may not be. If we allow the number of segments to increase indefinitely, we no longer have a finite number of design variables, but a *design function,* which specifies the variation of the rod diameter with the distance measured along the rod.

2.1.8. Design Objective

The design objective specifies the sense in which the design is to be optimal. Minimizations of the weight or cost of a structure are frequently encountered design objectives. Consider, for example, the minimization of the volume of material needed for a beam subject to certain behavioral constraints. The beam is to have a rectangular cross section, the height and breadth of which will be denoted by $H(x)$ and $B(x)$, x being the distance measured along the beam. The design objective, as in this example, usually is the minimization of an integral Γ of a quantity γ [here $B(x)H(x)$] over the extent of the structure (here the length of the beam). For the sake of uniformity of expression, we shall call Γ the *cost* of the structure, even if it is not expressed in dollars and cents. Similarly, γ will be called the *specific cost.*

2.2. Problem Types

With the terminology introduced above, the typical problem of structural optimization may be stated as follows: minimize the cost of a structure that is subject to given behavioral constraints and, possibly, also to geometrical and technological constraints.

Depending on the nature of the data and the behavioral constraints, problems of structural optimization may be classified as *deterministic or probabilistic.* For example, the intensity P of a load may be specified outright or by a probability distribution $f(P^*)$, which indicates the probability for P to exceed the value P^*. In the first case we may adopt the behavioral constraint that the structure should be on the verge of plastic collapse under the load λP, where λ is a given load factor. In the second case, however, we can only require that, for the given load factor λ, the probability of plastic collapse should not exceed a given value. *Only deterministic problems are treated in this book.*

Problems of optimal structural design may be grouped as follows:

Group A. In a problem of this group the structural system and its layout are specified in advance, and the optimization involves only cross-sectional dimensions.

Group B. In a problem of this group the structural system is specified in advance, but the layout, or perhaps only certain features of it, and the cross-sectional dimensions remain to be determined optimally.

Group C. In a problem of this group the structural system, the layout, and the cross-sectional dimensions are to be chosen optimally. As already indicated in Section 2.1.1, only problems of Groups A and B will be discussed in this book.

2.3. Elementary Problems of Structural Optimization

Some problems of structural optimization are readily solved without appeal to a formal theory. Consider, for example, the design of a single-purpose truss of a given statically determinate layout when the danger of buckling of bars may be disregarded and the total volume of a specified material needed for the bars is to be minimized, subject to the behavioral constraint that the magnitude of the compressive or tensile axial stresses should in no bar exceed the given values σ_c or σ_t. Because the truss is statically determinate, the axial forces in the bars do not depend on the values of the design variables—the cross-sectional areas of the bars. After the bar forces have been determined from equilibrium conditions, the cross-sectional area of each bar is therefore found by dividing its axial force by σ_c or σ_t, depending on whether the bar is stressed in compression or tension. While a design with larger cross-sectional areas would also satisfy the behavioral constraint, it would obviously require a greater amount of structural material and, hence, not be optimal. The optimal statically determinate single-purpose truss is thus a *fully stressed structure*; that is, each of its members is stressed right up to the allowable limit.

Next consider a statically determinate truss of given layout that must carry two or more alternative sets of loads and is subject to the same behavioral constraint as the single-purpose truss just discussed. To arrive at an optimal design we determine the bar forces *for each loading* from the conditions of equilibrium and the cross-sectional areas the bars would need to have, in view of the behavioral constraint, if the truss had to carry only the considered loading. For each bar we finally adopt the largest cross-sectional area obtained for it in this way. The resulting truss is a *fully stressed structure* in the sense that each of its members is stressed right up to the allowable limit by at least one loading. That this truss is optimal may be shown in the manner used above.

2.4. Empirical Optimality Criteria

The traditional approach to structural optimization consists of developing a sufficient number of alternative designs and choosing the best of these. It is only natural that, having found a nearly optimal design in this laborious manner, the designer should look for characteristic features of the chosen design that would have enabled him to obtain this design without exploring numerous alternatives. Consequently, the field of structural optimization abounds with empirical optimality criteria of uncertain validity.

Another source of questionable optimality criteria is the generalization of well-established criteria to problems beyond their proven range of validity. Consider, for example, the minimum-weight design of an elastic truss of given layout for alternative loadings when an upper bound (called *allowable stress*) is set on the absolute value of the stress in any bar. Although we suppose that the truss is statically indeterminate if all bars provided by the given layout are used, the optimal truss may be statically determinate because some of these bars may be omitted.

It is widely believed that under these conditions the truss of minimum weight is *statically indeterminate* and *fully stressed* (i.e., in each bar the allowable stress is reached under at least one loading). To test the validity of this belief, Chern and Prager (1972) investigated the minimum-weight design of the truss in Fig. 2.1 for the alternative loads P' and P'', where P' is horizontal and of unit magnitude and P'' has the horizontal and vertical components H and V. Figure 2.2 shows the findings of these authors. The optimal truss is statically indeterminate only if the *load point*, with coordinates H and V, lies in one of the shaded regions. The optimal truss is statically indeterminate and fully stressed only if the load point lies in one of the shaded regions marked F. If all positions of the load point in the square of Fig. 2.2 are equally likely, this means that, for $0 \leq H \leq 3$ and $0 \leq V \leq 3$, the optimal truss is statically indeterminate in only 29.2% of the cases and indeterminate and fully stressed in only 7.2% of the cases. For load points

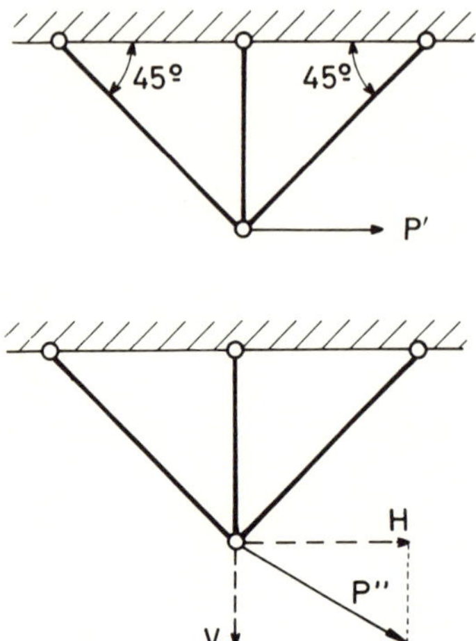

Figure 2.1. Three-bar elastic truss subjected to alternative loadings P' and P''.

above the shaded regions, the vertical bar is missing in the optimal truss; for load points below the shaded regions, the right-hand bar is missing. These two possibilities account for 43.0% and 27.8% of the cases.

It would, however, be premature to conclude from these results that the fully stressed three-bar truss is a poor design when it is not optimal. The numbers in Fig. 2.3 show the structural efficiency of the fully stressed three-bar truss for values of H and V that are integer multiples of 0.25. (The efficiency of a structure is customarily expressed in percentage and indicates the quotient of the absolute minimum of structural weight by the weight of the considered structure.) Fully stressed three-bar trusses exist

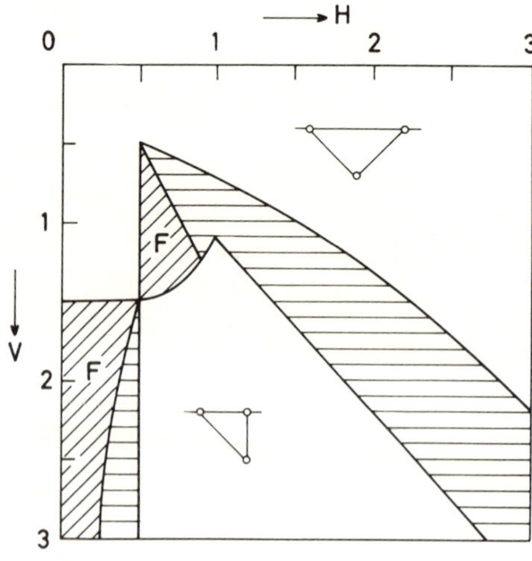

Figure 2.2. Domains of solution types: shaded zones for three-bar truss, F for fully stressed three-bar truss.

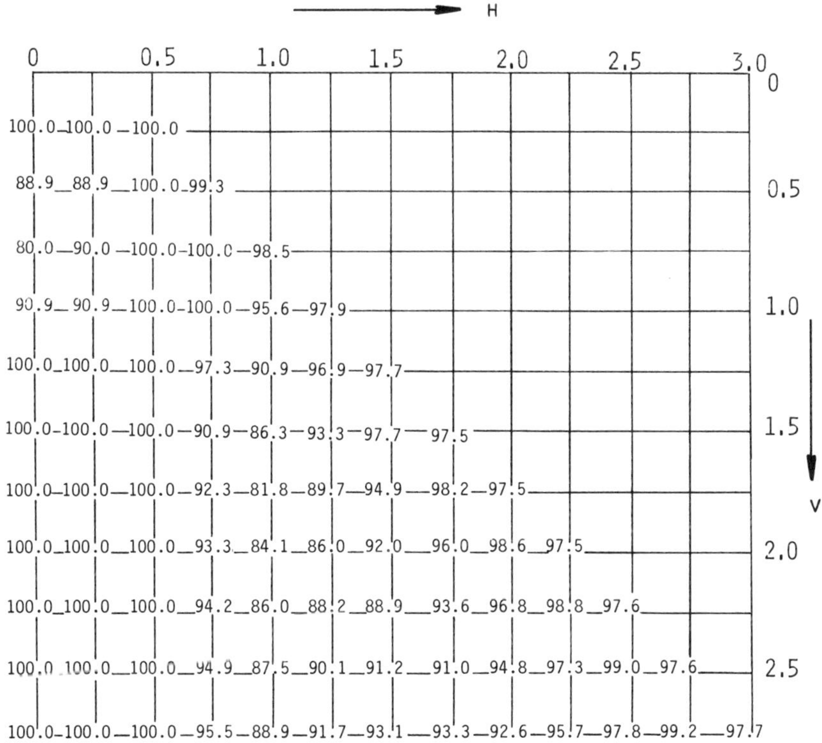

Figure 2.3. Efficiency of fully stressed three-bar truss.

only for $V \geq 0.5$ and $H \leq V$. For $H < 0.5$ the vertical bar is fully stressed by the load P'', and for $H > 0.5$ the right-hand bar is fully stressed by the load P'. In each case the two other bars are fully stressed by the other load. For $H = 0.5$ the two oblique bars are fully stressed by the load P'; the left-hand and vertical bars, by the load P''. For 88 of the combinations of H and V considered in Fig. 2.3, a fully stressed three-bar truss does exist. Its efficiency exceeds 95% for 54 of these combinations and drops as low as 80% for only one combination.

2.5. Ill-Posed Problems

Structural optimization is a field in which apparently reasonable problems often prove to be ill posed. The following examples illustrate this statement. The beams of the first two examples are to consist of a given elastic material and to have rectangular cross sections of given constant

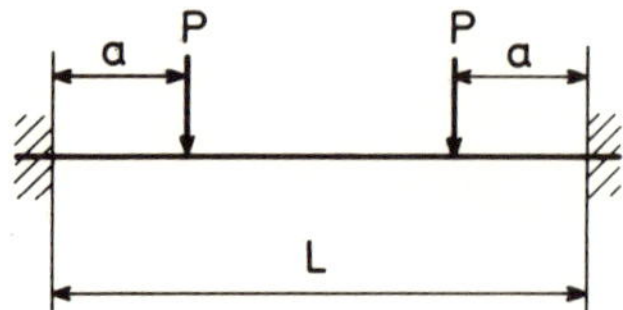

Figure 2.4. Doubly built-in beam.

height and continuously varying width. In both cases the design objective is the minimization of the beam volume.

Example A. The beam in Fig. 2.4 is subject to the behavioral constraint that the deflections of the loaded sections should not exceed a given value. If the length a is sufficiently small in comparison to the span L, the optimal design consists of two short cantilevers, at whose tips the loads are acting. However, the gap between the tips of these cantilevers may not be acceptable. To exclude the unacceptable solution we must improve the formulation of the problem by adding a technological constraint that sets a positive lower bound on the width of the beam.

Example B. The propped cantilever in Fig. 2.5 is loaded by a couple C at the simply supported end and by a vertical force P near the built-in end. It is subject to the behavioral constraint that the deflection at mid-span should not exceed a given absolute value. As Barnett (1961) has pointed out, the volume of a design satisfying this constraint can be made arbitrarily small. Indeed, in view of the fact that by themselves C and P produce mid-span deflections of opposite signs, there exist designs with vanishing deflection at mid-span. A design of this kind obviously satisfies the behavioral constraint. If its width is everywhere reduced in the same proportion, the resulting beam of smaller volume still satisfies the behavioral constraint because its mid-span deflection vanishes. Since the limit of this process of volume reduction—that is, a beam of zero volume—will not be acceptable, there is no minimum volume design for this problem. To avoid this difficulty we may set an upper bound on the absolute value of the deflection at any point, rather than just the absolute value of the mid-span deflection.

Example C. This example concerns the plastic limit design of an annular circular sandwich plate, the outer radius of which is twice the inner

Figure 2.5. Optimal propped cantilever of vanishing volume for assigned mid-span elastic deflection.

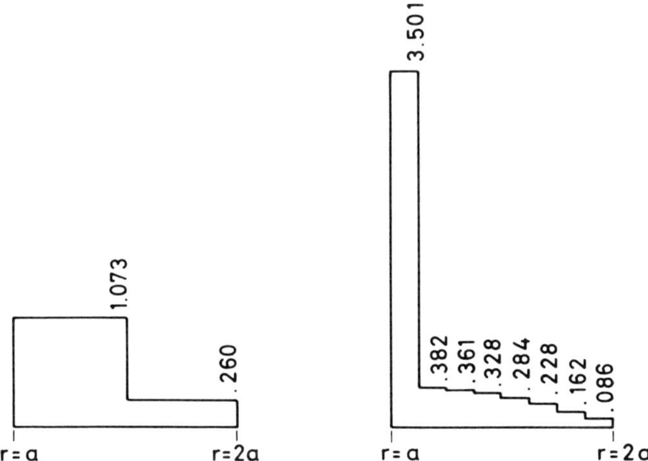

Figure 2.6. Optimal designs of an annular plate: two-ring and eight-ring solutions.

radius. The plate is simply supported at the outer edge and free at the inner edge. Its core has a given constant thickness H. The cover sheets consist of a material with tensile and compressive yield stresses $\pm\sigma_0$, and their continuously varying thickness T is to be determined in such a manner that plastic collapse occurs at a given value p of the uniformly distributed load, while the plate is as light as possible. Sheu and Prager (1969) investigated the variation of the optimal thickness T when it is required to be constant over each of n rings of equal width. Figure 2.6, which shows the variation of $\sigma_0 HT/pa^2$ for $n = 2$ and $n = 8$, suggests that the thickness at the inner edge increases rapidly with n and may become infinite as $n \to \infty$, that is, for continuously varying thickness. This edge effect was first pointed out by Megarefs (1968). The corresponding optimal design is not acceptable because it violates the basic assumption that $T \ll H$, but it indicates that in formulating the problem we overlooked the possibility that the optimal design might have a reinforcing beam along the inner edge.

3

Optimal Design of Beams and Frames

3.1. Optimality Condition for Prescribed Load Factor at Plastic Collapse

We treat the optimal plastic design of a beam carrying a distributed transverse load $p(x)$ along its span, where x denotes the position of a typical cross section. A design for the beam is the prescription of a yield-moment (fully plastic moment) distribution along the beam. Piecewise constant designs will be considered first; continuous designs will be discussed at the end of this section. The following assumptions are made concerning layout, constraints, and specific cost.

3.1.1. Layout

The beam span is divided into subspans or segments by specifying the positions of supports and sections where the yield moment may change value. Denote these positions by x_i, $i = 0, 1, 2, \ldots, N$; the length of the ith segment is $l_i = x_i - x_{i-1}$, $i = 1, 2, \ldots, N$. The type of each support is given; this constrains the class of kinematically admissible deflection rates. A design for the beam is specified by giving a piecewise constant yield-moment distribution with value Y_i, $i = 1, 2, \ldots, N$, in the ith segment.

3.1.2. Technological Constraint

Nonnegative bounds Y_i^- and Y_i^+ are prescribed for the yield moment of the typical segment:

$$0 \leq Y_i^- \leq Y_i \leq Y_i^+ \qquad (3.1)$$

3.1.3. Behavioral Constraint

The beam is to be on the verge of plastic collapse under the load $\Lambda p(x)$, where $\Lambda > 0$ is a given load factor.

3.1.4. Specific Cost

The specific cost (for definition see Section 2.1.8) is assumed to be a continuously differentiable, monotonically increasing, *convex* function $\gamma(Y)$ of the yield moment Y. If Y^* and Y are any two values of the yield moment from the interval $\{Y^-, Y^+\}$, it follows from the convexity of $\gamma(Y)$ that

$$\gamma(Y^*) - \gamma(Y) \geq (Y^* - Y)\gamma_Y(Y), \tag{3.2}$$

where the subscript Y indicates differentiation with respect to Y, and $\gamma_Y(Y) \geq 0$ (see Fig. 3.1). Note that increase of the yield moment from Y to $Y + dY$ entails an increase, $\gamma_Y\, dY$, of the specific cost. The derivative γ_Y will therefore be called the *marginal specific cost.*

For the beam let $v(x)$ be a kinematically admissible rate of deflection that has been normalized for the load $p(x)$. Thus,

$$\int p(x)v(x)\, dx = 1, \tag{3.3}$$

where the integration is extended over the entire length of the beam. While $v(x)$ is continuous, its first derivative $v'(x)$ is discontinuous at yield hinges. If there is a yield hinge at the interface of two segments, say at $x = x_i$, we must state whether we envision this hinge to be at $x = x_i - 0$ or $x = x_i + 0$ in order to avoid ambiguities in the evaluation of the power of dissipation.

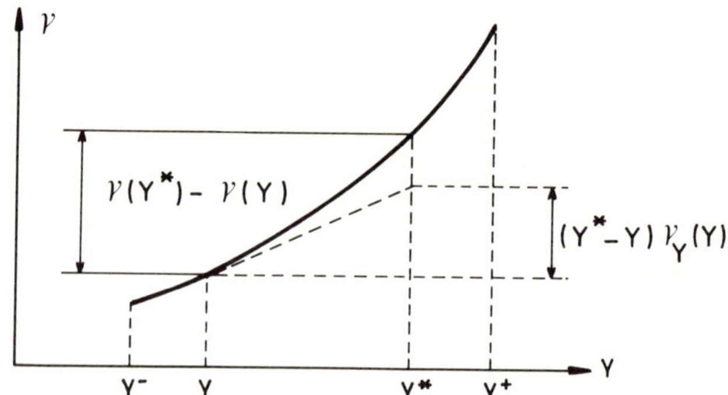

Figure 3.1. Convex specific cost function.

Statements of this kind form an essential part of the specification of the kinematically admissible rate of deflection $v(x)$.

For the $v(x)$ considered we write the power dissipated in the ith segment in the form

$$D_i = Y_i k_i[v(x)], \tag{3.4}$$

where k_i is a functional of $v(x)$ that represents the value of this power per unit yield moment. Note that, from its physical meaning, k_i must be nonnegative. If the ith segment is free from yield hinges,

$$k_i = \int_{x_{i-1}}^{x_i} |v''(x)|\, dx. \tag{3.5a}$$

If, on the other hand, this segment contains yield hinges with the relative angular velocities $\omega_{i\alpha}$ at $x = x_{i\alpha}$ $(\alpha = 1, 2, \ldots)$, then

$$k_i = \int_{x_{i-1}}^{x_i} |v''(x)|\, dx + \sum_{\alpha} |\omega_{i\alpha}|, \tag{3.5b}$$

where the summation includes all yield hinges of the ith segment, and the integral is to be interpreted as the sum of the integrals over the intervals into which these hinges divide the segment.

Because $v(x)$ has been normalized for the load $p(x)$, the load factor for plastic collapse of the design Y_i (that is, the design specified by the yield moments Y_i, $i = 1, 2, \ldots$) is given by

$$\Lambda = \min_{v(x)} \sum_i Y_i k_i[v(x)], \tag{3.6}$$

where the minimum is taken over all normalized, kinematically admissible rates of deflection, and the sum includes all segments of the beam. If the load factor Λ^* of the design Y_i^* has *at least* the value Λ, we have

$$\Lambda \leq \Lambda^* = \min_{v^*(x)} \sum_i Y_i^* k_i[v^*(x)] \leq \sum_i Y_i^* k_i[v(x)], \tag{3.7}$$

because the normalized, kinematically admissible rates of deflection $v^*(x)$ include $v(x)$.

For $v(x)$ we now choose the normalized rate of deflection of a collapse mechanism of the design Y_i under the load $\Lambda p(x)$. The symbol $\min_{v(x)}$ in Eq. (3.6) may then be omitted, and comparison of the resulting equation with Eq. (3.7) furnishes

$$\sum_i (Y_i^* - Y_i) k_i[v(x)] \geq 0. \tag{3.8}$$

We now use this inequality to show that the cost $\Gamma^* = \Sigma_i Y_i^* l_i$ of the design Y_i^* cannot be less than the cost $\Gamma = \Sigma_i Y_i l_i$ of the design Y_i if the normalized rate of deflection $v(x)$ of the considered collapse mechanism of the latter design under the load $\Lambda p(s)$ satisfies the following relations for all i:

$$k_i[v(x)] \left\{ \begin{matrix} \leq \\ = \\ \geq \end{matrix} \right\} c^2 l_i \gamma_Y(Y_i) \quad \text{if} \left\{ \begin{matrix} Y = Y^-, \\ Y^- < Y < Y^+, \\ Y = Y^+, \end{matrix} \right. \tag{3.9}$$

where $c^2 > 0$ has the same value for all segments. Indeed, the inequality of Eq. (3.8) may be written as

$$0 \leq c^2 \sum_i (Y_i^* - Y_i) l_i \gamma_Y(Y_i) - \sum_i (Y_i^* - Y_i)\{c^2 l_i \gamma_Y(Y_i) - k_i[v(x)]\}. \tag{3.10}$$

Using the convexity inequality of Eq. (3.2) and the definitions of the costs Γ^* and Γ in Eq. (3.10) yields

$$0 \leq c^2(\Gamma^* - \Gamma) - \sum_i (Y_i^* - Y_i)\{c^2 l_i \gamma_Y(Y_i) - k_i[v(x)]\}. \tag{3.11}$$

If the conditions of Eq. (3.9) are satisfied, the ith segment does not furnish a contribution to the sum in Eq. (3.11) if $Y_i^- < Y_i < Y_i^+$. On the other hand, if $Y_i = Y_i^-$, the difference $Y_i^* - Y_i$, the bracketed term in Eq. (3.11), and the contribution of the ith segment to the sum in Eq. (3.11) are nonnegative. Finally, if $Y_i = Y_i^+$, the difference $Y_i^* - Y_i$ and the bracketed term are nonpositive, and the contribution of segment i to the sum in Eq. (3.11) is nonnegative. It follows from these remarks that Eq. (3.11) is equivalent to $\Gamma^* - \Gamma \geq 0$. Since we did not have to assume that Y_i^* and Y_i were neighboring designs, the conditions of Eq. (3.9) are seen to be *sufficient* for the *global* optimality of Y_i.

To prove that the conditions of Eq. (3.9) are also *necessary* for the optimality of the design Y_i with normalized collapse mechanism $v(x)$ under the load $\Lambda p(x)$, we now consider a *neighboring* design $Y_i + \delta Y_i$, whose normalized collapse mechanism under $\Lambda p(s)$ is denoted by $v(x) + \delta v(x)$. The powers dissipated per unit yield moment in the ith segment of the two designs are denoted by k_i and $k_i + \delta k_i$. Because the load factor for plastic collapse is to have the same value Λ for the two designs, we have

$$\Lambda = \sum_i (Y_i + \delta Y_i)(k_i + \delta k_i)$$

or, to within higher-order terms,

$$\Lambda = \sum_i Y_i(k_i + \delta k_i) + \sum_i k_i \delta Y_i. \tag{3.12}$$

The typical term in the first sum in Eq. (3.12) represents the power dissipated in the ith segment of the design Y_i for a normalized, kinematically admissible rate of deflection that is near the normalized, kinematically admissible rate of deflection of the collapse mechanism of this design. On account of the minimum property, Eq. (3.6), of the load factor, the first sum in Eq. (3.12) thus has the value Λ, and Eq. (3.12) reduces to

$$\sum_i k_i \delta Y_i = 0. \tag{3.13}$$

We now assume that the design Y_i corresponds to a *local* minimum of cost. Because of the inequality constraints in Eq. (3.1), this does not imply that the difference $\delta\Gamma$ of the costs of designs $Y_i + \delta Y_i$ and Y_i vanishes, but only that

$$\delta\Gamma \equiv \sum_i l_i \gamma(Y_i) \delta Y_i \geq 0. \tag{3.14}$$

The changes δY_i of the yield moments must satisfy Eq. (3.13) and the inequality of Eq. (3.14). Using a positive Lagrange multiplier c^2 on the latter, we may combine them into the inequality

$$\sum_i [c^2 l_i \gamma(Y_i) - k_i] \delta Y_i \geq 0, \tag{3.15}$$

in which the changes δY_i may be regarded as *independent* variables.

Depending on whether Y_i has a value at the lower bound Y_i^-, between Y_i^- and Y_i^+, or at the upper bound Y_i^+, the change δY_i is nonnegative, or indeterminate sign, or nonpositive, respectively. Accordingly, Eq. (3.15) requires that expression in the braces be nonnegative, zero, or nonpositive, respectively, and these conditions are equivalent to the conditions of Eq. (3.9), which have thus been shown to be *necessary* for *local* optimality. Note that the convexity of the specific cost function $\gamma(Y)$ has been used in the sufficiency proof but not in the necessity proof.

To give a mechanical interpretation of the optimality condition of Eq. (3.9), we note that $\int_{x_{i-1}}^{x_i} \gamma_Y(Y_i)\, dx = l_i \gamma_Y(Y_i)$ is the *marginal cost of the ith segment.* Since $c^2v(x)$ is also a collapse mechanism of the beam, though no longer a normalized one, and since $k_i[c^2v(x)] = c^2k_i[v(x)]$, the optimality condition of Eq. (3.9) states that *the optimal beam admits a collapse mechanism such that for each segment, the power of dissipation per unit yield moment is not greater than, equal to, or smaller than the marginal cost, depending on whether the yield moment of the segment has a value at the lower bound, between bounds, or at the upper bound.*

For the sake of greater clarity we have explicitly taken care of the technological constraint in Eq. (3.1) in our discussion. It is worth noting, however, that this constraint may also be handled by stipulating a γ versus Y diagram of the form shown in Fig. 3.2, according to which nothing is saved by using yield moments smaller than Y^-, while yield moments in excess of Y^+ are ruled out by the vertical rise of the specific cost at $Y = Y^+$. This function $\gamma(Y)$ is convex but not continuously differentiable, as assumed in the preceding discussion. If, however, we use the small intervals $(Y^-, Y^- + \varepsilon)$ and $(Y^+ - \varepsilon, Y^+)$ to round out the corners of the diagram in Fig. 3.2, we may apply the arguments leading to the optimality condition in Eq. (3.9) without having to bother about the bounds on the yield moment. The optimal beam therefore admits a normalized collapse mechanism satisfying $k_i = c^2 l_i \gamma_Y(Y_i)$. Since $\gamma_Y(Y) < \gamma_Y(Y^- + \varepsilon)$ in $(Y^-, Y^- + \varepsilon)$ and $\gamma_Y(Y) > \gamma_Y(Y^+ - \varepsilon)$ in $(Y^+ - \varepsilon, Y^+)$, we again obtain the optimality condition, Eq. (3.9), when we let ε tend to zero. Similarly it can be shown that this condition must be supplemented by

$$c^2\gamma_Y(\bar{Y} - 0) \leq k_i \leq c^2\gamma_Y(\bar{Y} + 0) \tag{3.16}$$

if we allow the function $\gamma_Y(Y)$ to have a discontinuity at $Y = \bar{Y}$.

When the variation of the yield moment along the beam is specified by a continuous function $Y(x)$, a necessary and sufficient condition for

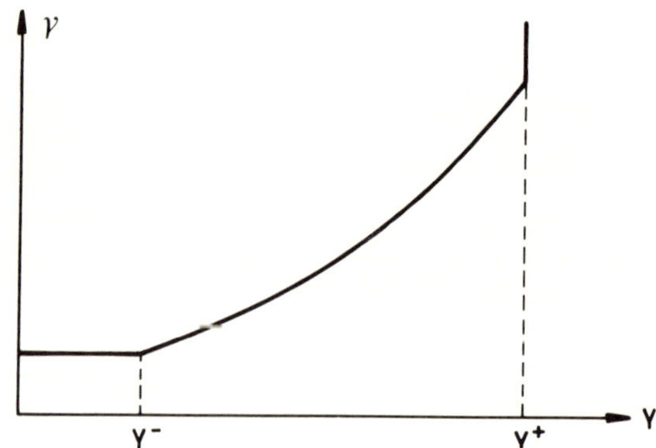

Figure 3.2. Convex specific cost function with bounds on design variable Y.

global optimality may be obtained from Eq. (3.9) by regarding the beam as consisting of infinitely many segments of infinitesimal length dx. Since the power of dissipation in such a segment is $|v''(x)|\,Y(x)\,dx$, the optimal beam admits a normalized collapse mechanism satisfying

$$|v''(x)|\left\{\begin{matrix} \leq \\ = \\ \geq \end{matrix}\right\} c^2\gamma_Y\{Y(x)\} \qquad \text{if} \left\{\begin{matrix} Y = Y^-, \\ Y^- < Y < Y^+, \\ Y = Y^+, \end{matrix}\right. \tag{3.17}$$

Instead of being regarded as a limiting case of a beam with piecewise constant yield moment, the beam with continuously varying yield moment may also be treated directly. Consider the designs $Y(x)$ and $Y^*(x)$ with load factors Λ and $\Lambda^* \geq \Lambda$ for plastic collapse under the given loading. Let $v(x)$ be a collapse mechanism for Y that is normalized according to Eq. (3.3). We have

$$\Lambda = \int Y|v''|\,dx, \tag{3.18}$$

where the integration is extended over the entire beam. Furthermore, it follows from the kinematic theorem of limit analysis that

$$\Lambda^* \leq \int Y^*|v''|\,dx.$$

Accordingly,

$$\int (Y^* - Y)|v''|\,dx \geq 0, \tag{3.19}$$

or

$$c^2 \int (Y^* - Y)\gamma_Y(Y)\,dx - \int (Y^* - Y)(c^2\gamma_Y(Y) - |v''|)\,dx \geq 0. \tag{3.20}$$

In view of the convexity of $\gamma(Y)$, the first term of Eq. (3.20) is not larger than $c^2(\Gamma^* - \Gamma)$. Now, if

$$|v''|\left\{\begin{matrix} \leq \\ = \\ \geq \end{matrix}\right\} c^2\gamma_Y(Y) \qquad \text{if} \left\{\begin{matrix} Y = Y^-, \\ Y^- < Y < Y^+, \\ Y = Y^+, \end{matrix}\right. \tag{3.21}$$

then the second integral in Eq. (3.20) vanishes for $Y^- < Y < Y^+$ and is positive for $Y = Y^-$ or $Y = Y^+$. Accordingly, $\Gamma^* \geq \Gamma$; that is, the design Y with a collapse mechanism satisfying Eq. (3.21) is optimal when a lower bound Λ is prescribed for the load factor at plastic collapse under the given loading.

It is important to note that the function $v(x)$ specifying a collapse mechanism of the optimal design not only must be kinematically admissible and satisfy Eq. (3.21) but also must have continuous first derivative. Indeed, if there is a discontinuity of v' at $x = x_1$, the term $Y(x_1)|v'(x_1+0) - v'(x_1-0)|$ must be added on the right of Eq. (3.18), and Eq. (3.19) must be modified in a similar manner. The reasoning that established the optimality of the design Y can then no longer be applied to this end.

In the following, a kinematically admissible collapse mechanism $v(x)$ with continuous angular velocity $v'(x)$ is called *kinematically admissible for optimality.* Note that this property excludes yield hinges with finite relative angular velocity. In particular, v' must vanish at a built-in end of the beam.

When the specific cost function is $\gamma = \alpha Y$, the cost of the beam may be computed from the integral $\int pv\, dx$, where $v = v(x)$ is a collapse mechanism. Indeed, for the collapse mechanism,

$$\int pv\, dx = \int Y|v''|\, dx.$$

Since the optimality condition requires $|v''| = c^2\alpha$, we have

$$\int pv\, dx = c^2\alpha \int Y\, dx = c^2\Gamma.$$

The collapse mechanism may thus be regarded as an influence function of the loads for the cost.

3.2. Optimal Plastic Design for Given Collapse Load

3.2.1. Linear Cost Function; No Explicit Bounds on Y

As a first example, consider the propped cantilever of Fig. 3.3a, which has a rectangular cross section of constant height h and variable breadth $b(x)$. The beam is made of a rigid, perfectly plastic material with tensile and compressive yield stresses $\pm\sigma_0$, and its load factor for plastic collapse under a given downward loading is Λ. No explicit bounds are set on the yield moment

$$Y(x) = \frac{\sigma_0 h^2 b(x)}{4}, \tag{3.22}$$

which is, however, nonnegative. The volume of the beam is to be minimized. The specific cost thus equals the volume $hb(x)$ per unit length of the beam.

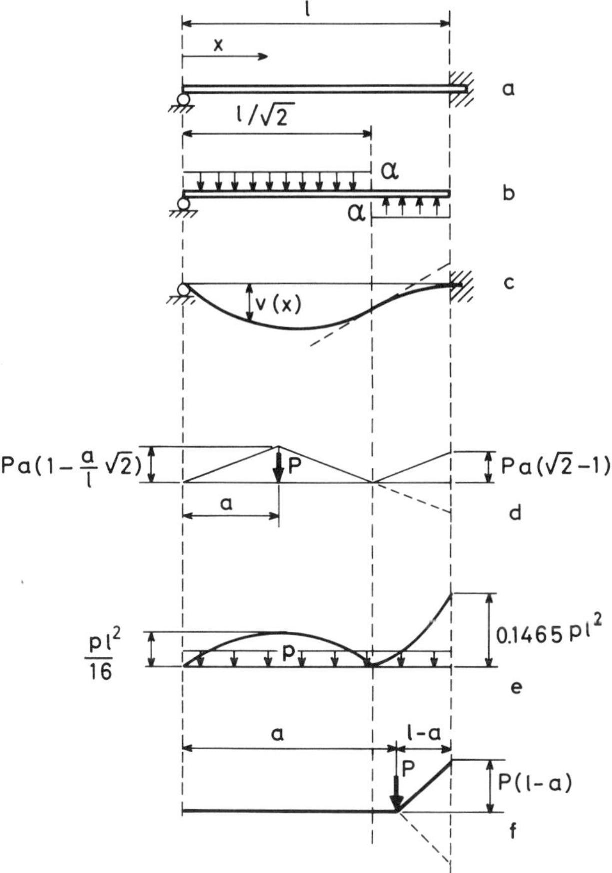

Figure 3.3. (a) Propped cantilever. (b) Curvature distribution on conjugate beam. (c) Collapse mechanism of optimal beam. (d), (e), (f) Examples of optimal designs.

In view of Eq. (3.22) we have

$$\gamma = 4Y/\sigma_0 h, \tag{3.23}$$

and the marginal specific cost γ_Y is a positive constant that will be denoted by α:

$$\alpha = 4/\sigma_0 h > 0. \tag{3.24}$$

The optimality condition, Eq. (3.17), thus reduces to

$$|\kappa(x)| \begin{Bmatrix} \leq \\ = \end{Bmatrix} c^2\alpha \qquad \text{if} \begin{cases} Y(x) = 0, \\ Y(x) > 0, \end{cases} \tag{3.25}$$

where $\kappa(x) = -v''(x)$ is the rate of curvature of the collapse mechanism of the optimal beam. Since a collapse mechanism is only defined to within a constant positive factor, the value of the positive constant $c^2\alpha$ in Eq. (3.25) is irrelevant.

For any downward loading the collapse mechanism of the beam tends to transform the centerline into a curve that, seen from above, is concave near the simply supported end and convex near the built-in end. The optimality condition, Eq. (3.25), thus requires a constant negative rate of curvature in some interval $(0, x_1)$ of x and a constant positive rate of curvature of the same absolute value in (x_1, l). The rotational equilibrium of the conjugate beam (Fig. 3.3b), which is simply supported at $x = 0$, free at $x = l$, and loaded with the rates of curvature, then furnishes the abscissa $x_1 = l/\sqrt{2}$ of the point of counterflexure. Figure 3.3c shows the collapse mechanism of the optimal beam.

Since the product of the bending moment $M(x)$ and the curvature $\kappa(x)$ is the nonnegative specific power of dissipation, and since the bending moment is continuous, we have $M(x_1) = 0$. As far as the bending moment is concerned, the optimal beam thus behaves as if it had a hinge at $x = x_1$. Note, however, that this does not mean that we may have a discontinuity of the angular velocity v' at $x = x_1$, because the collapse mechanism must be kinematically admissible for optimality.

It is worth emphasizing that the collapse mechanism in Fig. 2.3c does not depend on the loading as long as all loads act downward. For any loading of this kind we may therefore envision the optimal beam as having a hinge at $x = l/\sqrt{2}$. The bending moment $M(x)$ of the optimal beam thus is statically determinate, and the optimal design is specified by $Y(x) = |M(x)|$ or $b(x) = 4|M(x)|/\sigma_0 h^2$.

Figure 3.3d and 3.3e show the bending moments for a concentrated load P at $x = a < l/\sqrt{2}$ and for a uniformly distributed load p. For a concentrated load at $x = a > l/\sqrt{2}$, the bending moment is shown in Fig. 3.3f. Accordingly, $Y(x) = 0$ for x in $[0, a)$, and the optimal beam is a cantilever with P as tip load.

The preceding discussion was greatly simplified by the convenient decoupling of the kinematics and statics of the optimal beam. Unfortunately, this is not a general feature of optimal plastic design of beams, as we shall see in the next section.

The optimal design discussed above could also be obtained by the following, purely static approach, which has been advocated by some authors. The bending moment of the propped cantilever at the abscissa x may be written as a function $M(x, x_1)$ of x and the abscissa x_1 at which the bending moment changes sign. The cost $\Gamma = \int |M(x, x_1)|\, dx$ then becomes a function of x_1, and the optimal value of x_1 follows from the equation $d\Gamma/dx_1 = 0$.

This static approach is readily extended to beams of higher degrees of redundancy. For downward loads acting on a doubly built-in beam, for instance, the bending moment will change sign at two abscissas, x_1 and x_2, which may be obtained from the equations $\partial\Gamma/\partial x_1 = 0$, $\partial\Gamma/\partial x_2 = 0$, where

$$\Gamma = \int |M(x, x_1, x_2)|\, dx.$$

The static approach, however, is more cumbersome than the kinematic approach used above, as has been stressed by Rozvany (1976, p. 41 and Section 9.3). Whereas the kinematic method once and for all furnishes $x_1 = l/\sqrt{2}$ for the propped cantilever regardless of the distribution of the downward loads, the static analysis would have to be repeated for each loading for which a propped cantilever is to be optimally designed. Moreover, the equation for x_1 furnished by the static approach may be difficult to solve.

In concluding this section, we point out that the optimality condition, Eq. (3.25), remains valid if the specific cost function has the general linear form $\gamma = \alpha Y + \beta$, where α and β are positive. As a first example, we mention a sandwich beam of rectangular cross section with a core of constant breadth and height, B and H, and identical cover plates whose variable thickness $T(x)$ is small compared with H. The core is not supposed to transmit axial stresses, and plastic deformation is possible only if the cover plates are under the axial stresses $\pm\sigma_0$. In view of the assumed smallness of T/H, the moment arm for these stresses is essentially H, and the yield moment is approximately $Y = \sigma_0 BHT(x)$; that is, it is proportional to the cross-sectional area of a cover plate. The weight per unit length of the beam thus is of the form $\alpha Y + \beta$, where β is the contribution of the core to the weight per unit length. If the weight of a beam of length l is to be minimized, the *fixed* weight βl of the core may be disregarded; thus, the optimality condition again has the form of Eq. (3.25).

As a second example with specific cost $\gamma = \alpha Y + \beta$, we mention an I-beam with a web of constant breadth and height, b and H, and identical flanges of constant breadth B and variable thickness $T(x)$, which is small compared with H. The contribution of the web to the yield moment is to be neglected, and the volume of the beam is to be minimized. In the linear expression for the specific cost, α stands for $2/\sigma_0 H$ and β for the cross-sectional area of the web.

3.2.2. Linear Cost Function; Positive Lower Bound on Y

To avoid unrealistic designs with too small breadths in some parts of the beam, let us set a lower bound Y^- on the yield moment. Instead of

Eq. (3.25), we then have the following optimality condition:

$$|\kappa(x)|\begin{Bmatrix} \leq \\ = \end{Bmatrix}c^2\alpha \qquad \text{if}\begin{cases} Y(x) = Y^-, \\ Y(x) \geq Y^-. \end{cases} \tag{3.26}$$

In the case of a propped cantilever loaded downwards, in contrast to Eq. (3.25), the optimality condition of Eq. (3.26) does not allow us to discuss the kinematics of the optimal beam independently of its statics.

Any cross section at which the bending moment M vanishes belongs to a segment of the beam in which $Y(x) = Y^- > |M(x)|$, and hence $\kappa(x) = 0$. As shown in Fig. 3.4a, the rate of curvature of the collapse mechanism of the optimal beam is therefore of the form

$$\kappa = \begin{cases} 0 & \text{for } 0 \leq x \leq x_1 \text{ and } x_2 \leq x \leq x_3, \\ c^2\alpha & \text{for } x_1 < x < x_2, \\ -c^2\alpha & \text{for } x_3 < x \leq 1. \end{cases} \tag{3.27}$$

The rotational equilibrium of the conjugate beam loaded with these rates of curvature (Fig. 3.4b) furnishes the kinematic condition

$$\xi_3^2 + \xi_2^2 - \xi_1^2 = 1, \tag{3.28}$$

where $\xi_i = x_i/l$ $(i = 1, 2, 3)$. To obtain further equations for ξ_1, ξ_2, ξ_3, we must turn to statics and stipulate that

$$M(\xi_1 l) = M(\xi_2 l) = Y^-, \qquad M(\xi_3 l) = -Y^-. \tag{3.29}$$

The four conditions in Eqs. (3.28) and (3.29) will enable us to evaluate ξ_1, ξ_2, ξ_3 and the reaction R at the simply supported end of the beam.

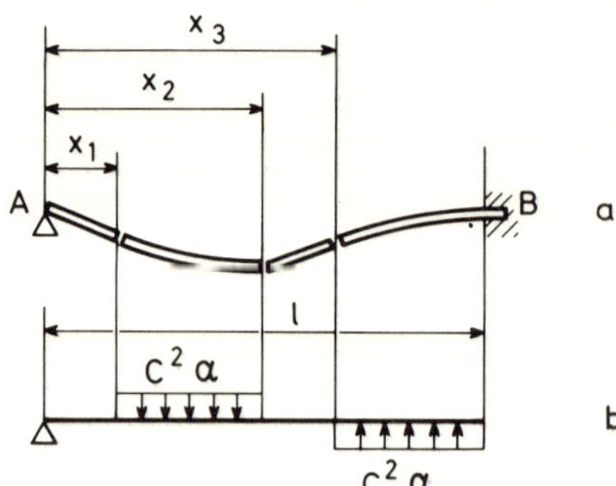

Figure 3.4. Propped cantilver with lower bound on Y.

Example A. Considering a uniformly distributed load p, we write the conditions in Eq. (3.29) as

$$Rl\xi_1 - \frac{pl^2\xi_1^2}{2} = Y^-, \tag{3.30}$$

$$Rl\xi_2 - \frac{pl^2\xi_2^2}{2} = Y^- \tag{3.31}$$

$$Rl\xi_3 - \frac{pl^2\xi_2^2}{2} = -Y^-. \tag{3.32}$$

Elimination of R between Eqs. (3.30) and (3.31) and between Eqs. (3.30) and (3.32) yields

$$\xi_1\xi_2 = \frac{2Y^-}{pl^2}, \tag{3.33}$$

$$\frac{\xi_1\xi_3(\xi_3 - \xi_1)}{\xi_1 + \xi_3} = \frac{2Y^-}{pl^2}. \tag{3.34}$$

Finally, subtraction of Eq. (3.33) from Eq. (3.34) furnishes

$$\xi_3^2 - \xi_1\xi_2 - \xi_2\xi_3 - \xi_3\xi_1 = 0. \tag{3.35}$$

For $Y^- = 0$ we have $\xi_1 = 0$, and, hence, by Eqs. (3.35) and (3.28), $\xi_2 = \xi_3 = l/\sqrt{2}$, which furnishes the collapse mechanism of Fig. 3.3c.

The other limiting case occurs when yielding is restricted to two cross sections; that is, when $\xi_1 = \xi_2$ and $\xi_3 = 1$. Equation (3.28) is then identically fulfilled, and Eqs. (3.35) and (3.33) yield $\xi_1^2 + 2\xi_1 - 1 = 0$, or $\xi = \sqrt{2} - 1$, and

$$\frac{2Y^-}{pl^2} = (\sqrt{2} - 1)^2 = 0.1716;$$

hence,

$$p = \frac{2(3 + 2\sqrt{2})Y^-}{l^2} = \frac{11.66Y^-}{l^2}$$

as the collapse load of a uniformly loaded, propped cantilever of span l and constant yield moment Y^-.

Table 3.1. Values of ξ_1, ξ_2, ξ_3 for Given Values of $2Y^-/pl^2$

$2Y^-/pl^2$	0	0.025	0.050	0.075	0.100	0.125	0.150	0.1716
ξ_1	0	0.0372	0.0788	0.1260	0.1800	0.2452	0.3253	0.4142
ξ_2	0.7010	0.6710	0.6340	0.5952	0.5539	0.5096	0.4610	0.4142
ξ_3	0.7071	0.7420	0.7772	0.8134	0.8518	0.894	0.9451	1.0000

For values of $2Y^-/pl^2$ between 0 and 0.1716, Eqs. (3.28), (3.33), and (3.35) were solved numerically to obtain the values in Table 3.1.

We thus have the following design procedure:

If $2Y^-/pl^2 \geq 0.1716$, the optimal design is $Y(x) = Y^-$.

If $2Y^-/pl^2 < 0.1716$, determine ξ_1, ξ_2, ξ_3 by interpolation in Table 3.1, and determine the reaction R from, say, Eq. (3.31), using $x_i = \xi_i l$.

The bending moment then is $M(x) = Rx - px^2/2$, and the optimal design is

$$Y(x) = \begin{cases} Y^- & \text{for } 0 \leq x \leq \xi_1 l \text{ and } \xi_2 l \leq x \leq \xi_3 l, \\ |M(x)| & \text{for } \xi_1 l < x < \xi_2 l \text{ and } \xi_3 l < x \leq l. \end{cases}$$

Remark. When confronted with the problem of optimal plastic design with a lower bound $Y^- > 0$ on the yield moment, one might envision the following procedure: disregarding the lower bound on yield moment, determine the optimal design $Y(x)$ and replace all $Y(x) \leq Y^-$ by Y^-. That this procedure does not furnish the correct result is shown by Fig. 3.5, which refers to a uniformly loaded, propped cantilever with $Y^- = 0.0858\, pl^2$. Whereas the optimal design has yield moment Y^- everywhere, the simple procedure above furnishes the design whose yield moment is represented

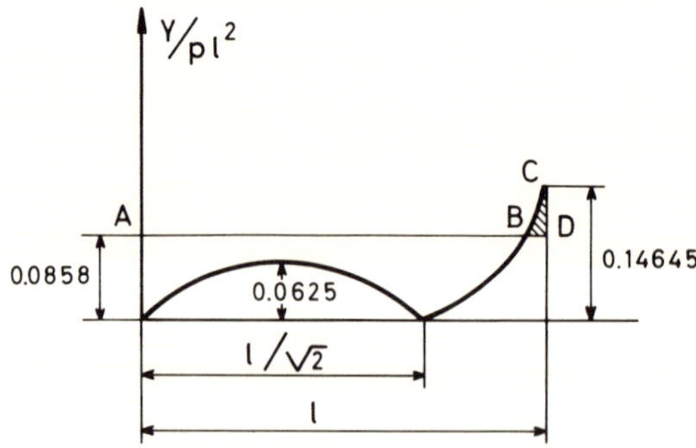

Figure 3.5. Optimal designs of a propped cantilever under uniform load.

by the line *ABC*. The cost of this design exceeds that of the optimal design by an amount represented by the shaded area *BCD*.

Example B. A uniformly loaded beam with built-in ends provides a slightly simpler example. The collapse mechanism, which is symmetric with respect to the center of the span, exhibits yielding zones of common length a near the ends and a central yielding zone of length $2b$ (Fig. 3.6a). The conjugate beam (Fig. 3.6b) is free at both ends. When it is loaded by the rates of curvature $-c^2\alpha$ in the end zones and $c^2\alpha$ in the central zone, its vertical equilibrium requires that $a = b$, and its rotational equilibrium is assured by symmetry. The bending moment for the beam in Fig. 3.6a is

$$M(x) = M(0) + px(l - x)/2.$$

The conditions $M(a) = -Y^-$ and $M(l/2 - a) = Y^-$ then furnish

$$M(0) = -(pl^2/16)(1 + 4\bar{a} - 8\bar{a}^2)$$

and

$$\bar{a} = a/l = \tfrac{1}{4}(1 - p_0/p).$$

Here $p_0 = 16Y^-/l^2$ is the collapse load of the uniformly loaded, prismatic beam with yield moment Y^-.

We thus have the following design procedure:

If $p \leq p_0$, the optimal design is $Y(x) = Y^-$.

If $p > p_0$, evaluate $\bar{a}$, $M(0)$, and $M(x)$ and take $Y(x) = Y^-$ for x in $[\bar{a}l, (\frac{1}{2} - \bar{a})l]$ and $[(\frac{1}{2} + \bar{a})l, (1 - \bar{a})l]$, and $Y(x) = |M(x)|$ elsewhere.

Note that for $Y^- = 0$, we have $a = b = \frac{1}{4}$ for any downward loading, and segments with negative and positive rates of curvature are no longer separated by rigid segments.

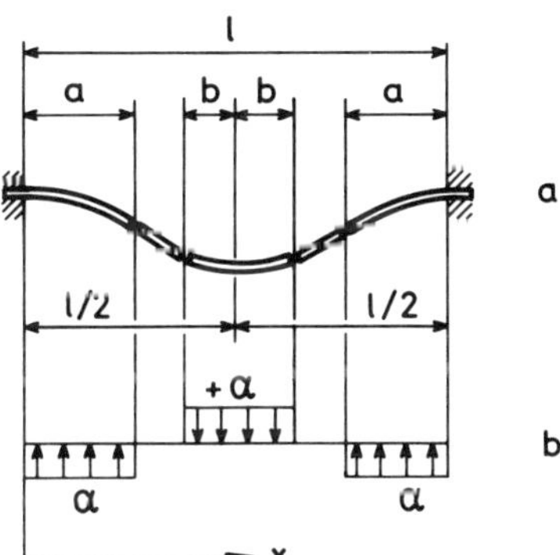

Figure 3.6. (a) Optimal collapse mechanism of a doubly built-in beam. (b) Curvature distribution on conjugate beam.

3.2.3. Parabolic Cost Function; No Explicit Bounds on Y

For the specific cost function $\gamma = \alpha Y^2/2$, where α is a positive constant, and in the absence of explicit bounds on Y, the optimality condition in Eq. (3.17) requires the collapse mechanism of the optimal beam to have a rate of curvature $\kappa(x)$ satisfying

$$|\kappa(x)| = \alpha c^2 Y(x). \tag{3.36}$$

Because a collapse mechanism is only defined to within a constant positive factor, we may set $\alpha c^2 = 1$ in Eq. (3.36). Now the relation between yield moment $Y(x)$ and bending moment is

$$Y(x) = |M(x)| = M(x) \operatorname{sgn} M(x) = M(x) \operatorname{sgn} \kappa(x).$$

Accordingly, Eq. (3.36) may be written as $\kappa(x) = M(x)$. The rate of deflection of the collapse mechanism of the optimal beam is therefore proportional to the deflection of an elastic beam of constant bending stiffness. [This is a special case of a general method of optimal plastic design for convex cost functions developed by Marcal and Prager (1964) and Prager and Shield (1967).]

For a uniformly distributed load p, an elastic beam of constant bending stiffness that is simply supported at $x = 0$ and built in at $x = l$ (Fig. 3.7a) has its point of counterflexure at $x = 3l/4$ (Fig. 3.7b) and a bending moment of $M(x) = px(3l - 4x)/8$. The optimal plastic design of a beam with collapse load p that is supported in this manner is thus $Y(x) = px|3l - 4x|/8$ if the specific cost is proportional to Y^2 (Fig. 3.7).

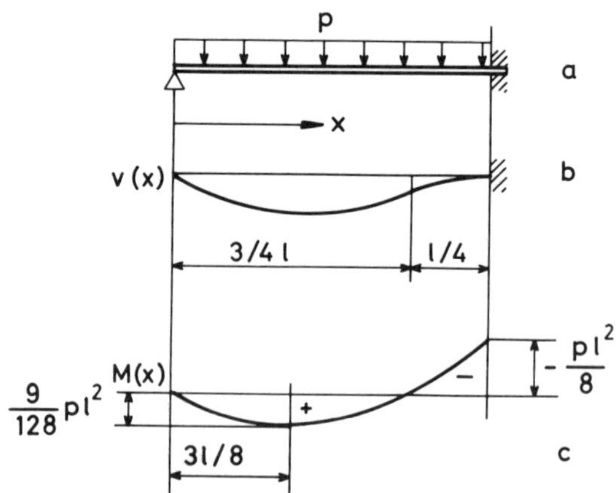

Figure 3.7. (a), (b), (c) Optimal design of a uniformly loaded propped cantilever with parabolic cost function.

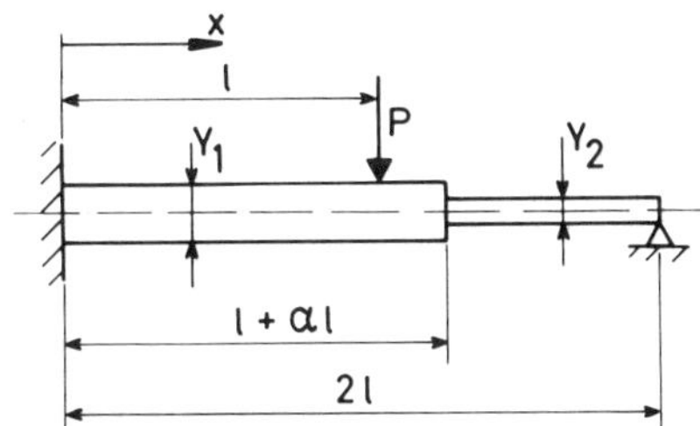

Figure 3.8. Propped cantilever with two yield moments: concave cost function, no bound on Y.

3.2.4. Concave Cost Function; No Explicit Bounds on Y

The beam in Fig. 3.8, which is built in at $x = 0$ and simply supported at $x = 2l$, is to be on the verge of plastic collapse under a mid-span load of the given intensity P. The rectangular cross section is to have the given breadth b over the entire span and constant heights h_1 and h_2 in $[0, (1 + \alpha)l]$ and $[(1 + \alpha)l, 2l]$, which are to be chosen to minimize the volume of the beam. The constant α has a given value in $[0, 1]$.

Since $Y = \sigma_0 bh^2/4$ and $\gamma = bh$, we have

$$\gamma = 2k\sqrt{Y}, \qquad k = \sqrt{b}/\sigma_0. \tag{3.37}$$

The specific cost γ thus is a concave function of Y. The optimality condition of Eq. (3.9), which is necessary and sufficient for global optimality when $\gamma(Y)$ is convex, is now only necessary for local optimality, as the following discussion will show.

We first consider the collapse mechanism in Fig. 3.9a that has three hinges and, hence, two degrees of freedom. The relative counterclockwise angular velocities at the hinges will be denoted by ω_1, ω_2, and ω_3. In view

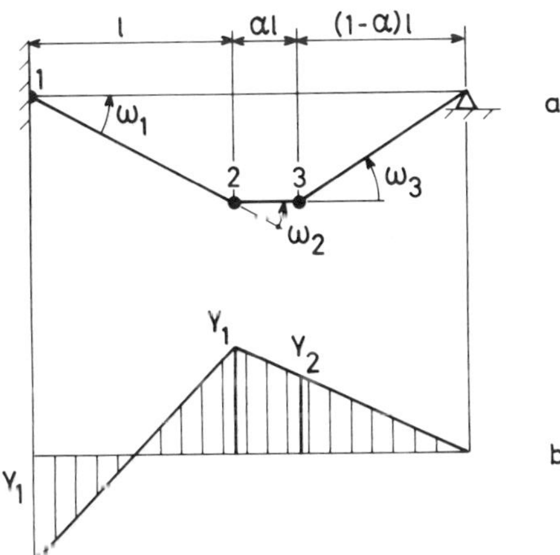

Figure 3.9. (a) Three-hinge collapse mechanism. (b) Corresponding bending moment diagram.

of Eq. (3.5b), and since the positive constant c^2 in the optimality condition of Eq. (3.9) may be given an arbitrary value, this condition yields

$$|\omega_1| + |\omega_2| = \frac{1+\alpha}{\sqrt{Y_1}}, \qquad |\omega_3| = \frac{1-\alpha}{\sqrt{Y_2}}. \tag{3.38}$$

The bending moments in Fig. 3.9b correspond to this collapse mechanism and furnish

$$Y_1 = \frac{Pl}{3}, \qquad Y_2 = \frac{Pl(1-\alpha)}{3}. \tag{3.39}$$

If the conjugate beam, which is free at $x = 0$ and simply supported at $x = 2l$, is subjected to concentrated loads equal to the relative angular velocities at the plastic hinges, its moment equilibrium requires that

$$2\omega_1 + \omega_2 + (1-\alpha)\omega_3 = 0. \tag{3.40}$$

Finally, the sign of the relative angular velocity at each hinge must agree with the sign of the bending moment. Accordingly, the absolute values of the relative angular velocities in Eq. (3.38) may be replaced by $-\omega_1$, ω_2, ω_3, and Eq. (3.40) shows that

$$0 \leq \omega_2 \leq -2\omega_1. \tag{3.41}$$

We now substitute Eq. (3.39) into Eq. (3.38), eliminate Pl between the resulting equations, and then substitute ω_3 from Eq. (3.40). Thus,

$$\omega_2 = \frac{\omega_1(1-2A)}{1+A}, \qquad \text{where } A = \frac{1+\alpha}{(1-\alpha)^{3/2}}. \tag{3.42}$$

Because A ranges from 1 to ∞ as α varies from 0 to 1, ω_2, as given by Eq. (3.42), ranges from $-\omega_1/2$ to $-2\omega_1$, and Eq. (3.41) remains satisfied. For any value of α in $[0, 1]$ there thus exists a mechanism of the considered type that satisfies the necessary condition of optimality, but this is not the only mechanism satisfying this condition.

Indeed, for the mechanism in Fig. 3.10, the relations in Eqs. (3.38) to (3.41) must be replaced by

$$|\omega_1| = -\omega_1 = \frac{1+\alpha}{\sqrt{Y_1}}, \qquad |\omega_2| = \omega_2 = \frac{1-\alpha}{\sqrt{Y_2}}, \tag{3.43}$$

$$Y_2 = \frac{(1-\alpha)(Pl - Y_1)}{2}, \tag{3.44}$$

$$2\omega_1 + (1-\alpha)\omega_2 = 0, \tag{3.45}$$

which yield

$$Y_1 = \frac{2(1+\alpha)^2 Pl}{D}, \quad Y_2 = \frac{(1-\alpha)^4 Pl}{2D}, \qquad \text{where } D = (1-\alpha)^3 + 2(1+\alpha)^2. \tag{3.46}$$

The bending moments in Fig. 3.10b are only admissible if the bending moment under the load, which equals $Y_2/(1-\alpha)$, does not exceed Y_1—that is, if

$$\frac{(1-\alpha)^3}{2} \leq 2(1+\alpha)^2. \tag{3.47}$$

Since the inequality is satisfied for all α in [0, 1], this mechanism also satisfies the necessary condition of optimality.

To continue the discussion of the problem, we shall compare the costs Γ_1 and Γ_2 of the two designs when $\alpha = 0.5$. According to Eqs. (3.37) and (3.39), we then have

$$\Gamma_1 = 2kl(1.5\sqrt{Y_1} + 0.5\sqrt{Y_2}) = 2.14030\ kl\sqrt{Pl}. \tag{3.48}$$

For the second design it follows from Eq. (3.46) that

$$\Gamma_2 = 2kl(1.5\sqrt{Y_1} + 0.5\sqrt{Y_2}) = 3.04138\ kl\sqrt{Pl}. \tag{3.49}$$

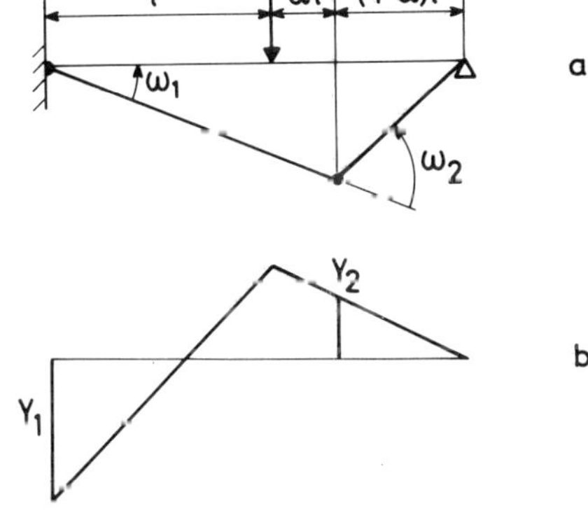

Figure 3.10. (a) Two-hinge yield mechanism. (b) Corresponding bending moment diagram.

Thus the first design is optimal. To find it, we only had to consider the two designs satisfying the optimality condition. It is, however, instructive to discuss the variation of the cost Γ with the ratio $r = Y_1/Y_2$. We note that the first design above corresponds to $r = 2$; the second, to $r = 144$.

For $\alpha = 0.5$ and $r < 2$, no plastic hinge can form at $x = 1.5l$ because the bending moment at $x = l$ would then exceed Y_1. We thus have hinges at $x = 0$ and $x = l$ only, and

$$Y_1 = Pl/3 = rY_2. \tag{3.50}$$

The corresponding cost,

$$\Gamma = (\sqrt{3} + 1/\sqrt{3r})kl\sqrt{Pl}, \tag{3.51}$$

increases indefinitely as r tends to zero. Note that for $r = 2$ the cost in Eq. (3.51) equals the cost Γ_1 from Eq. (3.48).

For $r > 2$ plastic hinges will form at $x = 0$ and $x = 1.5l$. Since Eq. (3.44) is then valid with $\alpha = 0.5$, substituting $Y_2 = Y_1/r$ into this equation yields

$$Y_1 = \frac{rPl}{4 + r} \tag{3.52}$$

and, hence,

$$\Gamma = \frac{kl\sqrt{Pl}(1 + 3\sqrt{r})}{(4 + r)^{1/2}}. \tag{3.53}$$

The necessary condition of optimality is satisfied for only one value of r in the range $(2, \infty)$: namely, $r = 144$, which corresponds to the ratio of Y_1 and Y_2 from Eq. (3.46) with $\alpha = 0.5$. Note that for $r = 2$ and $r = 144$, Eq. (3.53) furnishes the values Γ_1 and Γ_2 in (Eqs. 3.48) and (3.49).

Figure 2.9 shows the variation of the dimensionless quotient Γ/Γ_C with r, where $\Gamma_C = 4kl\sqrt{Pl}/3$ is the cost for $r = 1$; that is, the cost of the prismatic beam that is on the verge of plastic collapse under the load P. One finds

$$\Gamma/\Gamma_C = \begin{cases} 0.75 + \dfrac{0.25}{\sqrt{r}} & \text{for } 0 \le r \le 2, \\[2ex] \dfrac{\sqrt{3}(1 + 3\sqrt{r})}{4(4 + r)^{1/2}} & \text{for } 2 \le r < \infty. \end{cases} \tag{3.54}$$

As one sees from Fig. 3.11, the second design ($r = 144$) corresponds to a

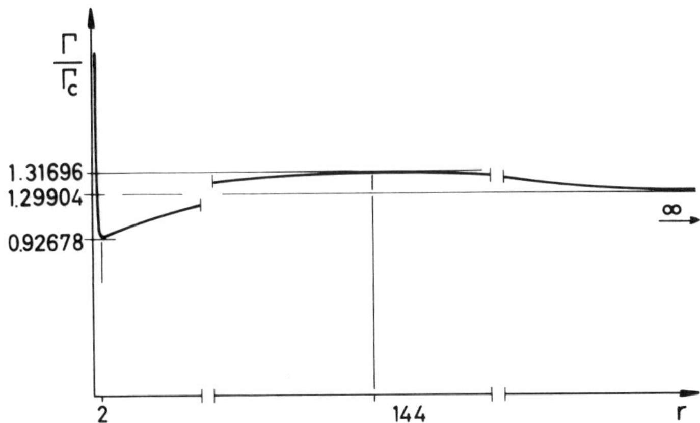

Figure 3.11. Dimensionless total cost versus ratio r of yield moments Y_1 and Y_2, $\alpha = 0.5$.

maximum of the cost. The necessary condition of optimality only assures stationary behavior of the cost in the neighborhood of the considered design and, hence, indicates maxima as well as minima. For $r \to \infty$ the quotient Γ/Γ_C tends toward the value 1.29904, which is only slightly smaller than the value $\Gamma/\Gamma_C = 1.31696$ for $r = 144$, but greater than $\Gamma/\Gamma_C = 0.92678$ for $r = 2$.

It is also of interest to examine the case $\alpha = 0$. The cost is now

$$\Gamma = 2kl(\sqrt{Y_1} + \sqrt{Y_2}). \tag{3.55}$$

There will be a negative plastic hinge at the built-in end and a positive hinge just left or right of mid-span according to whether $r < 1$ or $r > 1$. One readily finds

$$\begin{aligned} Y_1 &= \frac{Pl}{3}, \quad \Gamma = 2kl(Pl/3)^{1/2}\left(1 + \frac{1}{\sqrt{r}}\right) && \text{for } 0 \le r \le 1, \\ Y_1 &= \frac{Plr}{2+r}, \quad \Gamma = \frac{2kl\sqrt{Pl}(1+\sqrt{r})}{(2+r)^{1/2}} && \text{for } 1 \le r < \infty, \end{aligned} \tag{3.56}$$

The optimality condition is found to be satisfied for $r = 4$; the corresponding value of Γ from Eq. (3.56) is 2.44949 $kl\sqrt{Pl}$. This is again a maximum of the cost (see Fig. 3.12). For $r = 1$ we have a minimum, $\Gamma = 2.30940\ kl\sqrt{Pl}$, but for $r = \infty$—that is, for $Y_2 = 0$—we have an even smaller value: $\Gamma = 2kl\sqrt{Pl}$. Note that the local minimum at $r = 1$ is not analytic because Eq (3.56) shows that $d\Gamma/dr$ has different values for $r = 1^-$ and $r = 1^+$. Accordingly, our optimality condition does not indicate this minimum. It does,

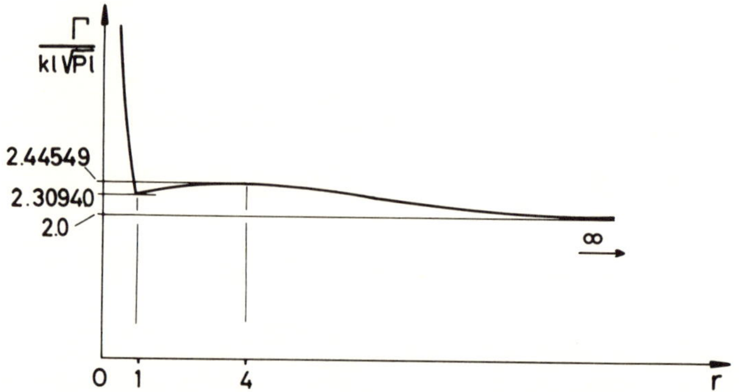

Figure 3.12. Dimensionless total cost versus ratio r of yield moments Y_1 and Y_2, $\alpha = 0$.

however, indicate the global minimum for $r \to \infty$, although this is by no means obvious. Indeed, the mechanism with $\omega_1 = -1$ at $x = 0$ and $\omega_2 = 2$ at $x = l^+$ furnishes a positive value for k_2, but $\gamma_Y(Y_2)$ is infinite because $Y_2 = 0$. The first inequality in Eq. (3.9) thus is satisfied.

Further examples of optimal design of beams with convex cost functions have been given by Megarefs and Hodge (1963), Martin and Ponter (1972), and Rozvany (1973a, 1976).

3.2.5. Beams and Frames Consisting of Prismatic Segments

Consider a beam or a frame with piecewise-constant cross section. For the sake of manufacturing simplicity, pleasing symmetry, or other reasons, some parts may be required to have the same cross section. These parts are then regarded as forming a single "segment" of the structure. The common yield moment of the parts forming the ith segment will be denoted by Y_i, and the sum of their lengths will be denoted by l_i.

In general, the magnitude of the bending moment in a structure of this kind will equal the yield moment only at a finite number of cross sections. Plastic flow is then restricted to these cross sections, which are called *yield hinges.* Between consecutive yield hinges, members remain straight. The relative angular velocities of the parts meeting at the αth yield hinge in segment i will be denoted by $\omega_{i\alpha}$. We assume these relative angular velocities correspond to *normalized* rates of deflection for which the power of the given loads is unity.

No explicit bounds will be introduced for the yield moments Y_i, but, because these are nonnegative, we have $Y_i^- = 0$.

The specific cost of the typical segment is to be equal to its yield moment; the marginal specific cost is thus unity. Since the power dissipated per unit yield moment in segment i is

$$k_i = \sum_{\alpha} |\omega_{i\alpha}|, \tag{3.57}$$

the optimality condition in Eq. (3.9) reduces to

$$\sum_{\alpha} |\omega_{i\alpha}| \left\{ \begin{matrix} = \\ \leq \end{matrix} \right\} c^2 l_i \qquad \text{if} \left\{ \begin{matrix} Y_i > 0, \\ Y_i = 0. \end{matrix} \right. \tag{3.58}$$

The condition in the upper line of Eq. (3.58) is due to Foulkes (1954), and a collapse mechanism satisfying this condition is called a *Foulkes mechanism.* As Smith (1974) has emphasized, the condition in the lower line of Eq. (3.58) does, however, become important when segments of the contemplated structure may be missing in the optimal design.

We will discuss the manner in which Foulkes derived his optimality condition because it reveals another aspect of the problem. For simplicity, let us take the prescribed collapse factor as unity—that is, demand that the structure is to be designed so that it will be on the verge of plastic collapse under the given loads. For any assumed collapse mechanism, the kinematic theorem of limit analysis then furnishes an inequality expressing the fact that the internal power of dissipation cannot be smaller than the power of the loads. Now the internal power is linear in the yield moments Y_i and so is the cost of the structure. We thus look for the minimum of a linear function of the design variables Y_i when these are subject to linear inequalities. In other words, we have a *linear programming problem* to solve.

Consider, for example, the portal frame in Fig. 3.13a. The two columns, with common yield moment Y_1, form segment 1 with length $l_1 = 2h$. The beam with length $l_2 = 2l$ is segment 2; its yield moment Y_2 may or may not be greater than Y_1. Under the actions of the loads shown in the figure,

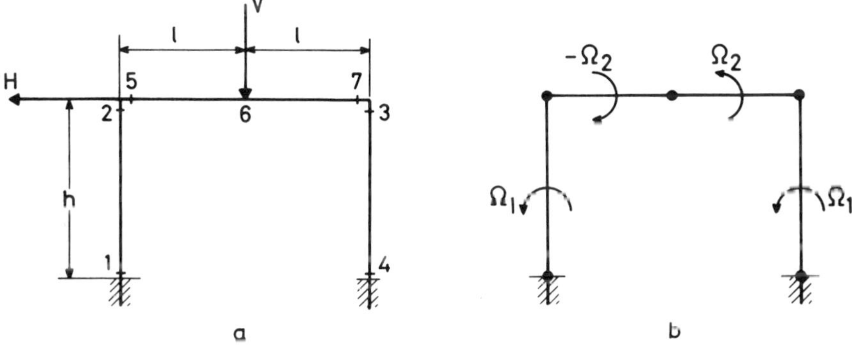

Figure 3.13. (a) Rectangular portal frame. (b) Possible hinge locations.

potential locations of yield hinges are the ends of the columns (sections 1–4) and the ends and the center of the beam (sections 5, 7, and 6).

A collapse mechanism of the frame may be specified by the common angular velocity Ω_1 of the columns and the angular velocity Ω_2 of the right half of the beam. For the given directions of the loads, these angular velocities will not be negative; that is, they will not correspond to rotations in the clockwise sense. The angular velocity of the left half of the beam is $-\Omega_2$. The *power of the loads* for this collapse mechanism is

$$\Pi = Hh\Omega_1 + Vl\Omega_2. \tag{3.59}$$

The relative angular velocities in the hinges are

$$\omega_1 = -\omega_4 = \Omega_1, \qquad \omega_6 = 2\Omega_2, \tag{3.60}$$

and

$$\omega_2 = -(\Omega_1 + \Omega_2), \quad \omega_3 = \Omega_1 - \Omega_2, \quad \omega_5 = \omega_7 = 0, \qquad \text{if } Y_1 < Y_2, \tag{3.61a}$$

or

$$\omega_2 = \omega_3 = 0, \quad \omega_5 = -(\Omega_1 + \Omega_2), \quad \omega_7 = \Omega_1 - \Omega_2, \qquad \text{if } Y_1 > Y_2. \tag{3.61b}$$

Since the *power of dissipation* in the hinges is

$$Y_1(|\omega_1| + |\omega_2| + |\omega_2| + |\omega_4|) + Y_2(|\omega_5| + |\omega_6| + |\omega_7|),$$

the kinematic theorem of limit analysis yields the inequalities

$$\left.\begin{matrix} Y_1(3\Omega_1 + \Omega_2 + |\Omega_1 - \Omega_2|) + 2Y_2\Omega_2 \\ 2Y_1\Omega_1 + Y_2(\Omega_1 + 3\Omega_2 + |\Omega_1 - \Omega_2|) \end{matrix}\right\} \geq Hh\Omega_1 + Vl\Omega_2 \qquad \text{if} \begin{cases} Y_1 < Y_2, \\ Y_1 > Y_2. \end{cases} \tag{3.62}$$

Depending on the values of the nonnegative parameters Ω_1 and Ω_2, each line of Eq. (3.62) represents an infinity of inequalities. All those furnished by the first line may, however, be obtained as nonnegative linear combinations of the following three inequalities, which respectively correspond to $\Omega_1 = 0$, $\Omega_2 = 0$, or $\Omega_1 = \Omega_2$:

$$2Y_1 + 2Y_2 \geq Vl, \quad 4Y_1 \geq Hh, \quad 4Y_1 + 2Y_2 \geq Hh + Vl, \qquad Y_1 < Y_2. \tag{3.63}$$

Indeed, for $\Omega_1 > \Omega_2$ the first line of Eq. (3.62) reduces to

$$4Y_1\Omega_1 + 2Y_2\Omega_2 \geq Hb\Omega_1 + Vl\Omega_2, \qquad (3.64)$$

which is the linear combination of the inequalities in Eq. (3.63) with the nonnegative coefficients 0, $\Omega_1 - \Omega_2$, and Ω_2. Similarly, for $\Omega_1 < \Omega_2$ the first line of Eq. (3.62) yields

$$2Y_1(\Omega_1 + \Omega_2) + 2Y_2\Omega_2 \geq Hh\Omega_1 + Vl\Omega_2, \qquad (3.65)$$

which is the linear combination of the inequalities in Eq. (3.63) with the nonnegative coefficients $\Omega_2 - \Omega_1$, 0, and Ω_1. All inequalities furnished by the second line of Eq. (3.62) may similarly be obtained as nonnegative linear combinations of the inequalities

$$4Y_2 \geq Vl, \quad 2Y_1 + 2Y_2 \geq Hh, \quad 2Y_1 + 4Y_2 \geq Hh + Vl, \qquad Y_1 > Y_2. \qquad (3.66)$$

Since the cost of the frame is proportional to $Y_1h + Y_2l$, the yield moments Y_1, Y_2 of the optimal frame minimize the linear form $\Gamma = Y_1h + Y_2l$ subject to the linear inequalities of Eqs. (3.63) and (3.66).

Let us, for instance, assume that $h/l = 1.25$ and $Hh/Vl = 0.3$. Dividing each inequality by Vl and the cost Γ by Vl^2, and setting $y_i = Y_i/l_i$ $(i = 1, 2)$, we have to minimize

$$\Gamma/Vl^2 = 1.25\, y_1 + y_2$$

subject to the inequalities

$$y_1 + y_2 \geq 0.5, \qquad (a)$$

$$y_1 \geq 0.075, \qquad (b)$$

$$2y_1 + y_2 \geq 0.65, \qquad (c)$$

$$y_2 \geq 0.25, \qquad (d)$$

$$y_1 + y_2 \geq 0.15, \qquad (e)$$

$$y_1 + 2y_2 \geq 0.65. \qquad (f)$$

Each of these inequalities restricts the *design point* with rectangular coordinates y_1, y_2 to a half-plane. In Fig. 3.14 the line bounding any one of these half-planes is marked by the letter appearing behind the corresponding inequality. Note that the region common to the half-planes, which is shaded in Fig. 3.14, is bounded by the lines *b*, *c*, *a*, and *d*.

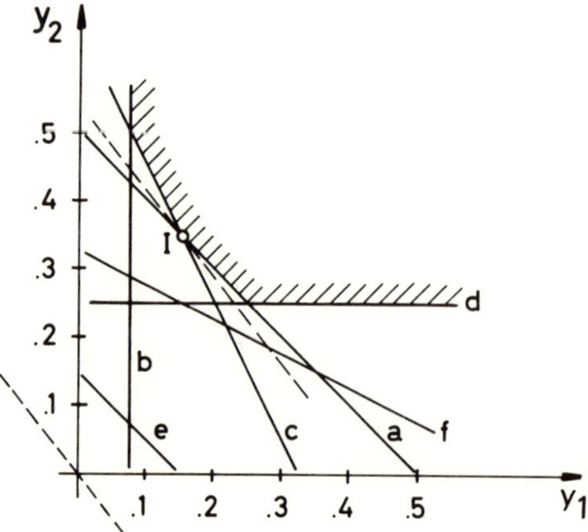

Figure 3.14. Design quarter-plane with acceptable (shaded) zone.

The dotted line through the origin in Fig. 3.14 corresponds to $\Gamma = 0$. As the value of Γ increases, the line undergoes a translation towards the right and first makes contact with the shaded region at the intersection I of lines a and c. This means that for the optimal design the inequalities in Eqs. (a) and (c) are fulfilled as equations. Their solution furnishes the optimal design $y_1 = 0.15$, $y_2 = 0.35$.

The equation of line a expresses the equality of power of the loads and power of dissipation in the hinges of a mechanism, which will be called mechanism a. Similarly, the equation of line b is derived from a mechanism b. The optimal design admits as collapse mechanisms not only mechanisms a and b, but also any nonnegative linear combination of them. For one of these combinations, the equality of power of loads and power of dissipation in the hinges is expressed by the equation of the dotted line through I, in which the coefficients of y_1 and y_2 have the ratio $h/l = 1.25$. In other words, the optimal design admits a collapse mechanism for which the rates of dissipation per unit yield moment in the hinges of the columns and the beam have the ratio h/l—that is, the ratio of the total lengths of members with yield moment Y_1 and Y_2. This statement, however, is Foulkes' theorem.

Figure 3.15 shows the *design chart* established by Foulkes (1954). The first quadrant, with abscissa h/l and ordinate Hh/Vl, is divided into regions A through G. For each region the table at the right gives the optimal values of Y_1 and Y_2.

To show how the boundaries of one of these regions and the formulas for the optimal values of Y_1 and Y_2 in this region are found, let us assume that

$$Y_1 < Y_2 \tag{3.67}$$

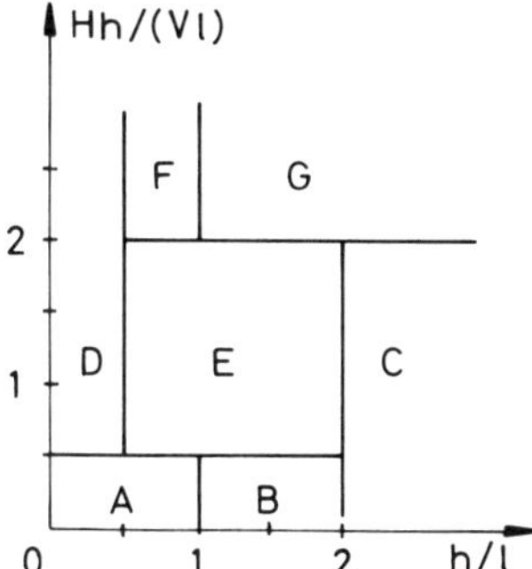

Figure 3.15. Graph of solution types. KEY: A: $Y_1 = Y_2 = Vl/4$; B: $Y_1 = Hh/2$, $Y_2 = (Vl - Hh)/2$; C: $Y_1 = Hh/4$; D: $Y_1 = Hh/2$, $Y_2 = Vl/4$; E: $Y_1 = Y_2 = (Hh + Vl)/6$; F: $Y_1 = (Hh - Vl)/2$, $Y_2 = Vl/2$; G: $Y_2 = Y_2 = Hh/4$.

and consider the mechanism specified by Eqs. (3.60) and (3.61a) with

$$0 \leq \Omega_1 \leq \Omega_2. \tag{3.68}$$

Foulkes' optimality condition, which then requires the coefficients of Y_1 and Y_2 in Eq. (3.65) to have the ratio h/l, yields $1 + \Omega_1/\Omega_2 = h/l$. In view of Eq. (3.68) this means that

$$1 \leq h/l \leq 2. \tag{3.69}$$

The relative angular velocities in the hinges of the considered mechanism are positive for hinges 1 and 6 and negative for hinges 2, 3, and 4. The bending moments thus satisfy

$$M_1 = -M_2 = -M_3 = -M_5 = -M_4 = -M_7 = Y_1, \quad M_6 = Y_2. \tag{3.70}$$

The shear forces in the left and right columns are $(m_2 - M_1)/h = -2Y_1/h$ and $(M_4 - M_3)/h = 0$. Horizontal equilibrium thus requires that

$$Y_1 = Hh/2. \tag{3.71}$$

The shear forces in the left and right halves of the beam are $(M_6 - M_5)/l = (Y_1 + Y_2)/l$ and $(M_7 - M_6)/l = -(Y_1 + Y_2)/l$. Vertical equilibrium thus requires that $2(Y_1 + Y_2) = Vl$. In view of Eq. (3.71) we therefore have

$$Y_2 = \frac{Vl - Hh}{2}. \tag{3.72}$$

The condition $Y_1 \geq 0$ and Eq. (3.67) finally yield

$$0 \leq Hh/Vl \leq \tfrac{1}{2}. \tag{3.73}$$

The inequalities in Eqs. (3.69) and (3.73) specify the region B in which the optimal design is given by Eqs. (3.71) and (3.72).

Note that since $Y_1 = Y_2$ in regions A, E, and G, Foulkes' optimality condition does not apply to these regions. To show how their boundaries and the optimal value of $Y_1 = Y_2 = Y$ are obtained, consider the mechanism specified by Eqs. (3.60) and (3.61a), with $\Omega_1 = \Omega_2 = \Omega$ and, hence,

$$\omega_1 = -\omega_4 = \Omega, \quad \omega_6 = -\omega_2 = 2\Omega, \quad \omega_3 = \omega_5 = \omega_7 = 0.$$

Equating power of the loads and power of dissipation in the hinges, we find

$$Y = \frac{Hh + Vl}{6}. \tag{3.74}$$

Since there is no hinge in the right corner, we have $-Y \leq M_3 = M_7 \leq Y$, while the other bending moments satisfy

$$M_1 = -M_2 = -M_5 = -M_4 = M_6 = Y. \tag{3.75}$$

We must now use the equations of equilibrium as above, first setting $M_3 = M_7 = -Y$ and then $M_3 = M_7 = Y$. In this way the equations $Hh/Vl = 1/2$ and $Hh/Vl = 2$ of the lower and upper boundaries of region E are obtained, in which Y is given by Eq. (3.74). The lateral boundaries of region E are the left boundary of region C and the right boundary of region D, because the considered case with $Y_1 = Y_2$, $\Omega_1 = \Omega_2$ represents a transition between the cases $Y_1 < Y_2, \Omega_1 > \Omega_2$ and $Y_1 > Y_2, \Omega_1 < \Omega_2$ that, respectively, furnish regions C and D.

On account of its practical importance, the subject of this subsection will be discussed in greater detail in Volume 2.

3.2.6. Cost of Supports

In optimizing a propped cantilever built in at $x = l$ and simply supported at $x = 0$, we have taken it for granted that a propped cantilever is in all circumstances superior to a simple cantilever. This tacit assumption, however, may not be justified when the cost of supporting the end $x = 0$ is taken into account.

For now let us assume that the support at $x = 0$ is provided by a short, rigid, perfectly plastic column for which there is no danger of buckling. The design of the structure then is specified by the axial yield force $\bar{Y}$ of the column and the variable yield moment $Y(x)$ of the beam. If the rate of deflection $v(x)$ represents a collapse mechanism for this design under the given distributed load $p(x)$, we have

$$\int pv\,dx = \int Y|v''|\,dx + \bar{Y}v(0), \tag{3.76}$$

where the integrations are extended over the span of the beam. On the other hand, if the design $\bar{Y}^*$, $Y^*(x)$ is not yet or, at most, just ready to collapse under the given load, it follows from the kinematic theorem of limit analysis that

$$\int pv\,dx \leq \int Y^*|v''|\,dx + \bar{Y}^*v(0). \tag{3.77}$$

Combining Eqs. (3.76) and (3.77), we obtain the fundamental inequality

$$\int (Y^* - Y)|v''|\,dx + (\bar{Y}^* - \bar{Y})v(0) \geq 0. \tag{3.78}$$

Let us now assume that the cost of the design $\bar{Y}$, $Y(x)$ is proportional to

$$\Gamma = \alpha \int Y\,dx + \beta\bar{Y}, \tag{3.79}$$

where α and β are known constants that, respectively, have the dimensions of a reciprocal and of a length. It then follows from Eq. (3.78) that the cost Γ^* of the design $\bar{Y}^*$, $Y^*(x)$ cannot be smaller than the cost of the design $\bar{Y}$, $Y(x)$ if

$$|v''| = \alpha, \qquad v(0) = \beta. \tag{3.80}$$

To study the implications of these conditions, which are sufficient for global optimality, we may again use the conjugate beam, which is simply supported at $x = 0$ and free at $x = l$ (Fig. 3.16). The rates of deflection of the original beam are represented by the bending moments of the conjugate beam when this is loaded by the rates of curvature α for $0 \leq x < x_1$ and $-\alpha$ for $x_1 < x \leq l$, and a couple β at $x = 0$, which corresponds to the rate

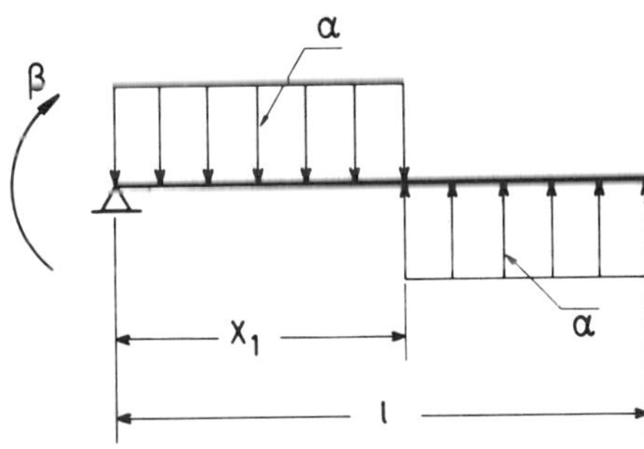

Figure 3.16. Conjugate beam of propped cantilever with costly simple support at left end.

of deflection $v(0)$ demanded by the second equation of Eq. (3.80). The rotational equilibrium of the conjugate beam then yields

$$x_1 = (l^2/2 - \beta/\alpha)^{1/2}. \tag{3.81}$$

The fact that the rate of curvature, and hence the bending moment, changes sign at $x = x_1$ makes the bending moment distribution $M(x)$ statically determinate. The optimal variation of the yield moment of the beam is given by $Y(x) = |M(x)|$, and the optimal yield force $\bar{Y}$ of the supporting column equals the reaction R required to make the bending moment vanish at $x = x_1$.

If we replace $\bar{Y}$ in Eq. (3.79) by R, we may abandon the concept of a yielding support at $x = 0$. The optimal design of the beam is not modified by this change, but the rate of deflection $v(x)$ stipulated by the optimality condition of Eq. (3.80) no longer represents a collapse mechanism of the beam with rigid support at $x = 0$; it simply serves to locate the section of vanishing bending moment.

Let us now consider a uniformly distributed load p. The condition of vanishing bending moment at $x = x_1$ is

$$Rx_1 - px_1^2/2 = 0,$$

which yields

$$R = px_1/2.$$

When $x_1 = 0$, that is, when $\beta/\alpha = l^2$, the support at $x = 0$ becomes useless, and for $\beta/\alpha > l^2$ is better omitted.

3.2.7. Partially Preassigned Yield Moments

The theory developed in Section 3.2.5 can be generalized to beams and frames for which *the yield moment is partially preassigned.* In each segment (x_{i-1}, x_i) we specify a design $Y_i(x)$ by giving its values Y_{ij} at cross sections x_{ij}, $j = 1, 2, \ldots, n_i$, and then prescribing it elsewhere in the segment as the linear combination

$$Y_i(x) = \sum_{j=1}^{n_i} Y_{ij}\phi_{ij}(x). \tag{3.82}$$

Here, the $\phi_{ij}(x)$ are given *shape functions* for the ith segment that must satisfy the conditions

$$\sum_j \phi_{ij}(x_{ij}) = 1 \qquad \text{for } j = 1, \ldots, n_i,$$

in order that $Y_i(x_{ij}) = Y_{ij}$. The design variables for the optimality problem are now the Y_{ij}; they must satisfy yield inequalities $0 \le Y_{ij}^- \le Y_{ij} \le Y_{ij}^+$.

The power of dissipation in segment i is now

$$\int_{x_{i-1}}^{x_i} Y_i(x)|v''(x)|\,dx = \int_{x_{i-1}}^{x_i} \left(\sum_j \phi_{ij}(x)\,Y_{ij}\right)|v''(x)|\,dx.$$

The dissipation per unit value of Y_{ij} is

$$\int_{x_{i-1}}^{x_i} \phi_{ij}(x)|v''(x)|\,dx$$

and here plays the role of k_i in prismatic members.

The fundamental inequality in Eq. (3.8) becomes

$$\sum_i \int_{x_{i-1}}^{x_i} \sum_j \phi_{ij}(x)(Y_{ij}^{\dagger} - Y_{ij})|v''(x)|\,dx \ge 0, \qquad (3.83)$$

and the assumption of convexity of the specific cost function is expressed, for every i, by

$$\gamma(Y_{ij}^*) - \gamma(Y_{ij}) \ge \sum_j (Y_{ij}^* - Y_{ij})\gamma_{Y_{ij}}(Y_{ij}), \qquad j = 1, 2, \ldots, n, \quad (3.84)$$

where the specific cost in each segment is a function of the design unknowns Y_{ij} for this segment. To obtain the total costs of the designs Y_{ij}^* and Y_{ij}, we integrate Eq. (3.84) from x_{i-1} to x_i and sum over all segments. We obtain

$$\Gamma(Y_{ij}^*) - \Gamma(Y_{ij}) \ge \sum_i \sum_j (Y_{ij}^* - Y_{ij}) \int_{x_{i-1}}^{x_i} \gamma_{Y_{ij}}(Y_{ij})\,dx. \qquad (3.85)$$

Comparing Eqs. (3.84) and (3.85), we obtain the following optimality condition when the Y_{ij} satisfy the strict inequalities $Y_{ij}^- < Y_{ij} < Y_{ij}^+$:

$$\int_{x_{i-1}}^{x_i} \phi_{ij}(x)|v''(x)|\,dx = c^2 \int_{x_{i-1}}^{x_i} \gamma_{Y_{ij}}(Y_{ij})\,dx. \qquad (3.86)$$

Because

$$c^2 \int_{x_{i-1}}^{x_i} \gamma_{Y_{ij}}(Y_{ij})\,dx = c^2 \frac{\partial}{\partial Y_{ij}} \int_{x_{i-1}}^{x_i} \gamma(Y_{ij})\,dx$$

by von Leibnitz's rule, the optimality condition requires the curvature field

$v''(x)$ to be such that, for all Y_{ij}, the power of dissipation in segment i for unit Y_{ij} be proportional, with nonnegative constant coefficient c^2, to the marginal cost of this segment computed from the Y_{ij}.

When the bounds Y_{ij}^- or Y_{ij}^+ are attained, the proportionality is replaced by an inequality, as found in Section 3.1. The proof is left to the reader. An example of application is given in Exercise 3.6.

Exercise 3.1. A beam of rectangular cross section of constant height and continuously varying breadth is built in at $x = 0$, simply supported at $x = l$, and free at $x = 1.5\,l$. Determine the abscissas $x_1 = (1 - \xi - \eta)l$ and $x_2 = (1 - \xi)l$ at which the rate of curvature of the collapse mechanism of the optimal design for a uniformly distributed load p changes sign. (Note that static, as well as kinematic, conditions must be used to solve the problem.)

Exercise 3.2. Assuming that a beam of the type considered in Exercise 3.1 carries a concentrated load P at $x = 1.5l$ instead of the uniformly distributed load p, show that the rate of curvature of the collapse mechanism of the optimal design changes sign at $x = l(2 - \sqrt{2})/2$.

Exercise 3.3. A propped cantilever has rectangular cross section with given nonuniform height $h(x)$ and unknown variable breadth $b(x)$. Find the optimality condition for minimum volume and apply it to the case $h = 1/[h_0^{-1} - (h_0^{-1} - h_1^{-1})x/l]$, where h_0 and h_1 are the heights at the simply supported and built-in ends ($x = 0$ and $x = l$, respectively). Treat the example with $l = 10$, $h_0 = 0.5$, and $h_1 = 1$.

Exercise 3.4. The propped cantilever considered in Section 3.2.6 is subjected to a uniform load p. Writing $\beta/\alpha = kl^2$, let the dimensionless parameter k take the values 0.0, 0.1, 0.2, 0.3, 0.4, and 0.5. Determine the corresponding values of the total cost and the percent contribution of the cost of the support to the total cost.

Exercise 3.5. The doubly built-in beam in Fig. 3.17a, whose yield moment varies as shown in Fig. 3.17b, has been designed for the concentrated central collapse load $6Y_0/L$. Show how to optimally strengthen the beam to double this collapse load. (The cost of an increase in yield moment is proportional to this increase.)

Exercise 3.6. A beam has rectangular cross section of uniform height and linearly varying width. It carries a uniformly distributed load p, is simply supported at $x = 0$, and built-in at $x = l$. Determine the yield moments Y_1 and Y_2 at $x = 0$ and $x = l$, respectively, to minimize the volume of the beam.

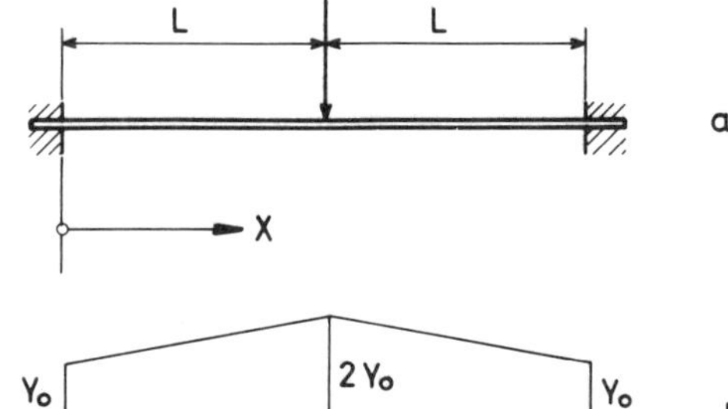

Figure 3.17. (a), (b) Built-in beam with partially preassigned yield moment distribution.

Exercise 3.7. Generalize the results in Section 3.2.1 to the case where the propped cantilever has a specific cost $\gamma = \alpha Y$ everywhere, except in a domain given by $a \le x \le b \le l/2$, where $\gamma = \beta Y$.

Exercise 3.8. Extend the solution of Exercise 3.7 to the case of concentrated cost C at some section $x = a$ (cost of a joint, for example).

3.3. Optimality Condition for Prescribed Elastic Compliance

We now consider the optimal design of an elastic beam carrying a transverse load $p(x)$. A design here is the prescription of a bending stiffness distribution along the beam; again, we consider piecewise-constant stiffness distributions first.

As in Section 3.1, the *layout* is prescribed by giving support locations and positions, x_i, $i = 0, 1, \ldots$, where bending stiffnesses may change. The lengths $l_i = x_i - x_{i-1}$, $i = 1, 2, \ldots$, are thus known. A design for the beam consists of the specification of a piecewise-constant bending stiffness with values $B_i = E_i I_i$, $i = 1, 2, \ldots$, for each segment. Here E_i is the material Young's modulus, and I_i is the second moment of cross-sectional area in the ith segment.

For *technological constraints* we introduce lower and upper bounds B_i^- and B_i^+ on the bending stiffness B_i of the typical segment.

The *specific cost* is assumed to be a monotonically increasing *convex* function $\gamma(B)$ of the bending stiffness; its derivative with respect to B will be denoted by $\gamma_B(B)$. In analogy with (3.2) we thus have the inequality

$$\gamma(B^*) - \gamma(B) \ge (B^* - B)\gamma_B(B), \tag{3.87}$$

where B^* and B are any two values from $[B^-, B^+]$.

The *elastic compliance* of the beam to the given load is defined as the work $\int p(x)v(x)\,dx$ of the given load $p(x)$ on the elastic deflection $v(x)$ it produces.† The compliance thus equals twice the *elastic strain energy* $\frac{1}{2}\int B(x)\kappa^2(x)\,dx$ stored in the deformed beam, $\kappa = -v''$ being the curvature of the deformed axis of the beam.

The smaller the compliance to a given load, the stiffer the beam behaves under this load. It thus makes sense to impose a *behavioral constraint* that sets an upper bound $\mathscr{E}$ on the elastic compliance. Since no other behavioral constraint is to be imposed here, the optimal beam will obviously have compliance $\mathscr{E}$ if the compliance constraint is at all relevant. (See the end of Section 3.4.2 for a case in which this constraint is not relevant.)

A kinematically admissible deflection $v(x)$ of the elastic beam is continuous, has continuous first derivative $v'(x)$ and piecewise-continuous second derivative $v''(x)$, and satisfies the kinematic conditions at the supports. The contribution of the ith segment to the compliance $\mathscr{E}$ of the beam with elastic deflection $v(x)$ is

$$\mathscr{E}_i = B_i e_i[v(x)], \tag{3.88}$$

where

$$e_i[v(x)] = \int_{x_{i-1}}^{x_i} v''^2(x)\,dx \tag{3.89}$$

is twice the strain energy per unit bending stiffness of this segment.

Let the designs B_i and B_i^*, which satisfy the technological constraints, have compliances $\mathscr{E}$ and $\mathscr{E} - \bar{\mathscr{E}}$ to the given load $p(x)$, where $\bar{\mathscr{E}}$ is nonnegative, and let $v(x)$ and $v^*(x)$ be their elastic deflections under this load. Thus,

$$\mathscr{E} = \sum_i B_i e_i\{v\} = \sum_i B_i^* e_i\{v_i^*\} + \bar{\mathscr{E}}. \tag{3.90}$$

The principle of minimum total potential energy, applied to the design B_i^* and the deflections v^* and v, yields

$$\sum_i B_i^* e_i\{v^*\} - 2(\mathscr{E} - \bar{\mathscr{E}}) \le \sum_i B_i^* e_i\{v\} - 2\mathscr{E}. \tag{3.91}$$

Substituting $\Sigma_i\, B_i^* e_i\{v^*\}$ from Eq. (3.90) into Eq. (3.91), we obtain

$$\sum_i (B_i^* - B_i) e_i\{v\} \ge \bar{\mathscr{E}} \ge 0. \tag{3.92}$$

† Although v is now being used to denote deflection, rather than rate of deflection, this should not cause confusion, because the rate of deflection does not play any role in the problems of elastic design considered here.

In the following, the nonnegativity of the left side of Eq. (3.92) plays exactly the same role that the nonnegativity of the left side of Eq. (3.8) played in the proof of the sufficiency of the optimality condition in Eq. (3.9). Replacing the quantities Y_i^*, Y_i, and k_i in this proof by B_i^*, B_i, and e_i, respectively, we thus show that the condition

$$e_i[v(x)]\begin{Bmatrix} \leq \\ = \\ \geq \end{Bmatrix} c^2 l_i \gamma_B(B_i) \quad \text{if} \begin{cases} B_i = B_i^-, \\ B_i^- < B_i < B_i^+, \\ B_i = B_i^+, \end{cases} \tag{3.93}$$

is sufficient for the global optimality of the design B_i.

The proof that Eq. (3.93) is also necessary for local optimality resembles the necessity proof for Eq. (3.9). Because the designs B_i and $B_i + \delta B_i$ are to have the same compliance to the load $p(x)$, we have

$$\sum_i (B_i\,\delta e_i + e_i\,\delta B_i) = 0, \qquad \int p\,\delta v\,dx = 0, \tag{3.94}$$

to within higher-order quantities. Furthermore, the first variation of the potential energy of the design B_i must vanish:

$$\delta\Pi = \tfrac{1}{2}\sum B_i \delta e_i - \int p\,\delta v\,dx = 0. \tag{3.95}$$

From Eqs. (3.94) and (3.95) there follows the equation

$$\Sigma\, e_i\,\delta B_i = 0, \tag{3.96}$$

which corresponds to Eq. (3.13). The remainder of the necessity proof is completely analogous to the proof in Section 3.1.

Since $l_i \gamma_B(B_i)$ in Eq. (3.93) is the marginal cost of the ith segment of the beam, this optimality condition states that *the given load produces a deflection of the optimal beam for which the strain energy stored per unit bending stiffness in any segment is a fixed positive multiple of an amount that is not greater than, equal to, or smaller than the marginal cost of this segment, depending on whether its bending stiffness has a value at the lower bound, between the bounds, or at the upper bound.*

When the variation of the bending stiffness along the beam is specified by a continuous function $B(x)$, the optimality condition in Eq. (3.93) reduces to

$$v''^2(x)\begin{Bmatrix} \leq \\ = \\ \geq \end{Bmatrix} c^2 \gamma_B[B(x)] \quad \text{if} \begin{cases} B(x) = B^-, \\ B^- < B(x) < B^+, \\ B(x) = B^+. \end{cases} \tag{3.97}$$

3.4. Optimal Elastic Design for Given Compliance

3.4.1. Linear Cost Function; No Explicit Bounds on *B*

Assume that

$$\gamma(B) = \alpha^2 B, \tag{3.98}$$

where α^2 is a positive constant. This proportionality between specific cost and bending stiffness holds, for example, for a beam with rectangular cross section of constant height h and variable breadth $b(x)$ whose volume is to be minimized. In the absence of explicit bounds on B, the optimality condition of Eq. (3.97) then reduces to

$$|\kappa(x)| \left\{ \begin{matrix} \leq \\ = \end{matrix} \right\} c\alpha \quad \text{if} \left\{ \begin{matrix} B(x) = 0, \\ B(x) > 0, \end{matrix} \right. \tag{3.99}$$

where $\kappa = -v''$ is the curvature of the optimal beam under the given loading. Comparing Eq. (3.99) with Eq. (3.25) shows that the deflection of the optimal elastic beam with given compliance is proportional to the rate of curvature of the optimal plastic beam for given collapse load, provided that the two beams are supported in the same manner and subjected to the same loading, and the specific costs are proportional to bending stiffness and yield moment.

For example, a propped elastic cantilever (Fig. 3.3a) subjected to *downward* loads will again have its point of counterflexure at $x = l\sqrt{2}$. Since the bending moment must vanish at this point, $M(x)$ for the optimal elastic beam becomes statically determinate and, hence, identical with the bending moment of the optimal plastic beam. Finally, we see that the optimal variation of bending stiffness for the elastic beam with given compliance is proportional to the optimal variation of yield moment for the plastic beam with given load factor at collapse. We show this by substituting the constant value of $|\kappa(x)|$ given by the optimality condition in Eq. (3.99), where $B(x) \neq 0$, into the general definition of bending stiffness, written in the form

$$B(x) = \frac{M(x)}{\kappa(x)} = \frac{|M(x)|}{|\kappa(x)|}; \tag{3.100}$$

thus the optimal $B(x)$ is proportional to $|M(x)| = Y(x)$, the optimal yield moment. Although this statement is made with reference to the propped cantilever of Fig. 3.3a, it is readily seen to be generally valid, provided that the two beams are supported in the same fashion and subjected to the same loading.

Whereas the value of the constant $c^2\alpha$ in the optimality condition of Eq. (3.25) is irrelevant, the value of the constant $c\alpha$ in Eq. (3.99) remains to be related to the prescribed compliance

$$\mathscr{E} = \int_0^l B\kappa^2\,dx = \int_0^l (B|\kappa|)|\kappa|\,dx. \tag{3.101}$$

Substituting $B|\kappa|$ from Eq. (3.100) and $|\kappa|$ from Eq. (3.99) into Eq. (3.101) furnishes

$$\mathscr{E} = c\alpha \int_0^l |M|\,dx = |\kappa| \int_0^l |M|\,dx, \tag{3.102}$$

where Eq. (3.99) has been used in the transition to the last term. Substituting $|\kappa|$ from Eq. (3.102) into Eq. (3.100) finally yields

$$B = |M| \int_0^l \frac{|M|\,dx}{\mathscr{E}}. \tag{3.103}$$

If the plastic beam is to have the load factor Λ for collapse under the load, whereas the elastic beam has the compliance $\mathscr{E}$, the optimal variation of yield moment is

$$Y = \Lambda |M|. \tag{3.104}$$

Equations (3.103) and (3.104), in which M denotes the same statically determinate bending moment for the given load, facilitate the transition from one kind of optimal design to the other.

3.4.2. Linear Cost Function; Positive Lower Bound on *B*

To avoid unrealistic designs a positive lower bound B^- may be set on the bending stiffness. For given elastic compliance $\mathscr{E}$, the beam of minimum cost then has a curvature $\kappa(x)$ under the given load that satisfies

$$|\kappa(x)| \left\{ \begin{matrix} \leq \\ = \end{matrix} \right\} c\alpha \qquad \text{if} \left\{ \begin{matrix} B(x) = B^- , \\ B(x) > B^- . \end{matrix} \right. \tag{3.105}$$

Consider, for instance, a beam that is built in at $x = 0$ and $x = l$ and carries a concentrated load p at $x = l/2$ (Fig. 3.18a). On account of the symmetry with respect to the center of the span, we need only consider the left half of the beam, in which the bending moment

$$M(x) = M(0) + Px/2 \tag{3.106}$$

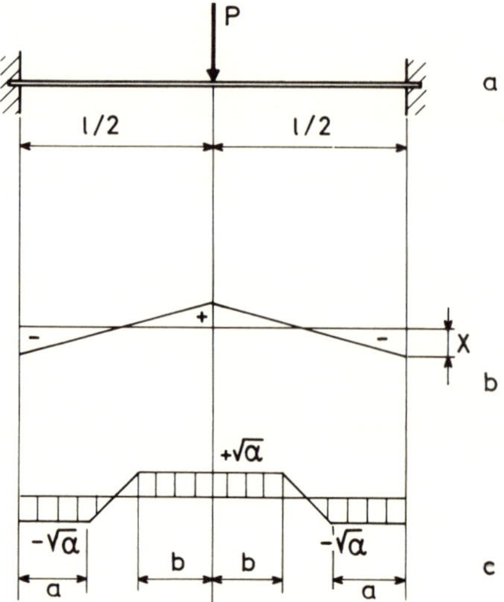

Figure 3.18. (a) Built-in beam. (b) Bending moment diagram. (c) Elastic curvatures of optimal beam with lower bound on stiffness.

has a single zero (Fig. 3.18b). In the neighborhood of this zero, say in the interval $[a, \frac{1}{2}l - b]$, the bending stiffness B will have the constant value B^-, and the curvature κ will be proportional to the bending moment, which varies linearly with x. In the intervals $[0, a]$ and $[\frac{1}{2}l - b, \frac{1}{2}l]$, however, the bending stiffness will exceed B^-, and the optimality condition of Eq. (3.105) will yield curvatures $\kappa = -c\alpha$ and $\kappa = c\alpha$, respectively. If we load the conjugate beam, which is free at both ends, with the curvatures (Fig. 3.18c), vertical equilibrium requires $b = a$, while rotational equilibrium is assured by symmetry. At $x = a$, we have $M = B^-\kappa$; that is,

$$M(0) + Pa/2 = -c\alpha B^-. \tag{3.107}$$

Similarly, at $x = \frac{1}{2}l - b = \frac{1}{2}l - a$,

$$M(0) + \frac{P(l-2a)}{4} = c\alpha B^-. \tag{3.108}$$

It follows from Eqs. (3.107) and (3.108) that

$$M(0) = -Pl/8, \tag{3.109}$$

$$a = \tfrac{1}{4}(l - 8c\alpha B^-/P). \tag{3.110}$$

Note that $M(l/2) = M(0) + Pl/4 = -M(0)$. The bending moment diagram for the left half of the beam is thus antisymmetric with respect to $x = l/4$, as is the curvature diagram (Fig. 3.18c).

We still must relate the length a to the prescribed elastic compliance, which may be written as $\mathscr{E} = \int_0^l M\kappa\, dx$. In view of the symmetry and antisymmetry properties of the M and κ diagrams, we have

$$\mathscr{E} = 4c\alpha \int_0^a |M|\, dx + B^- \int_a^{l/4} \kappa^2\, dx. \tag{3.111}$$

Using Eqs. (3.106) and (3.109), we see that the first integral in Eq. (3.111) has the value

$$\int_0^a |M|\, dx = \frac{Pa(l-2a)}{8}. \tag{3.112}$$

Over the interval $[a, l/4]$ the absolute value of the curvature decreases linearly from $c\alpha$ to zero. The second integral in Eq. (3.111) is thus

$$\int_a^{l/4} \kappa^2\, dx = \frac{c^2\alpha^2(l-4a)}{12}. \tag{3.113}$$

Substituting $c\alpha$ from Eq. (3.110) and using Eqs. (3.112) and (3.113) in Eq. (3.111) finally yields

$$(l-4a)(l^2+4al-8a^2) = 192B^-\mathscr{E}/P^2. \tag{3.114}$$

When this cubic equation has been solved for a, Eq. (3.110) furnishes

$$c\alpha = \frac{P(l-4a)}{8B^-}, \tag{3.115}$$

and the optimal design is

$$B = \begin{cases} \left|\dfrac{M}{\kappa}\right| = B^-\left|\dfrac{l-4x}{l-4a}\right| & \text{for } 0 \le x \le a \text{ and } \tfrac{1}{2}l - a \le x \le \tfrac{1}{2}l, \\ B^- & \text{for } a < x < \tfrac{1}{2}l - a. \end{cases} \tag{3.116}$$

This design, however, is only valid if Eq. (3.114) has a nonnegative root a. For $a = 0$ Eq. (3.114) yields

$$\mathscr{E} = P^2l^3/192\, B^-, \tag{3.117}$$

and the design in Eq. (3.116) has uniform bending stiffness B^-. Note that from these facts follows the well-known result that the central deflection is

$Pl^3/192\,B^-$ for a doubly built-in prismatic beam of bending stiffness B^- under a central load P.

We now recall that in Section 3.3 we introduced $\mathscr{C}$ as an upper bound on the elastic compliance and then concluded that the optimal beam would actually have compliance $\mathscr{C}$ if the compliance constraint was at all relevant. It ceases to be relevant when the given upper bound $\mathscr{C}$ exceeds the value of Eq. (3.117). The optimal design is then determined by the technological constraint $B \geq B^-$.

Exercise 3.9. Assuming that only a single cross-sectional dimension t is available to the designer, prove that $\gamma(B)$ will be convex if (i) $B(t)$ is concave and monotonically increasing, and (ii) $\gamma(t)$ is convex. Discuss the case $B = (pt + q)^n$, where $p > 0$ and $q \geq 0$ are constants.

3.5. Various Behavioral Constraints: Common Features and Optimality Conditions

The analogy between Sections 3.1 and 3.3 is obvious and has already been commented upon. We now analyze it in greater detail and extend it to other behavioral constraints.

We first note that the cross-sectional dimensions do not enter the formulation of the problem explicitly, but only through the yield moment Y or the bending stiffness B. The optimal design is obtained as a vector with components Y_i or B_i, or as a function $Y(x)$ or $B(x)$, and the designer is free to choose cross-sectional dimensions and other variables provided they furnish the required yield moment or bending stiffness. For example, if the prismatic segments of a beam do not have to consist of the same material, the designer may choose the material for the ith segment and then, taking account of the yield stress or Young's modulus of this material, determine the cross-sectional dimensions to obtain the value of Y_i or B_i that is required for optimality. It is therefore natural to regard the yield moment or the bending stiffness as the *design variable*. We note that the specific cost γ is assumed to be a *convex* function of the design variable.

For the sake of brevity, we will assume that the design variable is a nonnegative continuous function of x that is not bounded from above. The reader will readily make the modifications needed when the design variable is segmentwise constant or restricted to a given closed interval.

The behavioral constraint sets a bound on a scalar (load factor for plastic collapse or elastic compliance) that may be characterized as the minimum, over all kinematically admissible, normalized velocities $v(x)$ or all kinematically admissible deflections $v(x)$, of a functional $\int Y|v''|\,dx$ or

$\int Bv''^2\,dx$ whose integrand is the product of the design variable with an expression depending only on $v(x)$.

Two designs, $Y(x)$ and $Y^*(x)$ or $B(x)$ and $B^*(x)$, are now considered that satisfy the behavioral constraint, with the first satisfying it as an equality. When the minimum principle is applied to these designs and the field $v(x)$ that minimizes the functional for the first design, the fundamental inequality in Eq. (3.8) or Eq. (3.92) is obtained. In view of the assumed convexity of the specific cost function, this inequality furnishes a sufficient condition for the global optimality of the first design. This condition can also be shown to be necessary for a local optimum, and this proof makes no use of the convexity of the specific cost function.

We now review other behavioral constraints that may be treated in this manner.

3.5.1. Prescribed Lower Bound on Load Factor for Plastic Collapse in the Presence of Design-Dependent Loads

If, in addition to the given distributed load p, we must take account of the distributed design-dependent load $q(Y)$, where the function $q(Y)$ is convex, the argument leading to the optimality condition in Eq. (3.20) must be modified as follows.

If the rate of deflection $v(x)$ specifies a collapse mechanism of the design $Y(x)$, the load factor for plastic collapse of this design is

$$\Lambda = \frac{\int Y|v''|\,dx}{\int [p + q(Y)]v\,dx}. \tag{3.118}$$

If the design $Y^*(x)$ has load factor $\Lambda^* \geq \Lambda$, it follows from the kinematic theorem of limit analysis that

$$\Lambda \leq \Lambda^* \leq \frac{\int Y^*|v''|\,dx}{\int [p + q(Y^*)]v\,dx}. \tag{3.119}$$

Accordingly,

$$\int Y|v''|\,dx - \Lambda \int [p + q(Y)]v\,dx \leq \int Y^*|v''|\,dx - \Lambda \int [p + q(Y^*)]v\,dx, \tag{3.120}$$

or, in view of the convexity of the function $q(Y)$,

$$\int (Y^* - Y)[|v''| - \Lambda q_Y(Y)v]\,dx \geq 0, \tag{3.121}$$

where the subscript Y indicates differentiation with respect to Y. This fundamental inequality may be used in the same manner as Eq. (3.19) to show that the condition

$$|v''| - \Lambda q_Y(Y)v = c^2\gamma_Y(Y) \tag{3.122a}$$

assures the global optimality of the design $Y(x)$. We leave it to the reader to develop the corresponding argument for a beam consisting of prismatic elements (Exercise 3.9).

Note that the above discussion tacitly assumed that p as well as q is a load acting permanently on the beam. When the first is a service load whose intensity may have any value from 0 to p, and the second is a permanent load, such as the weight of the beam, the fact that at some cross sections the bending moments due to p and q may have opposite signs makes it necessary to demand that the load factor for plastic collapse have at least the given value Λ for the combined action of the permanent load and a service load of any intensity in the range $(0, p)$.

Note also that, in practical applications, the useful service load p is multiplied by a safety factor Λ larger than unity, wheras the beam's weight q per unit length, regarded as known with a much smaller uncertainty, may be affected with a load factor of unit magnitude. Hence, the plastic limit load is $\Lambda p + q$ instead of $\Lambda(p + q)$, as considered above. Assuming that the beam cannot collapse under its weight alone, the reader will easily show that the corresponding optimality condition is

$$|v''| - q_Y(Y)v = c^2\gamma_Y(Y). \tag{3.122b}$$

3.5.2. Prescribed Upper Bound on Elastic Compliance in the Presence of Design-Dependent Loads

With $q(B)$ as the distributed design-dependent load, the elastic compliance $\mathscr{C}$ of the beam with bending stiffness $B(x)$, which takes the deflection $v(x)$ under the load $p(x) + q[B(x)]$, is equal to twice the negative potential energy:

$$\mathscr{C} = 2\int\{p + q(B)\}v\,dx - \int Bv''^2\,dx. \tag{3.123}$$

If the design $B^*(x)$, which takes the deflection $v^*(x)$ under the load $p(x) + q[B^*(x)]$, has *at most* the compliance $\mathscr{C}$, we have

$$2\int[p + q(B^*)]v^*\,dx - \int B^*v^{*''2}\,dx \le \mathscr{C}. \tag{3.124}$$

According to the principle of minimum potential energy, the expression on the left of Eq. (3.124) is greater than the expression that would be obtained when v^* is replaced by v. Substituting Eq. (3.123) into Eq. (3.124) and using the convexity of the function $q(B)$ then furnish the fundamental inequality

$$\int (B^* - B)[v''^2 - 2q_B(B)v]\,dx \geq 0, \tag{3.125}$$

which shows that the condition

$$v''^2 - 2q_B(B)v = c^2\gamma_B(B) \tag{3.126}$$

assures the global optimality of the design $B(x)$.

3.5.3. Prescribed Upper Bound on Dynamic Elastic Compliance to Harmonically Varying Loads

As pointed out in Section 1.2.9, the minimum principle for steady-state forced vibrations is identical with the principle of minimum potential energy applied to the static loading consisting of the amplitude $p(x)$ of the given load and the amplitude $\omega^2\{m(x) + \mu[B(x)]\}v(x)$ of the inertia load. If the function $\mu(B)$ is convex, the optimality condition for the prescribed upper bound on dynamic elastic compliance to harmonically varying loads may therefore be derived in the same manner as that for the prescribed upper bound on static elastic compliance in the presence of design-dependent loads. Instead of p and q in Section 3.5.2, we now have $p + \omega^2 mv$ and $\omega^2\mu v$ as the design-independent and design-dependent loads, and instead of Eq. (3.126) we obtain the optimality condition

$$v''^2 - 2\omega^2\mu_B(B)v^2 = c^2\gamma_B(B). \tag{3.127}$$

3.5.4. Prescribed Lower Bound on Fundamental Natural Frequency

If the fundamental natural frequency of an elastic beam with bending stiffness $B(x)$ equals the prescribed lower bound ω, and if $v(x)$ is the fundamental mode of this beam, we have

$$\omega^2 = \frac{\int Bv''^2\,dx}{\int [m + \mu(B)]v^2\,dx}, \tag{3.128}$$

where $m(x)$ and $\mu[B(x)]$ are the design-independent and design-dependent masses per unit length. If, moreover, the fundamental natural frequency ω^* of the design $B^*(x)$ has *at least* the value ω, we have

$$\omega^2 \leq \omega^{*2} \leq \frac{\int B^* v''^2\,dx}{\int [m + \mu(B^*)]v^2\,dx} \tag{3.129}$$

on account of the minimum property of the Rayleigh quotient. For convex $\mu(B)$ it follows from Eqs. (3.128) and (3.129) that

$$\int (B^* - B)[v''^2 - \omega^2 \mu_B(B) v^2]\, dx \geq 0. \tag{3.130}$$

The condition

$$v''^2 - \omega^2 \mu_B(B) v^2 = c^2 \gamma_B(B) \tag{3.131}$$

therefore assures the global optimality of the design $B(x)$.

In general, the optimality condition in Eq. (3.131) contains two unknown functions, $B(x)$ and $v(x)$, between which the differential equation

$$(Bv'')'' - \omega^2(m + \mu)v = 0 \tag{3.132}$$

establishes a second relation. If, however, $\gamma = \mu = \alpha^2 B$, the optimality condition in Eq. (3.131) takes the form

$$v''^2 - \alpha^2 \omega^2 v^2 = \text{const.} \tag{3.133}$$

Since a natural mode is only determined to within a constant factor, the value of the constant on the right of Eq. (3.133) is not relevant and may be taken as unity. The resulting nonlinear differential equation and the kinematic conditions of support do not, however, specify a problem with a unique solution; we must add the condition that the fundamental mode has no zero except at the supports of the beam.

If a lower bound ω is prescribed, not for the fundamental (or first) natural frequency, but for the nth natural frequency of the beam, the optimality condition in Eq. (3.133) remains valid provided we now require $v(x)$ to have $n - 1$ zeros in addition to those at the supports.

3.5.5. Prescribed Lower Bound on Elastic Buckling Load

Using the Rayleigh quotient for the buckling load [see, for instance, Libove (1962)]

$$P = \frac{\int Bv''^2\, dx}{\int v'^2\, dx} \tag{3.134}$$

in a manner analogous to the quotient in Eq. (3.128), one readily obtains

$$\int (B^* - B) v''^2\, dx \geq 0 \tag{3.135}$$

in the place of Eq. (3.130). The condition

$$v''^2 = c^2 \gamma_B(B) \tag{3.136}$$

therefore assures the global optimality of the design $B(x)$.

When $\gamma = \alpha^2 B$, this optimality condition is equivalent to $v''^2 = 1$. For a simply supported beam it may, for instance, be satisfied by choosing either $v'' = 1$ for the entire span or $v'' = 1$ over one half and $v'' = -1$ over the other half of the span. The second choice, however, would not correspond to the lowest eigenvalue, which alone is physically relevant for buckling.

In deriving the optimality condition, Eq. (3.136), we have tacitly assumed that the buckling load p is a *single* eigenvalue. This means that, to within a constant factor, the buckling mode is unique. Olhoff and Rasmussen (1977) have drawn attention to the possibility that, for certain conditions of support, the buckling load of the optimal design may be a *double* eigenvalue. We then have linearly independent buckling modes $v_1(x)$ and $v_2(x)$, for each of which an inequality of the form in Eq. (3.135) holds. Accordingly,

$$\int (B^* - B)[\lambda v_1''^2 + (1 - \lambda) v_2''^2]\, dx \geqq 0, \qquad 0 \leqq \lambda \leqq 1. \tag{3.137}$$

The condition

$$\lambda v_1''^2 + (1 - \lambda) v_2''^2 = c^2 \gamma_B(B) \tag{3.138}$$

then assures the global optimality of the design $B(x)$.

It can be shown that the natural frequencies of a beam are always single eigenvalues. A portal frame, however, may have natural frequencies that are double eigenvalues. The optimality condition in Eq. (3.131) for the beam and the columns of the frame must then be modified in the same manner as Eq. (3.136).

3.5.6. Prescribed Upper Bound on Elastic Deflection at Specified Cross Section

To treat this behavioral constraint the principle of *stationary* mutual potential energy (Section 1.2.5) is used. Contrary to the principles used for the constraints discussed so far, this is not a *minimum* principle. Accordingly, it furnishes, in general, only a condition for *local* optimality—that is, a condition for the cost of a design to be, within higher-order quantities, equal to that of any neighboring design. Only for statically determinate beams does this condition assure *global* optimality.

Let $v(x)$ be the deflection of the design $B(x)$ caused by the given load $p(x)$, and let $\bar{v}(x)$ be its deflection caused by a unit concentrated load at the section $x = x_0$, for whose deflection the upper bound v_0 is prescribed. For the design $B^*(x)$ let $v^*(x)$ and $\bar{v}^*(x)$ denote the corresponding deflections. We shall assume that $v(x_0) = v_0$, whereas $v^*(x_0) \leq v_0$.

According to Eq. (1.26) the inequality $v(x_0) - v^*(x_0) \geq 0$ may be written as

$$U(v^*, \bar{v}^*; B^*) - U(v, \bar{v}; B) \geq 0. \tag{3.139}$$

Since the deflections v, $\bar{v}$ are kinematically admissible for the design $B^*(x)$, applying Eq. (1.27) to this design yields

$$U(v, \bar{v}; B^*) - U(v^*, \bar{v}^*; B^*) \geq \tfrac{1}{2}\int B^*(v^{*\prime\prime} - v'')(\bar{v}^{*\prime\prime} - \bar{v}'')\,dx. \tag{3.140}$$

Addition of Eqs. (3.139) and (3.140) and use of the definition of mutual potential energy then furnishes

$$\int (B^* - B)v''\bar{v}''\,dx \geq \int B^*(v^{*\prime\prime} - v'')(\bar{v}^{*\prime\prime} - \bar{v}'')\,dx. \tag{3.141}$$

Since the right-hand side of Eq. (3.141) cannot, in general, be shown to be nonnegative, this inequality cannot be used in the manner in which the fundamental inequalities have been used for the behavioral constraints discussed so far. If, however, we restrict the design $B^*(x)$ to the neighborhood of the design $B(x)$, we may write $B^* = B + \delta B$, $v^* = v + \delta v$, etc. The right-hand side of Eq. (3.141) then vanishes to the first order, and it follows from the convexity of $\gamma(B)$ that the condition

$$v''\bar{v}'' = c^2\gamma_B(B) \tag{3.142}$$

assures that the cost of the design $B(x)$ equals that of any neighboring design to within higher-order quantities. Note that Eq. (3.142) implies that $v''\bar{v}''$ and, hence, the product of the bending moments $M = -Bv''$ and $\bar{M} = -B\bar{v}''$ must be positive, which was first noted by Barnett (1961).

If the considered beam is *statically determinate*, the bending moments of the designs $B(x)$ and $B^*(x)$ under the given load $p(x)$ are identical; that is, $Bv'' = B^*v^{*\prime\prime}$ or

$$v^{*\prime\prime} - v'' = -\frac{(B^* - B)v''}{B^*}, \tag{3.143}$$

where $B^*(x)$ is no longer restricted to the neighborhood of $B(x)$. Substituting Eq. (3.143) and the corresponding expression of $\bar{v}^{*\prime\prime} - \bar{v}^{\prime\prime}$ into Eq. (3.141) and using the optimality condition in Eq. (3.142) then yield the relation

$$\Gamma^* - \Gamma \geq \int (B^* - B)\gamma_B(B)\,dx \geq \int \left[\frac{(B^* - B)^2\gamma_B(B)}{B^*}\right] dx \geq 0, \qquad (3.144)$$

which establishes the global optimality of the design $B(x)$ of a *statically determinate* beam satisfying the condition in Eq. (3.142). For an *indeterminate* beam, however, this condition assures only stationary behavior of the cost in the neighborhood of $B(x)$, and the global minimum of cost may occur for some other design satisfying the condition of Eq. (3.142).

3.5.7. Prescribed Upper Bound on Maximum Elastic Deflection

In general, the designer is not as interested in bounding the deflection at a given cross section as he is in bounding the *greatest* deflection of the beam under the given load. For a *statically determinate* beam the optimality condition in Eq. (3.142) may then be used to obtain the bending stiffness $B(x; x_0)$ of the optimal design whose deflection at $x = x_0$ equals the prescribed upper bound v_0 for the maximum deflection. For this design to be the solution of the present problem, its greatest deflection must occur at $x = x_0$, and this condition may be used to determine the value of x_0.

For a *statically indeterminate* beam this procedure tends to become complicated, because Eq. (3.142) assures only the stationary behavior of the cost in the neighborhood of the design $B(x; x_0)$, and even for a fixed x_0 several designs satisfying Eq. (3.142) may have to be investigated.

3.5.8. Prescribed Upper Bound on Maximum Elastic Bending Stress

This constraint, which is not related to one of the classical extremum principles of structural theory, may be treated as follows.

The bending moment $M(x)$ produces the maximum bending stress

$$\sigma_{\max}(x) = \frac{\frac{1}{2}E|M(x)|h(x)}{B(x)}, \qquad (3.145)$$

where $h(x)$ is the height of the cross section x, and E is Young's modulus of the beam material. Let the design $B(x)$ be *fully stressed* by the given load $p(x)$; that is, let $\sigma_{\max}$ attain the prescribed bound σ_0 at *all* cross sections, and assume that the design $B^*(x)$ nowhere has a bending stress exceeding this bound when it carries the load $p(x)$. We then have

$$2\sigma_0/E = |M|h/B \geq |M^*|h^*/B^*. \qquad (3.146)$$

The principle of virtual work applied to the given load $p(x)$ and the corresponding bending moments $M(x)$ of the design $B(x)$ as well as to the deflection $v(x)$ under this load and the corresponding curvature $\kappa(x) = -v''(x)$ of this design then furnishes

$$\int pv\,dx = \int M\kappa\,dx = \int |M||\kappa|\,dx, \tag{3.147}$$

because $M(x)$ and $\kappa(x)$ have the same sign. If we replace M by the bending moment $M^*(x)$ under load p for the design $B^*(x)$, keeping $v(x)$ and $\kappa(x)$ the same, then the principle of virtual work yields

$$\int pv\,dx = \int M^*\kappa\,dx \leq \int |M^*||\kappa|\,dx, \tag{3.147}$$

because $M^*(x)$ and $\kappa(x)$ need not have the same sign. It follows from the relations in Eq. (3.147) that

$$\int |M^*||\kappa|\,dx \geq \int |M||\kappa|\,dx. \tag{3.148}$$

Using $|M|$ and $|M^*|$ from Eq. (3.146) finally gives

$$\int (B^*/h^* - B/h)|\kappa|\,dx \geq 0, \tag{3.149}$$

and this relation may be used in the same manner as earlier fundamental inequalities. Indeed, if we now regard the specific cost as a function $\bar{\gamma}$ of the variable B/h and assume it to be convex and monotonically increasing, it follows from Eq. (3.149) that the *global* optimality of the design $B(x)$ is assured by the condition

$$|\kappa| = \bar{c}\bar{\gamma}_{B/h}(B/h), \tag{3.150}$$

where the subscript now indicates differentiation with respect to B/h, and the positive constant $\bar{c}$ must be evaluated from the condition that $\sigma_{\max}$ from Eq. (3.145) has the constant value σ_0.

Consider, for instance, a beam with rectangular cross section of given height $h(x)$ and a breadth $b(x)$ that is the choice of the designer. If the volume of the beam is to be minimized, we may set $\bar{\gamma} = (E/12)bh = (1/h)(B/h)$. With $\bar{\gamma}_{B/h} = 1/h$, the optimality condition of Eq. (3.150) then yields

$$|\bar{\kappa}| = \bar{c}/h, \tag{3.151}$$

where the curvature has been denoted by $\bar{\kappa}$ to distinguish this design from the one to be discussed next.

When a beam of this kind that is supported in the same manner and subject to the same loads is to be optimally designed for given compliance, we may set $\gamma = (E/12)bh = B/h^2$. With $\gamma_B = 1/h^2$, the optimality condition of Eq. (3.97) furnishes $\kappa^2 = c^2/h^2$ or

$$|\kappa| = c/h. \tag{3.152}$$

Except for a scale factor the two optimal designs thus have identical curvatures. Accordingly, the cross sections at which the curvatures change sign and the bending moments vanish are the same for the two beams. This means that the bending moments of the two beams are identical. Since Eq. (3.152) indicates that the second beam also has constant maximum stress, the breadths of the the two beams differ only by a constant factor.

Exercise 3.10. Consider a beam of rectangular cross section of variable breadth and given (nonuniform) height $h(x)$. Minimum volume is desired. Show that the optimality criterion is the same for

(i) assigned limit load
(ii) assigned elastic compliance
(iii) assigned maximum elastic bending stress

and compare the designs obtained in the three cases.

3.6. Examples of Optimal Single Purpose Structures for Various Behaviorial Constraints

3.6.1. Prescribed Load Factor for Plastic Collapse When Structural Weight is to Be Taken into Account

Consider a doubly built-in beam of span l that is to carry a permanent downward load $p(x)$ in addition to its own weight. The beam is to be optimally designed for a prescribed load factor Λ for plastic collapse. The load $p(x)$ is to be symmetric with respect to the center of the span. The optimal design and its weight $q(x)$ will then also be symmetric with respect to this center, and so will the bending moment $M(x)$ caused by p and q.

The bending moment M will be negative near the supports and positive in the central region. Accordingly, in the optimality condition in Eq. (3.122), $|v''|$ may be replaced by v'' near the supports and by $-v''$ in the central region.

When applying the optimality condition, we shall, for simplicity, suppose that $q_Y(Y)$ and $\gamma_Y(Y)$ are constants and denote them by α and β, respectively. We now take account of the fact that the positive constant c^2 in Eq. (3.122) is arbitrary, because a positive multiple of a collapse mechanism is also a collapse mechanism. Setting $c^2 = \Lambda\alpha/\beta$, we then obtain the optimality conditions

$$v'' - \Lambda\alpha(1 + v) = 0 \tag{3.153}$$

near the supports and

$$v'' + \Lambda\alpha(1 + v) = 0 \tag{3.154}$$

in the central region. Integrating Eq. (3.153), with the boundary conditions $v(0) = v'(0) = 0$, yields, near the support,

$$v(x) = \cosh(x\sqrt{\Lambda\alpha}) - 1, \qquad 0 \leq x \leq l/2 - b, \tag{3.155}$$

where the parameter b is to be determined. Integrating Eq. (3.154), with symmetry condition $v'(l/2) = 0$, yields, in the central region,

$$v(x) = A\cos[(l/2 - x)\sqrt{\Lambda\alpha}] - 1, \qquad l/2 - b \leq x \leq l/2. \tag{3.156}$$

The amplitude parameter A and the parameter b are determined from continuity conditions for v and v' at the interface $x = l/2 - b$ between the two regions. These continuity conditions,

$$\cosh[(l/2 - b)\sqrt{\Lambda\alpha}] = A\cos(b\sqrt{\Lambda\alpha}), \tag{3.157}$$

$$\sinh[(l/2 - b)\sqrt{\Lambda\alpha}] = A\sin(b\sqrt{\Lambda\alpha}), \tag{3.158}$$

furnish

$$\tanh[(l/2 - b)\sqrt{\Lambda\alpha}] = \tan(b\sqrt{\Lambda\alpha}). \tag{3.159}$$

Table 3.2 shows corresponding values of $l\sqrt{\Lambda\alpha}$ and b/l. Note that for $\alpha = 0$—that is, when the weight of the beam is neglected—the points of counterflexure are the outward quarter-chord points, as shown in Section 3.2.2.

Table 3.2. b/l as a Function of $l\sqrt{\Lambda\alpha}$

$l\sqrt{\Lambda\alpha}$	b/l
0.0	0.2500
1.0	0.2450
2.0	0.2295
4.0	0.1775

With the points of counterflexure known, the bending moment distribution $M(x)$ is statically determinate, and the optimal design is given by $Y(x) = |M(x)|$. Its cost is

$$\Gamma = \int \gamma \, dx = \beta \int Y \, dx = \beta \int |M| \, dx. \tag{3.160}$$

3.6.2. Prescribed Elastic Buckling Load

Consider a column that is simply supported at its ends $x = 0$ and $x = l$ and subject to the given compressive axial load P. Assume that the specific cost $\gamma(x)$ is proportional to the bending stiffness $B(x)$, and determine the function $B(x)$ to minimize the cost $\int_0^l B(x) \, dx$ and to yield an elastic buckling load that has at least the value P.

According to the optimality condition in Eq. (3.136), the optimal column admits a buckling mode $v(x)$ satisfying

$$v'' = -c. \tag{3.161}$$

Integrating this differential equation, with boundary conditions $v(0) = v(l) = 0$, furnishes

$$v = \frac{cx(l - x)}{2}. \tag{3.162}$$

Since the bending moment Pv equals $-Bv''$, we have $B = Px(l - x)/2$. This design with vanishing bending stiffness at the ends of the column is not practical. Let us therefore introduce a positive lower bound B_0 for B. The optimality condition above then applies only over a central region $\xi \leq x \leq l - \xi$, whereas, in the end regions $0 \leq x \leq \xi$ and $l - \xi \leq x \leq l$, we have bending moment

$$Pv = -B_0 v''. \tag{3.163}$$

Integrating this differential equation, with initial condition $v(0) = 0$, yields

$$v = \beta \sin \alpha x \qquad \text{for } 0 \leq x \leq \xi, \tag{3.164}$$

where β is a constant of integration and $\alpha = \sqrt{P/B_0}$. On the other hand, integrating Eq. (3.161), with symmetry condition $v'(l/2) = 0$ and continuity condition of v at $x = \xi$, gives

$$v = \beta \sin \alpha\xi + \frac{c(x - \xi)(l - x - \xi)}{2} \qquad \text{for } \xi \leq x \leq l - \xi \tag{3.165}$$

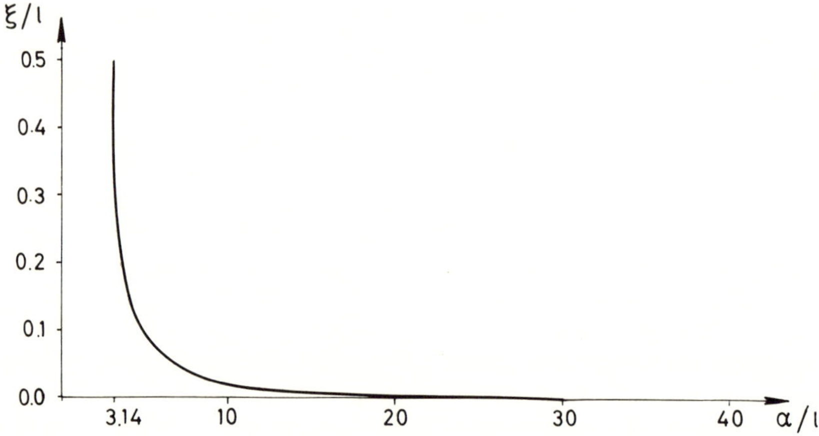

Figure 3.19. Minimum-volume design of simply supported beam for assigned buckling load P with lower bound B_0 on stiffness: dimensionless extent of constant stiffness zones versus $\alpha = (P/B_0)^{1/2}$.

From the continuity of the slope v' and the bending moment $-Bv''$ at $x = \xi$, we finally obtain the transcendental equation

$$\alpha\xi + \cot \alpha\xi - \alpha l/2 = 0 \tag{3.166}$$

for $\alpha\xi$. The smallest positive root of this equation furnishes the value of ξ.

Note that for $P = \pi^2 B_0/l^2$, that is, for $\alpha l = \pi$, the beam with constant stiffness B_0 has given buckling load P, and Eq. (3.166) furnishes $\xi = l/2$. As P increases beyond this value, ξ decreases. For $\alpha l = 4$, for instance, one finds $\alpha\xi = 0.6308$ and, hence, $\xi = 0.1577l$. The variation of ξ/l with αl for $\alpha l > \pi$ is shown in Fig. 3.19. The preceding discussion is due to Prager and Taylor (1968).

3.6.3. Optimal Elastic Design for Bounded Deflection

Let us consider a simply supported beam of span l with rectangular cross section of constant height h and variable breadth $b(x)$. It is subjected to a uniformly distributed load p. It is to be designed for minimum volume subject to the constraint of an upper bound kl on the central deflection $v(l/2)$. By symmetry of supports and loading we see that we may treat only one half of the beam. In order to avoid vanishing cross section at the supports, we will set a lower bound B_0 on the bending stiffness $B = Ebh^3/12$, where E is Young's modulus. The specific cost is $\gamma = bh$.

If we let $v'' = \kappa$ and $\bar{v}'' = \bar{\kappa}$, the optimality condition of Eq. (3.142) can be written as

$$\zeta \kappa \bar{\kappa} = 1, \qquad \zeta > 0, \tag{3.167}$$

where $B > B_0$. Indeed, in the right-hand side of Eq. (3.142), $\gamma_B(B) = 12/Eh^2$, and c^2 is an unknown constant, as ζ now is. We recall that κ is produced by the loading p; $\bar{\kappa}$, by a concentrated unit load at mid-span. With the origin of x at one support, we have

$$M = \kappa B = \frac{px(l-x)}{2}, \tag{3.168}$$

$$\bar{M} = \bar{\kappa} B = \frac{x}{2}, \qquad 0 \leq x \leq \frac{l}{2}. \tag{3.169}$$

In an unknown interval $0 \leq x \leq a$, we have $B(x) = B_0$, and the optimality condition becomes

$$\zeta \kappa \bar{\kappa} \leq 1. \tag{3.170}$$

On the other hand, in the interval $a \leq x \leq l/2$, Eq. (3.167) holds and gives

$$B(x) = (M\bar{M}\zeta)^{1/2}. \tag{3.171}$$

The constant ζ is obtained from the condition that the central deflection is equal to kl; that is,

$$\int_0^{l/2} \frac{M\bar{M}}{B}\, dx = \frac{kl}{2},$$

or, more explicitly,

$$\int_0^a \frac{M\bar{M}}{B_0}\, dx + \int_a^{l/2} \left(\frac{M\bar{M}}{\zeta}\right)^{1/2} dx = \frac{kl}{2}, \tag{3.172}$$

where the expression of Eq. (3.171) for $B(x)$ has been used in the second integral. Substituting the expressions in Eqs. (3.168) and (3.169) for M and $\bar{M}$, respectively, in Eq. (3.172) and formally integrating, we obtain

$$\frac{p(a^3 l/3 - a^4/4)}{4B_0} + \frac{p^{1/2}[(3a + 2l)(l - a)^{3/2} - 7(l/2)^{5/2}]}{15\zeta^{1/2}} = kl/2. \tag{3.173}$$

The unknown abscissa a is related to ζ by the condition of continuity of $B(x)$ at $x = a$:

$$B_0 = [M(a)\bar{M}(a)\zeta]^{1/2},$$

which gives

$$B_0^2 = \frac{\zeta p a^2(l - a)}{4}. \tag{3.174}$$

It is easily verified, using Eqs. (3.168), (3.169), and (3.174), that the optimality condition in Eq. (3.170) is satisfied in the interval $0 \le x \le a$. Hence, the desired optimal design is

$$B(x) = B_0, \qquad 0 \le x \le a,$$

$$B(x) = (M\bar{M}\zeta)^{1/2}, \qquad a \le x \le l/2,$$

where M and $\bar{M}$ are given by Eqs. (3.168) and (3.169), and ζ and a are solutions of the system of Eqs. (3.173) and (3.174).

When $B_0 = 0$, the first integral in Eq. (3.172) vanishes, a is equal to zero, Eq. (3.174) disappears, and Eq. (3.172) gives

$$\zeta = 0.102 p l^3/k^2. \tag{3.175}$$

On the other hand, setting $a = l/2$ in Eq. (3.173) gives

$$B_0^* = 5pl^3/384k. \tag{3.176}$$

When the lower bound imposed on B is at least equal to the value given by Eq. (3.176), the prismatic beam with bending rigidity B_0 is the optimal design.

Other cases of statically determinate beams could be treated in a similar manner for absolute optimality. When the beam is redundant, we recall that only local stationarity of the cost is achieved. Typical solutions can be found, for example, in Shield and Prager (1970), Barnett (1961), Huang and Tang (1969), Chern (1971), and, also, Huang (1971), Cinquini (1979), and Haug (1981), who treated limitation of maximum deflection.

3.7. General Formulation of Optimality Conditions for Multipurpose Beams and Frames

3.7.1. Optimal Plastic Design for Alternative Limit Loadings

Consider the optimal design of a beam for two given alternative limit loadings $p_1(x)$ and $p_2(x)$. Let us assume that both loadings are relevant for the optimal design, since otherwise the problem becomes one of optimal design for the single relevant loading. For simplicity let us restrict the discussion to the case of continuously varying yield moment in the given fixed range $[Y^-, Y^+]$.

Comparing the optimal design $Y(x)$ to another design $Y^*(x)$ capable of carrying the given loadings, and denoting the rates of deflection of collapse mechanisms of the first design under these loadings by $v_1(x)$ and $v_2(x)$, we may write a fundamental inequality of the form of Eq. (3.20) for each of these two mechanisms. Linear combination of these inequalities with the nonnegative multipliers α and $1 - \alpha$, where $0 \leq \alpha \leq 1$, and transformation of the resulting inequality in the manner used for Eq. (3.20) show that

$$\alpha|v_1''| + (1 - \alpha)|v_2''| \begin{Bmatrix} \leq \\ = \\ \geq \end{Bmatrix} c^2\gamma_Y \qquad \text{if} \begin{cases} Y = Y^-, \\ Y^- < Y < Y^+, \\ Y = Y^+, \end{cases} \tag{3.177}$$

is a sufficient condition for the global optimality of the design $Y(x)$.

Since a positive multiple of a yield mechanism is again a yield mechanism, we may drop the factors α and $1 - \alpha$ and the constant c^2 from the optimality condition in Eq. (3.177). The simplified condition states that an optimal design admits yield mechanisms for the two loadings for which the sum of the absolute rates of curvature is at any section at most equal to, at least equal to, or equal to the value of γ_Y at this section, depending on whether the yield moment there equals Y^- or Y^+ or lies between these values. When there are no explicit bounds on the yield moment, this optimality condition may be regarded as a special case of an optimality condition given by Shield (1963) for the optimal plastic design of plates.

If n, rather than two, alternative loadings are to be considered , the optimal design admits collapse mechanisms with rates of deflection $v_1(x)$, $v_2(x), \ldots, v_n(x)$ satisfying

$$\sum_{i=1}^{n} |v_i''| \begin{Bmatrix} \leq \\ = \\ \geq \end{Bmatrix} \gamma_Y \qquad \text{if} \begin{cases} Y = Y^-, \\ Y^- < Y < Y^+, \\ Y = Y^+. \end{cases} \tag{3.178}$$

Let us now consider a movable loading $p(x, \xi)$, where the variable ξ, which specifies the position of the loading, may vary from ξ_1 to $\xi_2 > \xi_1$. If all positions of the loading are relevant, a design is globally optimal if it admits, for each position ξ of the loading, a collapse mechanism with rates of deflection $v(x, \xi)$ such that

$$\int_{\xi_1}^{\xi_2} |v''(x, \xi)|\, d\xi \begin{Bmatrix} \leq \\ = \\ \geq \end{Bmatrix} \gamma_Y(Y(x)) \quad \text{if} \begin{cases} Y = Y^-, \\ Y^- < Y < Y^+, \\ Y = Y^+. \end{cases} \tag{3.179}$$

This optimality condition, due to Save and Prager (1963) for a linear cost function and to Save (1972) for a convex cost function, is readily extended to the case where certain positions of the loading are not relevant by setting $v(x, \xi) \equiv 0$ for such positions ξ. The condition can also be shown to be necessary for concave as well as convex cost functions.

3.7.2. Optimal Design for Various Behavioral Constraints or Alternative Loadings or Both

In Section 3.5, where various behavioral constraints are considered, we remarked that the fundamental inequality from which the optimality criterion is derived is always of the form

$$\int (D^* - D)\, dx \geq 0 \tag{3.180}$$

(see Eqs. (3.8), (3.92), and similar relevant relations in Section 3.5). In this inequality D is the specific energy, or power, associated with the behavioral constraint considered for any loading, where D^* and D are computed from the same displacement field $v(x)$ or collapse mechanism $\dot{v}(x)$,† but for two different designs. The superscript $*$ indicates that the corresponding design satisfies the behavioral constraint, with possibly some safety margin, whereas the other design satisfies this constraint *strictly.* Four important cases and the forms of D occurring in them are

(1) assigned limit load in the absence of design-dependent forces, where $D = Y|\dot{v}''|$;
(2) assigned limit load in the presence of design-dependent forces, where $D = Y\{|\dot{v}''| - \Lambda q_Y(Y)\dot{v}\}$;
(3) assigned elastic static compliance, where $D = Bv''^2/2$; and
(4) assigned elastic deflection at a given point, where $D = Bv''\bar{v}''$.

† We recall that when displacements and velocities must be considered simultaneously we use a dot to denote differentiation with respect to time or to some monotonically increasing function of time.

Consider the optimal design of a beam under the imposition of the combination of n behavioral constraints and loadings, each of which has a specific energy D_i for which the relation in Eq. (3.180) holds. Now, all these energies D_i, $i = 1, 2, \ldots, n$, are proportional to the design variables Y or B (if, as often accepted, the design-dependent forces are also proportional to Y or B). Hence, the assumed convexity of $\gamma(Y)$ or $\gamma(B)$ is equivalent to the convexity of the specific cost $\gamma(D_1, D_2, \ldots)$; that is,

$$\gamma^* - \gamma \geq \sum_{i=1}^{n} (D_i^* - D_i)\frac{\partial \gamma}{\partial D_i}. \tag{3.181}$$

Because an equality of the type in Eq. (3.180) can be written for every behavioral constraint (or loading case), we can derive the fundamental inequality

$$\sum_{i=1}^{n} \int \alpha_i(D_i^* - D_i)\, dx \geq 0. \tag{3.182}$$

Comparing Eqs. (3.181) and (3.182) shows that a sufficient condition for $\Gamma^* - \Gamma = \int (\gamma^* - \gamma)\, dx \geq 0$ is

$$\gamma_{D_i} = \alpha_i, \qquad i = 1, 2, \ldots, n, \quad \alpha_i > 0, \tag{3.183}$$

where the subscript D_i indicates differentiation with respect to D_i.

The very general condition in Eq. (3.183), due to Save (1975), must be particularized for each specific problem. For example, let a beam be optimally designed for some assigned limit load and for assigned maximum elastic compliance under another loading. We then have

$$D_1 = Y|\dot{\kappa}| \quad \text{and} \quad D_2 = B\kappa^2/2.$$

If the cross section is rectangular with given constant height h and variable breadth $b(x)$, and if minimum volume is desired, we have

$$Y = \frac{bh^2\sigma_0}{4}, \quad B = \frac{Ebh^3}{12}, \quad \gamma = bh, \quad \text{and} \quad \frac{\partial \gamma}{\partial b} = h.$$

Then, because

$$\frac{\partial \gamma}{\partial b} = \frac{\partial \gamma}{\partial D_1} \cdot \frac{\partial D_1}{\partial b} + \frac{\partial \gamma}{\partial D_2} \cdot \frac{\partial D_2}{\partial b},$$

we obtain, using the optimality conditions in Eq. (3.183),

$$h = \frac{\alpha_1 h^2 \sigma_0 |\dot{\kappa}|}{4} + \frac{\alpha_2 E h^3 \kappa^2}{24}.$$

Because a collapse mechanism is defined except for a positive scalar factor. we can set $\alpha_1 h \sigma_0 / 4 = 1$ and define $\beta = \alpha_2 E h^2 / 24$. Then we may write the optimality condition as

$$|\dot{\kappa}| + \beta \kappa^2 = 1. \tag{3.184}$$

Equation (3.184) and other analogous conditions derived from Eq. (3.183) will be applied in Section 3.8. After an optimality condition such as Eq. (3.184) has been obtained, it can be generalized to take into account upper and lower bounds on Y or B or both, and also to deal with stepwise variations of Y and B, design-dependent forces, movable loads, and a general convex cost function. It must be emphasized that when the two designs considered in Eq. (3.184) are not arbitrarily different, but differ only by an infinitely small variation, the resulting optimality criterion is that the cost be stationary or possibly have a local minimum.

3.8. Examples of Optimal Design of Multipurpose Structures

3.8.1. Prescribed Limit Load for Multiple Loadings and Movable Loads

We hereafter consider designing a beam or a frame for a finite set of equally possible loadings, for *each* of which the load factor at plastic collapse is assigned separately. This loading situation is called *multiple loadings* or *alternative loadings*, whereas movable loads imply an infinite set of equally possible loadings obtained by the application of a given load system anywhere in some subspan of the beam or frame.

Let a propped cantilever be optimality designed for *two alternative limit loads* P' and P'', with respective abscissas a and b, as shown in Fig. 3.20a. The specific cost γ is supposed to be proportional to the yield moment Y.

We first assume that both loads are to influence the design and that the corresponding bending moment diagrams at collapse are as shown in Fig. 3.20a. Hence, the optimal design $Y = |M(x)|$ will be of the type drawn in Fig. 3.20. The collapse mechanism will exhibit (1) under load P', a constant positive rate of curvature $\dot{\kappa} = c^2$ in the interval $[0, x_1]$ and a negative rate of curvature $\dot{\kappa} = -c^2$ in $[x_2, x_3]$, the two intervals being separated by rigid regions; (2) under load P'', $\dot{\kappa} = c^2$ in $[x_2, c_3]$, $\dot{\kappa} = -c^2$ in $[x_3, l]$, and $\dot{\kappa} = 0$ elsewhere.

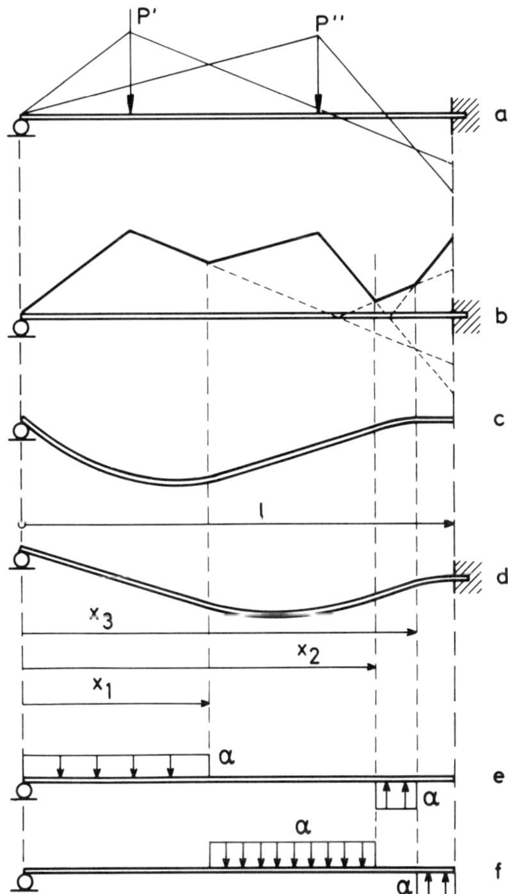

Figure 3.20. (a) Beam, alternative loads, and bending moment diagrams. (b) Envelope of bending moment diagrams. (c) Collapse mechanism for load P'. (d) Collapse mechanism for load P''. (e) Conjugate beam for loading P'. (f) Conjugate beam for loading P''.

If we denote by R' and R'' the redundant reactions of the simple support under loads P' and P'', respectively, we determine the five unknowns x_1, x_2, x_3, R', and R'' from the conditions that the bending moments $M(P', x)$ and $M(P'', x)$ are equal for $x = x_1$ and $x = x_3$, that they are opposite at $x = x_2$, and that the conjugate beams loaded with the rates of curvature of the collapse mechanisms (Fig. 3.20e, f) are in rotational equilibrium.

Since

$$M(P', x) = R'x, \qquad 0 \leq x \leq a,$$

$$M(P', x) = R'x - P'(x - a), \qquad a \leq x \leq l,$$

and

$$M(P'', x) = R''x, \qquad 0 \le x \le b,$$

$$M(P'', x) = R''x - P''(x - b), \qquad b \le x \le l,$$

the first three conditions give

$$x_1 = \frac{P'a}{P' + R'' - R'}, \tag{3.185}$$

$$x_2 = \frac{P'a + P''b}{P' + P'' - R' - R''}, \tag{3.186}$$

$$x_3 = \frac{P'a - P''b}{P' - P'' + R'' - R'}. \tag{3.187}$$

The two kinematic conditions are

$$x_1^2 + x_2^2 = x_3^2, \tag{3.188}$$

$$x_2^2 - x_1^2 = l^2 - x_3^2. \tag{3.189}$$

Adding Eqs. (3.188) and (3.189) immediately gives

$$x_2 = l/\sqrt{2}. \tag{3.190}$$

Eliminating $R'' - R$ from Eqs. (3.185) and (3.187) gives

$$(P'a - P''b)x_1 + P''x_1x_3 - P'ax_3 = 0.$$

If both sides of this equation are divided by $P''l^2$ and the dimensionless variables

$$p \equiv \frac{P'}{P''}, \quad \alpha \equiv \frac{a}{l}, \quad \beta \equiv \frac{b}{l}, \quad \zeta_1 \equiv \frac{x_1}{l}, \quad \zeta_2 \equiv \frac{x_2}{l}, \quad \zeta_3 \equiv \frac{x_3}{l}$$

are introduced, this equation becomes

$$(p\alpha - \beta)\zeta_1 - p\alpha\zeta_3 + \zeta_1\zeta_3 = 0. \tag{3.191}$$

Subtracting Eqs. (3.188) and (3.189), we get

$$\zeta_3^2 - \zeta_1^2 = \tfrac{1}{2}. \tag{3.192}$$

Substituting ζ_3 from Eq. (3.191) in Eq. (3.192) furnishes a fourth-degree equation in ζ_1 that can be solved numerically. The only acceptable root must satisfy the inequalities

$$\alpha \leq \zeta_1 \leq \beta \leq 1/\sqrt{2}.$$

The corresponding value of ζ_3 is obtained from Eq. (3.191). Equations (3.186) and (3.185), or (3.187), then give R' and R'', and the optimal design is defined by

$$Y(x) = \begin{cases} M(P', x) & \text{for } 0 \leq x \leq x_1, \\ M(P'', x) & \text{for } x_1 \leq x \leq x_2, \\ -M(P', x) & \text{for } x_2 \leq x \leq x_3, \\ -M(P'', x) & \text{for } x_3 \leq x \leq l. \end{cases}$$

The range of applicability of this solution is most readily obtained by considering that both loads are active in the optimal design when the optimal design for a single load is unable to support the other load. When P' acts alone, we apply the method of Section 3.2 and find the optimal design to be $Y(x) = |M(x)|$, with

$$M(x, P') = P'(1 - \sqrt{2}\, a/l)x, \qquad 0 \leq x \leq a,$$

$$M(x, P') = P'a(1 - \sqrt{2}\, x/l), \qquad a \leq x \leq l.$$

Similarly, for P'' acting alone we have $Y(x) = |M(x)|$, with

$$M(x, P'') = P''(1 - \sqrt{2}\, b/l)x, \qquad 0 \leq x \leq b,$$

$$M(x, P'') = P''b(1 - \sqrt{2}\, x/l), \qquad b \leq x \leq l.$$

P' alone is relevant as long as $M(b, P'') \leq M(b, P')$. This condition gives

$$P'a - P''b \geq 0 \tag{a}$$

P'' alone is relevant as long as $M(a, P) \leq M(a, P'')$. From this condition we obtain

$$P'' - P' \geq \frac{\sqrt{2}(P''b - P'a)}{l}.$$

But, because Eq. (a) is not satisfied, $P''b - P'a > 0$, and we conclude that we must have

$$\frac{P'' - P'}{P''b - P'a} \geq \frac{\sqrt{2}}{l}. \tag{b}$$

In dimensionless variables, Eqs. (a) and (b) are written

$$p \geq \frac{\beta}{\alpha}, \tag{a$'$}$$

$$p \leq \frac{1 - \sqrt{2}\,\beta}{1 - \sqrt{2}\,\alpha}, \tag{b$'$}$$

taking into account that when (b′) is valid (a′) is not and, hence, $p(\alpha/\beta) - 1 < 0$.

Because $\alpha < \beta$, we have

$$\frac{1 - \sqrt{2}\,\beta}{1 - \sqrt{2}\,\alpha} < \frac{\beta}{\alpha},$$

and Eqs. (a′) and (b′) define some real finite interval for p. Examples of solutions for values of the parameters p, α, and β for which both loads are active are given in Table 3.3. Note that ζ_2 is always equal to $1/\sqrt{2} = 0.7071$.

Table 3.3. Propped Cantilever with Two Concentrated Loads: Dimensionless Parameters for Optimal Design

α	β	β/α	$\dfrac{1-\sqrt{2}\beta}{1-\sqrt{2}\alpha}$	p	ζ_1	ζ_3
0.100	0.200	2.000	0.835	1.418	0.153	0.723
0.100	0.300	3.000	0.671	1.835	0.217	0.740
0.100	0.400	4.000	0.506	2.253	0.291	0.765
0.100	0.500	5.000	0.341	2.671	0.375	0.800
0.100	0.600	6.000	0.176	3.088	0.469	0.849
0.200	0.300	1.500	0.803	1.151	0.253	0.751
0.200	0.400	2.000	0.606	1.303	0.317	0.775
0.200	0.500	2.500	0.408	1.454	0.391	0.808
0.200	0.600	3.000	0.211	1.606	0.476	0.852
0.300	0.400	1.333	0.754	1.044	0.351	0.789
0.300	0.500	1.667	0.509	1.088	0.413	0.819
0.300	0.600	2.000	0.263	1.132	0.486	0.858
0.400	0.500	1.250	0.674	0.962	0.445	0.835
0.400	0.600	1.500	0.349	0.924	0.502	0.867
0.500	0.600	1.200	0.517	0.859	0.531	0.884

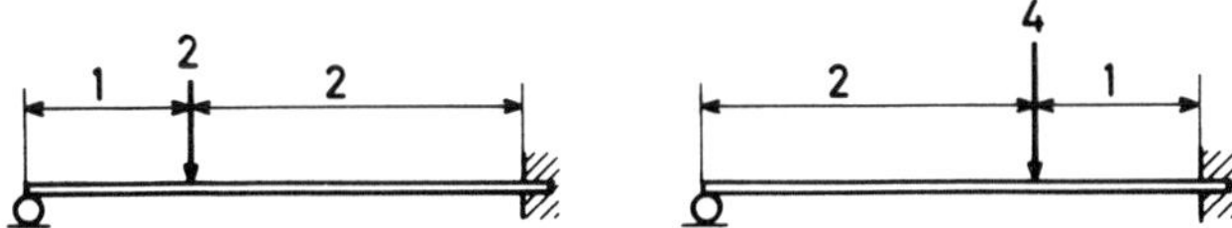

Figure 3.21. Alternative loadings.

Similar problems were treated by Mayeda and Prager (1967): They consider the propped cantilever subjected to three, equal, alternative, concentrated loads located at one-, two-, and three-fifths of the span from the simple support, respectively. It is to be noted that, with alternative loadings, intermediate cross sections with vanishing yield moment disappear when at least two loadings are active in the design. Of course, bounds on the yield moment can be introduced, as for the doubly built-in beam subjected to alternative, uniformly distributed and central, concentrated loads, also treated in Mayeda and Prager (1967).

If there are only two alternative loadings with parameters P' and P'', and if the unbounded yield moment is the same for both positive and negative bendings, the optimal design, corresponding to a specific cost proportional to the yield moment, can also be obtained from the *Hemp–Nagtegaal–Prager superposition principle*† [see Hemp (1968), Nagtegaal and Prager (1973), and Nagtegaal (1973)]. According to this principle, it suffices to construct the optimal designs Y^+ and Y^- for the separate loadings $P^+ = (P' + P'')/2$ and $P^- = (P' - P'')/2$ [or, equivalently, $(P'' - P')/2$] and add the two designs. We refer the reader to the original proof [Nagtegaal (1973)] and simply illustrate the application of that principle to our example of the propped cantilever with two concentrated loads when $p = \frac{1}{2}$, $\alpha = \frac{1}{3}$, and $\beta/\frac{2}{3}$, For simplicity, let $l = 3$, $P' = 2$, and $P'' = 4$, as shown in Fig. 3.21.

The loading P^+ is shown in Fig. 3.22, and the corresponding optimal design is a direct application of the method developed in section 3.2. With $M = Y = 0$ in $x = l/\sqrt{2} = 2.1213$, we obtain $R^+ = 0.6430$, $M_A^+ = 0.6430 = Y_A^+$, $M_B^+ = 0.2860 = Y_B^+$, and $M_C^+ = -2.0711 = -Y_C^+$.

The loading P^- is shown in Fig. 3.23a. The bending moment vanishes at $x = 0$ and also at x_1 and x_2, such that $1 < x_1 < 2$ and $2 < x_2 < 3$. The equilibrium of the conjugate beam loaded with constant rates of curvature

† A generalization to more than two alternative loads can be found in Rozvany and Hill (1978).

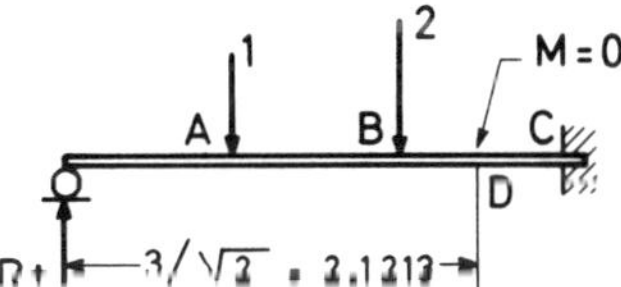

Figure 3.22. Loading P^+ for application of superposition.

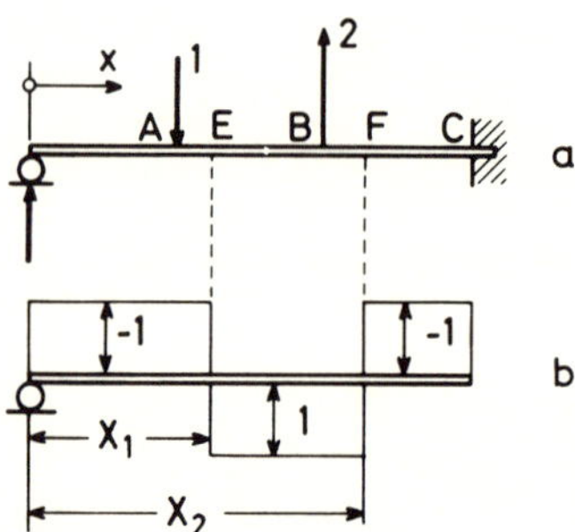

Figure 3.23. (a) Loading P^- for application of superposition. (b) Conjugate beam for loading P^-.

(Fig. 3.23b) gives

$$x_2^2 - x_1^2 = 4.5. \tag{c}$$

Vanishing moments at $x = x_1$ and $x = x_2$ imply that

$$R^- = 1 - 1/x_1 \tag{d}$$

and

$$R^- = -1 + 3/x_2. \tag{e}$$

The solution of system (c)-(e) is $x_1 = 1.271$, $x_2 = 2.4729$, $R^- = 0.2132$, from which we obtain $M_A^- = Y_A^- = 0.2132$, $M_B^- = -0.5736 = -Y_B^-$, $M_C^- = -0.6397 = -Y_C^-$. By superposition the following design is found:

$$Y_A = 0.8562, \qquad Y_E = Y_A^+ - 0.271(Y_A^+ - Y_B^+) = 0.5463,$$

$$Y_B = 0.8596, \qquad Y_D = Y_B^- - \frac{0.1213}{0.4729} Y_B^- = 0.4265,$$

$$Y_F = Y_C^+ \frac{0.4729 - 0.1213}{3 - 2.1213} = 0.8287,$$

$$Y_C = 2.7108.$$

This solution is easily verified to be the optimal design for the alternative loadings by setting $M' = 2$, $P'' = 4$, $l = 3$, $x_1 = 1.271$, $x_2 = 2.1213$, $x_3 = 2.4729$, or, equivalently, $\zeta_1 = 0.4237$, $\zeta_2 = 0.07071$, and $\zeta_3 = 0.8243$, in Eqs. (3.185)-(3.187), (3.191), and (3.192), and in the expressions for $Y(x)$.

We are now able to deal with the problem of *one movable load P*—that is, a statically applied limit load of magnitude P that can occupy any

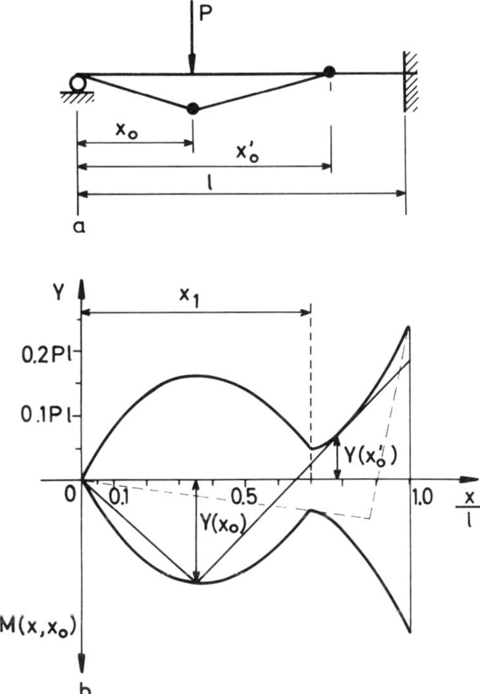

Figure 3.24. (a) Collapse mechanism for movable load. (b) Optimal design and bending moment diagrams.

location on the beam.† According to our optimality condition, Eq. (3.179), we must use a *family* of collapse mechanisms. We shall assume that when the load acts at a cross section with abscissa x_0 smaller than some unknown value x_1 a positive plastic hinge develops under the load, whereas a negative plastic hinge appears at some abscissa $x_0' > x_1$ (Fig. 3.24a). When the positive hinge moves from $x_0 = 0$ to $x_0 = x_1$ with the load, the negative hinge travels from $x_0' = x_1$ to $x_0' = l$. Hence, the family of mechanisms does not involve any superposition of hinges at any cross section. It is convenient for the analysis if we regard the hinges as narrow regions of widths dx_0 and dx_0'. We satisfy the optimality condition, Eq. (3.179), by setting $\dot{\kappa} = 1$ at x_0 and $\dot{\kappa} = -1$ at x_0' over dx_0 and dx_0', respectively, for every collapse mechanism. Note that no integration over x_0 (which here plays the role of ξ) is to be done in Eq. (3.179) because of the absence of superposition of hinges.

When the conjugate beam is loaded with the given rates of curvature, we have

$$x_0\, dx_0 = x_0'\, dx_0'. \tag{3.193}$$

† In opposition to the case of *moving* loads, neither dynamic effects nor cycling effects are considered here.

Upon integration, with boundary conditions $x_0' = x_1$ for $x_0 = 0$ and $x_0' = l$ for $x_0 = x_1$, we obtain $x_1 = l/\sqrt{2}$ and

$$x_0' = (x_0^2 + l^2/2)^{1/2}. \tag{3.194}$$

Equation (3.194) completely defines the family of collapse mechanisms. When $x_1 < x_0 \le l$, the beam is assumed to remain rigid.

Having completed the kinematical part of the solution, we undertake the statical part. As the point of application x_0 of the load varies, we require the bending moment diagram $M = M(x, x_0)$ to just touch the curve $Y(x)$ at $x = x_0$ and to be the envelope of the curve $Y(x)$ in $x_1 \le x \le l$, always being tangent to that curve at $x = x_0'$, as shown in Fig. 3.24b. These conditions are

$$M(x_0, x_0) = Y(x_0), \tag{3.195}$$

$$M(x_0', x_0) = -Y(x_0'), \tag{3.196}$$

$$\frac{\partial M(x, x_0)}{\partial x_0} = 0 \qquad \text{at } x = x_0'. \tag{3.197}$$

The three unknown functions are $Y(x_0)$, $Y(x_0')$ [with x_0' related to x_0 by Eq. (3.194)], and a redundant $R(x_0)$, which can be eliminated in the course of the solution.

The equation of the right-hand branch of the bending moment diagram is

$$M(x, x_0) = Y(x_0) - (x - x_0)[P - Y(x_0)/x_0], \qquad \text{with } 0 \le x_0 \le l/\sqrt{2},\ x_0 \le x \le l. \tag{3.198}$$

Substituting Eq. (3.198) in Eq. (3.197) gives

$$x_0' \cdot \frac{d}{dx_0}\left[\frac{Y(x_0)}{x_0}\right] + P = 0, \qquad 0 \le x_0 \le \frac{l}{\sqrt{2}}.$$

Using Eq. (3.194), we integrate this equation to obtain

$$Y(x_0) = x_0\{C - \ln(1/l)[x_0 + (x_0^2 + l^2/2)^{1/2}]\}, \qquad 0 \le x_0 \le l/2. \tag{3.199}$$

The integration constant C is determined from the continuity of the design $Y(x)$ at $x = x_1$. Indeed, we first substitute $Y(x_0)$ from Eq. (3.199) into Eq.

(3.198) and apply Eq. (3.196) to obtain

$$Y(x_0') = -x_0'\{C - \ln(1/l)[x_0 + (x_0^2 + l^2/2)^{1/2}]\} + P(x_0' - x_0). \quad (3.200)$$

Using Eq. (3.194), we eliminate x_0 from Eq. (3.200) and obtain

$$Y(x_0') = -x_0'\{C - \ln(1/l)[(x_0'^2 - l^2/2)^{1/2} + x_0']\}$$
$$+ P[x_0' - (x_0'^2 - l^2/2)^{1/2}], \qquad l/\sqrt{2} \leq x_0' \leq l. \quad (3.201)$$

From the condition $Y(x_0) = Y(x_0')$ when $x_0 = x_0' = l/\sqrt{2}$, we finally have

$$C = \tfrac{1}{2}[1 + \ln \tfrac{1}{2}(1 + \sqrt{2})]. \quad (3.202)$$

Substituting Eq. (3.202) in Eqs. (3.199) and (3.201) furnishes the design shown in Fig. 3.24b.

It remains to show that this design can remain rigid for $l/\sqrt{2} \leq x_0 \leq l$. This is achieved by finding an attainable, statically admissible, bending moment diagram for each of these values of x_0. Such a diagram is shown as a dotted line in Fig. 3.24b.

If we note that $x_1 = l/\sqrt{2}$ is the abscissa of the section with vanishing yield moment in the problems of Section 3.2.1, we can now very easily deal with the simultaneous action of the movable load and any system of fixed loads of Section 3.2.1. Indeed, because the equations of equilibrium are linear, simple addition of the bending moment diagrams for the separate loadings gives a bending moment diagram in equilibrium with the simultaneously applied loads. This diagram is attainable for a design obtained by simple addition of the respective optimal designs for the separate loads because we assumed Y to be the design variable (or proportional to it). Moreover, the beam constructed in this manner is at collapse with the mechanisms corresponding to the case of the movable load alone, because positive hinges of that family of mechanisms strictly cover the region of positive curvature rate of the mechanism for fixed loads; negative hinges and negative rate of curvature are handled similarly. Hence, the condition of Eq. (3.179) is fulfilled, and we conclude that we have obtained the optimal design for simultaneous fixed and moving loads. This *superposition method*, due to Save and Prager (1963) and extended to shells by Save and Shield (1966), holds when the mechanism for fixed loads is "compatible" with the family of mechanisms for the movable load in the sense given above. Various other problems of beams subjected to movable, concentrated, or distributed loads have been treated along the same line by Save and Prager (1963), Lamblin and Save (1971), and Lamblin (1972).

3.8.2. Prescribed Minimum Limit Load and Maximum Elastic Compliance

We first recall the findings of Section 3.4.1. Suppose the boundary conditions are given, there are no bounds on B and Y, and the specific cost is proportional to B and Y. Then if the loadings for both behavioral constraints differ only by a positive scalar factor, so do the optimal designs for the separate behavioral constraints.

Hence, when the two behavioral constraints must be regarded as alternative design constraints, it suffices to compare the two optimal designs for the separate constraints; the most costly of the two will necessarily satisfy the other behavioral constraint.

To be confronted with a "mixed" design where both behavioral constraints will influence the design simultaneously, we must deal either with two different loadings or with bounds on B and Y. Consider, for example, a simply supported beam carrying a uniformly distributed service load p under which the compliance must not exceed a given value $\mathscr{C}_0$. In addition, it must be able to support a central concentrated load F at plastic collapse. The cross section is rectangular with constant height h and variable width $b(x)$, and minimum volume is required. The bending moment diagrams are

$$M(p, x) = \frac{px(l-x)}{2}, \tag{3.203}$$

$$\begin{aligned} M(F, x) &= \frac{Fx}{2}, \qquad & 0 \le x \le \frac{l}{2}, \\ M(F, x) &= \frac{Fl}{2} - \frac{Fx}{2}, \qquad & \frac{l}{2} \le x \le l. \end{aligned} \tag{3.204}$$

The optimality condition in Eq. (3.184) applies directly:

$$|\dot{\kappa}| + \beta\kappa^2 = 1. \tag{3.205}$$

When subjected to the central load F, the beam will collapse with a central plastic region of length $l - 2a$ and with two elastic regions of length "a" near the supports. Hence, in the plastic region,

$$M(F, x) = Y, \qquad a \le x \le l - a, \tag{3.206}$$

or

$$b(x) = C_1 M(F, x), \qquad a \le x \le l - a, \tag{3.207}$$

with

$$C_1 \equiv 4/h^2\sigma_0. \tag{3.208}$$

In the elastic regions,

$$\beta\kappa^2 = 1, \quad 0 \le x \le a, \quad \text{and} \quad l - a \le x \le l \tag{3.209}$$

because $\dot{\kappa} = 0$ in Eq. (3.205) in these two intervals. If we use Hooke's law $\kappa = M(p, x)/B$ in Eq. (3.209) and let

$$C_2 \equiv 12/Eh^3, \tag{3.210}$$

we obtain

$$b(x) = C_2 M(p, x)\sqrt{\beta}, \qquad 0 \le x \le a, l - a \le x \le l, \tag{3.211}$$

Assigning the value $\mathscr{E}_0$ to the compliance of the design $B(x)$, we have

$$\int_0^l \frac{M^2(p, x)}{B(x)}\, dx = \mathscr{E}_0. \tag{3.212}$$

If we take into account the symmetry with respect to the center of the beam and use the relations in Eqs. (3.207) and (3.211), Eq. (3.212) becomes

$$2\int_0^a \frac{M(p, x)}{\sqrt{\beta}}\, dx + 2\int_a^{l/2} \frac{C_2 M^2(p, x)}{C_1 M(F, x)}\, dx = \mathscr{E}_0. \tag{3.213}$$

Substituting the expressions in Eqs. (3.203) and (3.204) for $M(p, x)$ and $M(F, x)$ in Eq. (3.203) and integrating, we obtain

$$\sqrt{\beta} = \frac{pa^2(l/2 - a/3)}{\mathscr{E}_0 - AC_2p^2/C_1F}, \tag{3.214}$$

where

$$A \equiv 0.0729l^4 - 0.5a^2l^2 - 0.25a^4 + 0.666a^3l.$$

The unknown a is obtained by setting $b(a)$ from Eq. (3.207) equal to $b(a)$ given by Eq. (3.211):

$$p(l - a) - C_1F/C_2\sqrt{\beta}. \tag{3.215}$$

When a and β are obtained from Eqs. (3.214) and (3.215), the design is given by Eqs. (3.207) and (3.211).

It remains to verify the optimality condition in Eq. (3.205) for $a \le x \le l/2$. Because $\dot{\kappa}$ is positive but arbitrary, we must simply have $\beta\kappa^2 < 1$. Since $\kappa = M(p, x)/B$ and $B = C_1 M(F, x)/C_2$, this condition becomes $(l - x)/(l - a) < 1$, which is obviously satisfied in the considered interval.

For the present solution to be valid, we must have $0 \le a \le l/2$, which, from Eqs. (3.214) and (3.215) requires that

$$0.0417 \le FC_1\mathscr{E}_0/p^2C_2l^4 \le 0.0729.$$

For smaller values of the parameter $FC_1\mathscr{E}_0/p^2C_2l^4$, the optimal elastic design with compliance $\mathscr{E}_0$ supports a limit load larger than F and is the solution. For larger values the optimal plastic design for F has an elastic compliance smaller than $\mathscr{E}_0$ [see Save and Igic (1982)].

3.8.3. Prescribed Limit Load and Elastic Deflection at a Given Section

Consider again the simply supported beam of Section 3.8.2. Design codes often require the maximum elastic deflection produced by service loads $p(x)$ to not be larger than some given bound k, and the load factor λ applied to the service loads at collapse to be not smaller than some assigned value. In addition, shearing forces at the supports require that a lower bound b_0 be set on the breadth $b(x)$. When the load p is distributed, symmetry makes the maximum deflection occur at mid-span and allows us to consider only one half of the beam, which we take to be the left half, $0 \le x \le l/2$.

The specific energies (or powers) associated with the two behavioral constraints are

$$D_1 = \beta\kappa\bar{\kappa}/2 \tag{3.216}$$

and

$$D_2 = Y|\dot{\kappa}|, \tag{3.217}$$

where κ and $\bar{\kappa}$ are the elastic curvatures under p and a central, concentrated unit load, respectively, and $\dot{\kappa}$ is the rate of plastic curvature at collapse. The bending moment distribution is

$$M(\lambda p, x) = \lambda M(p, x) = \frac{\lambda p x(l - x)}{2}. \tag{3.218}$$

Using the same procedure as we did to obtain the condition of Eq. (3.184), we find the optimality criterion

$$\zeta\kappa\bar{\kappa} + |\dot{\kappa}| \leq 1, \tag{3.219}$$

where equality holds for $b(x) > b_0$ and inequality holds for $b(x) = b_0$.

In the situation where both behavioral constraints are relevant, we assume that (1) the lower bound b_0 is reached in $0 \leq x \leq a_1$; (2) at collapse, plastic regions extend over the interval $a_1 \leq x \leq a_2 \leq l/2$. Note that, though it might be thought intuitively that collapse would exhibit a central plastic region, this assumption would render the condition in Eq. (3.219) impossible to satisfy, whereas with assumption (2) it turns out that Eq. (3.219) is actually satisfied. Hence, the optimal design is given by

$$b(x) = b_0 \qquad \text{for } 0 \leq x \leq a_1; \tag{3.220}$$

$$Y(x) = M(\lambda p, x) = \frac{\lambda p x(l-x)}{2}, \quad \text{or}$$

$$b(x) = C_1 \lambda M(p, x), \qquad \text{for } a_1 \leq x \leq a_2, \tag{3.221}$$

with C_1 given by Eq. (3.208), because this interval is at collapse under λp;

$$B(x) = (M\bar{M}\xi)^{1/2}, \quad \text{or} \quad b(x) = C_2\left[\frac{\zeta p x^2(l-x)}{4}\right]^{1/2}, \qquad a_2 \leq x \leq l/2, \tag{3.222}$$

with C_2 given by Eq. (3.210), because the optimality condition of Eq. (3.219) holds in this interval, with $\dot{\kappa} = 0$, equality holds, $\kappa = M/B$, and $\bar{\kappa} = \bar{M}/B$, where

$$\bar{M} = x/2 \tag{3.223}$$

is the bending moment under a concentrated unit load at mid-span.

The unknowns a_1, a_2, and ζ are obtained from the following conditions:

(i) continuity of $b(x)$ at $x = a_1$:

$$b_0 = \frac{C_1 \lambda p a_1(l-a_1)}{2}; \tag{3.224}$$

(ii) continuity of $b(x)$ at $x = a_2$:

$$\frac{C_1 \lambda p a_2(l-a_2)}{2} = \left[\frac{\zeta p a_2^2(l-a_2)}{4}\right]^{1/2} C_2; \tag{3.225}$$

(iii) central deflection equal to κl:

$$\int_0^{l/2} \frac{M\bar{M}}{B_0}\,dx = \frac{kl}{2}$$

or, explicitly,

$$\int_0^{a_1} \frac{M\bar{M}}{B_0}\,dx + \int_{a_1}^{a_2} \frac{C_2}{C_1\lambda}\bar{M}\,dx + \int_{a_2}^{l/2} \left(\frac{M\bar{M}}{\zeta}\right)^{1/2} dx = \frac{kl}{2}. \tag{3.226}$$

Keeping in mind the condition $0 \leq a_1 \leq a_2 \leq l/2$, we extract the relevant root a_1 from Eq. (3.224), solve Eq. (3.225) for ζ to get

$$\zeta = C_1^2\lambda^2\frac{p(l-a_2)}{C_2^2}, \tag{3.227}$$

and substitute this expression for ζ in Eq. (3.226) to determine a_2.

It remains to show that the optimality condition, already used for $a_2 \leq x \leq l/2$, is also satisfied for $0 \leq x \leq a_2$. In the interval $0 \leq x \leq a_1$, we have $B = B_0 = b_0h^2E/12$ and $\dot{\kappa} = 0$. Hence, using Eq. (3.224), we obtain

$$\zeta\kappa\bar{\kappa} = \frac{\xi M\bar{M}}{B_0^2} = \frac{x^2(l-a_2)(l-x)}{a_1^2(l-a_1)^2}.$$

Because $x^2(l-x)$ is a monotonic increasing function from $x = 0$ to $x = 2l/3$, and $a_1 \leq l/2$, we have $\zeta\kappa\bar{\kappa} < 1$. In the interval $a_1 \leq x \leq a_2$, we have $|\dot{\kappa}| > 0$, and the optimality condition also requires $\zeta\kappa\bar{\kappa} < 1$. In the interval considered, the left-hand side of this inequality turns out to be equal to $(l-a_2)/(l-x)$, which obviously does not exceed unity because $x \leq a_2$.

If, in the course of the solution, the continued inequality $0 \leq a_1 \leq a_2 \leq l/2$ is not satisfied, it means that the optimal design for one behavioral constraint is able to satisfy the other more than strictly. For simplicity of the discussion, assume $B_0 = 0$, and let $a_1 = 0$, $a_2 = a$. As soon as a vanishes, the optimal elastic design with central deflection kl is able to support at least a load λp at collapse. On the other hand, as soon as a attains $l/2$, the optimal plastic design for λp exhibits an elastic central deflection smaller than kl. These two conditions give the range in which a mixed design is optimal: namely,

$$8.00000 \leq lC_2/k\lambda C_1 \leq 9.83525. \tag{3.228}$$

Other loading cases, support conditions, and types of cross sections for statically determinate beams can be treated in a similar manner [see Igič (1979)].

3.9. Optimization of the Layout

3.9.1. Optimality Criteria

In this section we assume that some freedom has been left to the designer in the definition of the layout: namely, the *segmentation* when the structure consists of prismatic bars, and *the location of some supports.* We derive hereafter some criteria for optimal segmentation and optimal location of supports from purely mechanical arguments. For a more general and rigorous (but more abstract) approach, we refer the reader to Chapter 7, devoted to Lagrange multiplier methods, as well as to original papers by Rozvany (1973b, 1974, 1975), Mroz and Rozvany (1975), Masur (1974, 1975a, b), Olhoff and Taylor (1978), and Dems and Mroz (1980).

Let us denote by x_0 the abscissa of a segmentation in a beam or frame made of prismatic segments, the yield moments of which have been selected for minimum cost under assigned limit load. The optimal value of the abscissa x_0 will be obtained by replacing the discontinuity in the yield moment at x_0 by a steep but continuous variation on a small interval of magnitude ε on each side of the considered section and letting ε tend to zero. In this small interval the optimality condition $|\dot{\kappa}| = \gamma_Y$ must be satisfied, implying that $Y = |M|$. The average values $|\dot{\kappa}|^a$ and γ_Y^a are equal, and the increase Δr in the rate of rotation on the interval is

$$\Delta r = \gamma_Y^a 2\varepsilon \cdot \operatorname{sgn} M(x_0)$$

because of the flow rule. But, if ΔM is the variation of M from $x_0 - \varepsilon$ to $x_0 + \varepsilon$ and V is the average shear force, we have $2\varepsilon V = \Delta M$, with $\Delta M = \Delta Y \cdot \operatorname{sgn} M(x_0)$. Hence, we obtain $\Delta r = \gamma_Y^a \cdot \Delta Y / V$, or $\Delta r = \Delta\gamma / V$. When ε tends to zero, Δr tends to the localized rotation velocity $\omega(x_0)$. The local optimality condition is then

$$\omega(x_0) = \Delta\gamma / V(x_0). \tag{3.229}$$

When the behavioral constraint is a given upper bound on the elastic compliance, the corresponding condition turns out to be [see Rozvany (1976, p. 158)]:

$$\gamma[B(x_0^+)] - \gamma[B(x_0^-)] + \frac{kM^2(x_0^+)}{B(x_0^+)} - \frac{kM^2(x_0^-)}{B(x_0^-)} = 0, \tag{3.230}$$

where k is a constant and the superscripts + and − indicate sections on either side of x_0.

We now consider that the location x_0 of a support can be chosen in some given range. We shall regard x_0 as optimal for a given design if a small variation of x_0 leaves the measure of the behavior stationary. Let us first study the case of an assigned limit load $p(x)$. For a given design that collapses in a given mechanism, the limit load will remain stationary in a slight motion of the support if the power of the corresponding reaction R vanishes; that is, if

$$R \cdot v'(x_0) = 0, \tag{3.231}$$

where $v(x)$ is the rate of deflection of the collapse mechanism. If $v(x)$ is regarded as an elastic deflection, the same condition, Eq. (3.231), is necessary for stationariness of the compliance with respect to the support location. It applies also to optimal location of intermediate supports of columns subjected to assigned buckling load [see Rozvany and Mroz (1977a) and Olhoff and Taylor (1978)]. Since the buckling mode with intermediate supports corresponds, as a rule, to $v'(x_0) \neq 0$, the optimally located supports must furnish vanishing reactions; in other words, they must be placed at the nodes of a higher-order buckling mode.

3.9.2. Optimal Segmentation—Plastic Design

Let us consider the example of a propped cantilever built in at $x = 0$, simply supported at $x = 2l$, and carrying a transverse concentrated load P at $x = l$. Assume that the beam is made of two segments with lengths βl and $2l - \beta l$ and yield moments Y_1 and Y_2, respectively (Fig. 3.25a). The total cost is taken to be

$$\Gamma = \beta l Y_1 + (2 - \beta) l Y_2. \tag{3.232}$$

Depending on whether $Y_1 > Y_2$ or $Y_1 < Y_2$, the two possible mechanisms and the corresponding bending moment diagrams at collapse are depicted in Fig. 3.25(b), and (c), respectively. The dissipations are

$$\omega Y_1 + (2 + 3k - \beta k)\omega Y_2 \qquad \text{for Fig. 3.25(b)},$$

$$(1 + k)\omega Y_1 + (2 - 2k + \beta k)\omega Y_2 \qquad \text{for Fig. 3.25(c)}.$$

The optimality condition of Eq. (3.58), where the arbitrary constant c^2 is given a unit value, furnishes the relations for the mechanisms of Fig. 3.25(b), (c):

$$\begin{aligned} &\omega = \beta l, && k\omega = \frac{(2 - 3\beta)l}{3 - \beta} && \text{for } \beta < \frac{2}{3}, \\ &\omega = \frac{(2 + \beta - \beta^2)l}{4 - \beta}, && k\omega = \frac{(3\beta - 2)l}{4 - \beta} && \text{for } \beta > \frac{2}{3}, \end{aligned} \tag{3.233}$$

since k and ω must be positive in both cases.

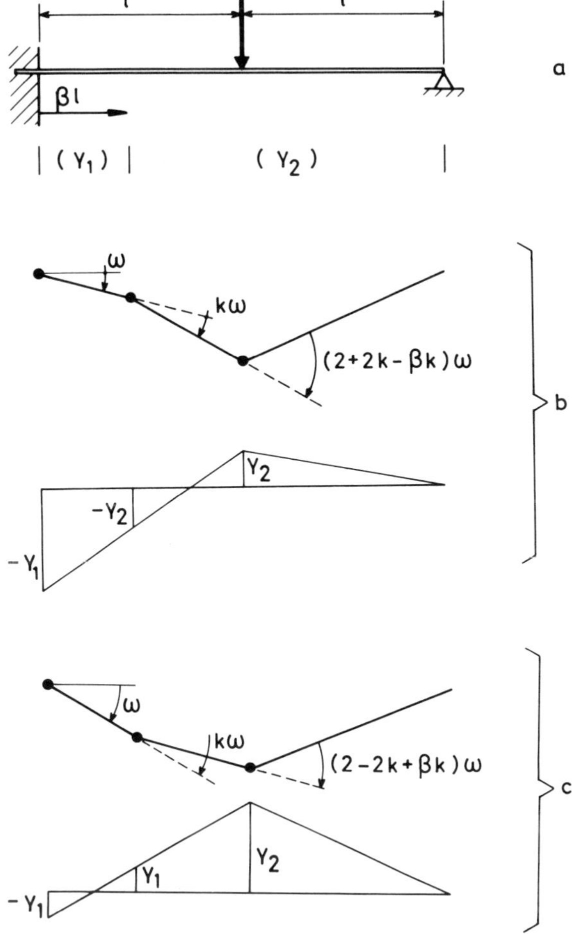

Figure 3.25. (a) Beam with piecewise-constant yield moments of unknown division abscissa βl. (b), (c) Collapse mechanisms and bending moment diagrams.

We can, as usual, express the bending moment $M(x)$ in terms of the load P and the redundant reaction R of the simple support. With the conditions [see Fig. 3.25(b), (c)]

$$M(0) = -Y_1, \qquad M(\beta l) = -M(l) = -Y_2 \qquad \text{for } \beta < \tfrac{2}{3},$$

and

$$M(0) = -Y_1 = -M(\beta l), \quad M(l) = Y_2 \qquad \text{for } \beta > \tfrac{2}{3},$$

we obtain the optimal designs

$$Y_1 = \frac{(1+\beta)Pl}{3-\beta}, \quad Y_2 = Rl = \frac{(1-\beta)Pl}{3-\beta} \quad \text{for } \beta < \frac{2}{3}, \tag{3.234a}$$

$$Y_1 = \frac{\beta Pl}{4-\beta}, \quad Y_2 = Rl = \frac{(2-\beta)Pl}{4-\beta} \quad \text{for } \beta > \frac{2}{3}. \tag{3.234b}$$

For these designs the cost is found to be

$$\Gamma(\beta) = \begin{cases} \dfrac{2(1-\beta+\beta^2)Pl^2}{3-\beta} & \text{for } \beta < \dfrac{2}{3}, \\ \dfrac{2(2-2\beta+\beta^2)Pl^2}{4-\beta} & \text{for } \beta > \dfrac{2}{3}. \end{cases} \tag{3.235}$$

When β tends to $\frac{2}{3}$ from below or above, the reactions R given by Eqs. (3.234a) and (3.234b) are $R = P/7$ and $R = 2P/5$. For $\beta = \frac{2}{3}$ the design $Y_1 = (P - 2R)l$, $Y_2 = Rl$ furnishes $\Gamma = 2Pl^2/3$ for any R satisfying $\frac{1}{7} \leq R/P \leq \frac{2}{5}$.

So far we have regarded β as given in advance. We now choose β from the optimality criterion in Eq. (3.229) to minimize $\Gamma(\beta)$. Assuming that $\beta_{\text{opt}} < \frac{2}{3}$, we use Eqs. (3.234a) to obtain $V = P - R = 2P/(3-\beta)$, whereas $\Delta\gamma = \Delta Y = Y_2 - Y_1 = 2\beta Pl/(3-\beta)$. The criterion for optimal segmentation thus gives

$$\omega(\beta l) = -\beta l.$$

Equating this value of $\omega(\beta l)$ to $-k\omega$, given by the first line of Eq. (3.233), yields $\beta_{\text{opt}} = 3 - \sqrt{7} = 0.35425$. It can be verified that this value of β actually minimizes $\Gamma(\beta)$ given by Eq. (3.235).

3.9.3. Optimal Location of Support—Plastic Design

Let us consider the beam of Fig. 3.26, built in at the left end and simply supported at some unknown abscissa x_0. Assume that the cost of the support is proportional to its reaction R. Hence, the cost of the beam is taken to be

$$\Gamma = \int_0^l \alpha Y \, dx + \beta R, \tag{3.236}$$

where l is the length of the beam.

We know from Section 3.2.6 that the collapse mechanism of the optimal design will have constant rates of curvature, and a downward velocity at the simple support of magnitude β. For optimal location of this support, we must add to the preceding conditions a vanishing slope at its abscissa.

Now consider the example of a uniformly distributed load p. The collapse mechanism is shown in Fig. 3.26b. Because the slope at section B must vanish, the region with positive curvature rate between A and B must have a length equal to the sum $(a + b)$ of the lengths of the adjacent regions with negative curvature rate, resulting in a vanishing shear force at section B for the conjugate beam of Fig. 3.26c.

Also, the moment at B of this conjugate beam must equal β, giving

$$a^2 - b^2 = \beta. \tag{3.237}$$

Now, as the shear force at the counterflexure points of the given beam is $p(a + b)/2$, we can equate the bending moment at B due to the left-hand forces to that created by the right-hand forces and obtain

$$l^2 + 4a^2 + 7ab + 2b^2 - 4al - 4bl = 0. \tag{3.238}$$

For every β, eqs. (3.237) and (3.238) give a and b. When the locations of

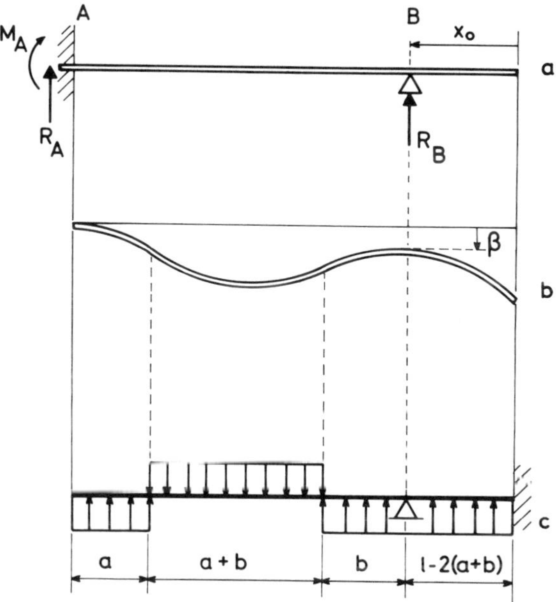

Figure 3.26. (a) Beam with support of unknown location, uniformly distributed load. (b) Collapse mechanism. (c) Conjugate beam.

the counterflexure points are known, the bending moment diagram is readily obtained, and the optimal design is $Y = |M|$, and the reaction is $R_B = p(2l - 3a - b)/2$. The cost can be computed to be $\Gamma = \int pv\, dx$. From Fig. 3.26b, constructed with the aid of Fig. 3.26c, where the constant rates of curvature are given the value α, we obtain

$$\Gamma = \alpha p\left[a^2 + \left(\frac{a^2 - b^2}{2}\right)2b + \frac{(l - 2a - 2b)^3}{6}\right] + \beta p(l - 2a - 2b)l^2.$$

For example, with $\beta/\alpha = \rho l^2$ (ρ is called the "cost ratio"), for $\rho = 0.05$ we find $a = 0.25472l$, $b = 0.12197l$, $\Gamma/\alpha = 0.045370pl^3$, and $R = 0.556939pl$. Hence, the cost of the support R is 61.38% of the total cost Γ.

The reader will verify that, whatever the magnitude of ρ, neither the abscissa $x_0 = l - 2(a + b)$ nor the reaction R_B vanish; the support at B is always present for all optimal designs. When the support at B is costless, $\rho = 0$, $a = b = 0.1745l$, $x_0 = l - 2(a + b) = 0.30217l$, and $\Gamma = 0.015218\alpha pl$. If this costless support were fixed at the right end, the minimal cost would be $0.048816\alpha pl$, that is, 3.21 times larger.

The stationary condition $v'(x_0) = 0$ may sometimes be useless, because the minimum cost may occur at a nonstationary value of $\Gamma(x_0)$. Consider, for example, the case of a concentrated force P acting at an intermediate section C between sections A and B of the beam considered above (Fig. 3.27a). The stress-free overhang to the right of section B will obviously have vanishing cross section, and the absolute value of its rate of curvature can be taken as any value smaller than α.

The mechanism of Fig. 3.27b does not satisfy the stationary condition $v'(x_0) = 0$, but the mechanism of Fig. 3.27c does. But in both cases the bending moment in the conjugate beam of Fig. 3.27d must be β. With $a + b = l'$ we obtain

$$a = l' - [l'^2/2 - \rho]^{1/2}, \tag{3.239}$$

$$b = [l'^2/2 - \rho]^{1/2}. \tag{3.240}$$

The reaction R_B at support B is such that $M(l - a) = 0$. With no load to the right of B, we get

$$R_B = \frac{P(c - a)}{b}. \tag{3.241}$$

Assuming $a < l'$ or, from Eq. (3.239), $\rho < l'^2/2$, we have $R_B = 0$ as soon

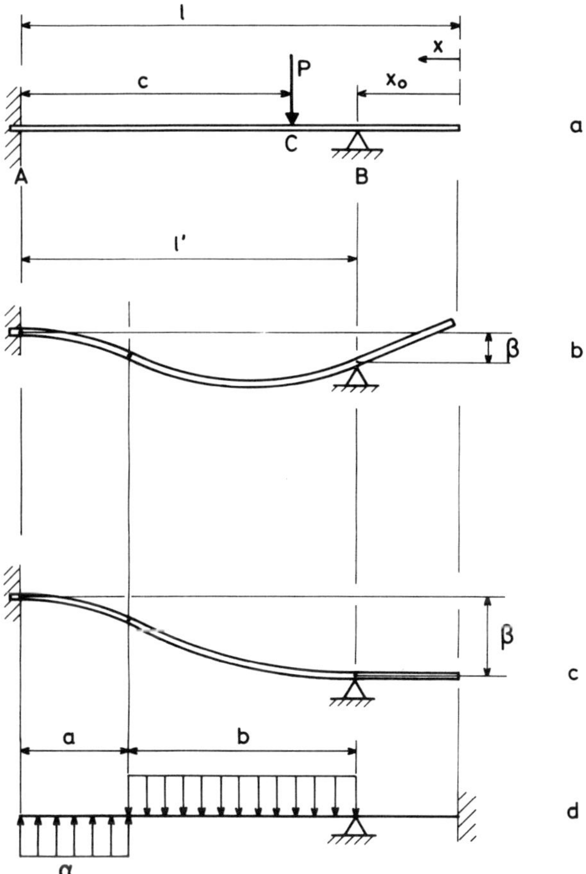

Figure 3.27. (a) Beam with support of unknown location, concentrated load. (b), (c) Collapse mechanisms. (d) Conjugate beam for collapse mechanism (c).

as $a = c$, or, again from Eq. (3.239),

$$\rho = 2l'c - c^2 - l'^2/2 \equiv A. \tag{3.242}$$

For larger values of ρ, R_B vanishes because $a = c$ and the support at B is useless and cost-free. If we take, for example, $l' = l$ and $c = l/2$, this particular value of the cost ratio is $\rho = l^2/4$.

First consider $\rho < A$ and l' given. Then Eq. (3.239) gives the abscissa of the counterflexure point, and the optimal design is $Y = |M|$. The cost is measured by $P \cdot v_C$, because $\kappa = \pm\alpha$ and $v_B = \beta$. In the conjugate beam,

$v_C = \alpha a(c - a/2) - (c - a)^2\alpha/2$. Hence

$$\Gamma = \alpha(2ac - a^2 - c^2/2). \tag{3.243}$$

Now x_0 or, equivalently, l' is regarded as a variable that satisfies $\rho < A$. This is achievable with sufficiently small ρ. When l' decreases from l to c, a varies according to Eq. (3.239), but $d\Gamma/da = 2\alpha(c - a)$ remains positive as long as $a < c$. Hence Γ is a continuously increasing function of a. But from Eq. (3.239) we see that a increases with l' if $\rho \leq l'^2/4$. Consequently, subject to $\rho < l'^2/4$, Γ decreases continuously with decreasing l', and the (nonstationary) minimum value of Γ is obtained for $l' = c$ (that gives $A = c^2/2$).

Thus, the design procedure will be as follows:

1. If $0 < \rho < c^2/4$ (small relative cost of support at B), then take $l' = c$. The load is directly balanced at the simple support, and no beam is needed. This optimal design was obvious for cost-free support ($\rho = 0$). Note that the stationary condition $v'(x_0) = 0$ does *not* apply.
2. If $c^2/4 \leq \rho$, then take $l' = 2\rho^{1/2}$, because the minimum of Γ is obtained for the smallest possible l' as long as $\rho < l'^2/4$. It must be verified that $R_B \neq 0$. This condition implies $\rho < c^2$. The choice $l' = 2\rho^{1/2}$ is thus justified when $c^2/4 \leq \rho \leq c^2$. Note that the minimum value of Γ is now a stationary value, contrary to the case where $\rho < c^2/4$. Indeed, $v'(x_0)$ gives $a = l'/2$ and, from Eq. (3.239), $l' = 2\rho^{1/2}$. When ρ reaches c^2, then $l' = 2c$ and $a = c$, but the reaction R_B vanishes, and the design is a cantilever of span c with cost $\alpha Pc^2/8$.
3. If $c^2 \leq \rho$, no right support is needed because $a > c$, $R_B = 0$, and the optimal design beam is again a cantilever of span c.

Note finally that an optimal solution with $l' > c$ never occurs. Indeed, with $l' > c$, equilibrium requires $R'_b > P$ and $Y = |M| \neq 0$ for $c \leq x \leq l$, except at the counterflexure point, whereas $l' = c$ gives $R'_B = P$ and $Y = 0$ everywhere.

4

Optimal Design of Trusses

4.1. Optimal Plastic Design of a Truss of Given Layout for a Single Loading

We consider a truss of given layout that is subject to a single set of loads. The bars of the truss are to consist of a given, rigid, perfectly plastic material with tensile and compressive yield stresses $\pm\sigma_0$. The truss is to have a given load factor for plastic collapse under the given type of loading, while the total volume of its prismatic bars is to be as small as possible. For the sake of simplicity, we disregard the possible buckling of bars in the treatment of the problem.

Denoting the length and cross-sectional area of the typical bar by l_i and A_i, respectively, we choose the design variables to be the *yield forces* $Y_i = \sigma_0 A_i$, and the design objective to be the minimization of the *cost*

$$\Gamma = \sum_i Y_i l_i, \tag{4.1}$$

which is proportional to the total volume of the bars. The specific cost is thus the function

$$\gamma(Y) = Y, \tag{4.2}$$

which satisfies the convexity inequality, Eq. (3.2), as equality.

Of the joints on the typical bar i, designate one as the *origin* and the other as the *terminal*, and denote by $\mathbf{U}_i$ the unit vector of the direction from the origin to the terminal. Denoting the typical joint by α, we define the *incidence matrix* $a_{\alpha i}$ as follows:

$$a_{\alpha i} = \begin{Bmatrix} -1 \\ 1 \\ 0 \end{Bmatrix} \text{depending on whether the joint } \alpha \text{ is} \begin{cases} \text{the origin of bar } i, \\ \text{the terminal of bar } i, \\ \text{not on bar } i. \end{cases} \tag{4.3}$$

A kinematically admissible rate of deformation of the truss is specified by velocities of the joints that satisfy the kinematic conditions at the supports. If $\mathbf{v}_\alpha$ is the velocity vector of joint α in a kinematically admissible rate of deformation of the truss, the corresponding axial strain rate ε_i of bar i is given by

$$\varepsilon_i = \sum_\alpha a_{\alpha i}\mathbf{v}_\alpha \cdot \frac{\mathbf{U}_i}{l_i}, \tag{4.4}$$

where the center dot indicates scalar product. Since the power of dissipation in bar i is

$$D_i = Y_i|\varepsilon_i|l_i, \tag{4.5}$$

the power of dissipation per unit yield force is

$$k_i = |\varepsilon_i|l_i. \tag{4.6}$$

In analogy with the discussion in Section 3.1, let bounds Y_i^- and Y_i^+ be set on the yield forces Y_i:

$$0 \le Y_i^- \le Y_i \le Y_i^+. \tag{4.7}$$

Proceeding exactly as in Section 3.1, one may then show that the optimal truss with the given layout admits a collapse mechanism for which

$$|\varepsilon_i| \begin{Bmatrix} \le \\ = \\ \ge \end{Bmatrix} c^2, \text{ depending on whether} \begin{cases} Y_i = Y_i^-, \\ Y_i^- < Y_i < Y_i^+, \\ Y_i = Y_i^+. \end{cases} \tag{4.8}$$

Here $c^2 > 0$ has the same value for all bars. In the same way as the condition in Eq. (3.9), the condition in Eq. (4.8) is sufficient for global optimality and necessary for local optimality.

To explore consequences of this optimality condition, consider the truss shown in Fig. 4.1a, which is to transmit the given load **P** from its point of application O to the rigid, vertical wall at the left of O by means of bars

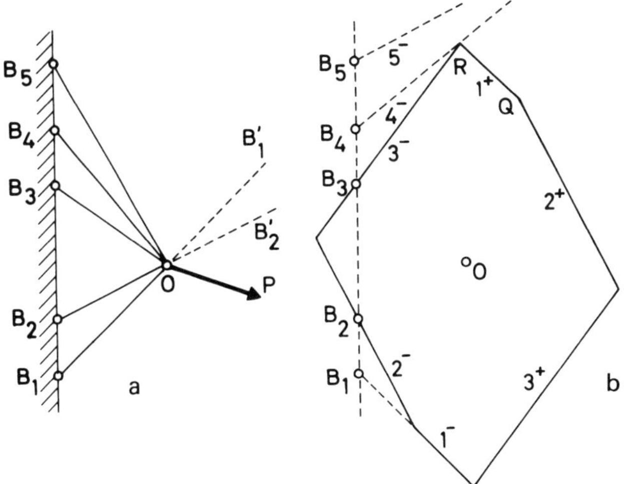

Figure 4.1. (a) Truss and loading. (b) Geometric representation of the optimality condition.

connecting O to some or all of the given points B_1, B_2, ..., B_5. The load **P** is to be the collapse load of the truss whose cost, Eq. (4.1), is to be minimized. The following geometric discussion of the problem is taken from Prager (1976).

To allow for the possibility that the optimal truss may not contain all bars of the given layout, we set $Y_i^- = 0$ for all bars. Moreover, to simplify the discussion, we set $Y_i^+ = \infty$ for all bars. Finally, we regard the point O as the terminal of each bar. In view of Eq. (4.4), the optimality condition of Eq. (4.8) may then be written as

$$|\mathbf{v} \cdot \mathbf{U}_i| \left\{ \begin{matrix} \leq \\ = \end{matrix} \right\} c^2 l_i, \text{ depending on whether } \left\{ \begin{matrix} Y_i = 0, \\ Y_i > 0. \end{matrix} \right. \tag{4.9}$$

For greater clarity the following construction, which could be carried out directly on Fig. 4.1a, is shown in a separate figure (Fig. 4.1b). For a given value of i the condition in Eq. (4.9) restricts the point V, whose radius vector from O is $\mathbf{V} = \mathbf{v}/c^2$, to lying in the strip

$$|\mathbf{V} \cdot \mathbf{U}_i| \leq l_i. \tag{4.10}$$

One boundary of this strip is the perpendicular to OB_i through B_i, which will be labelled i^-, because a point V on it implies a negative axial strain rate of bar i. The other boundary i^+ of the strip in Eq. (4.10) is symmetric to i^- with respect to O.

The intersection of the strips corresponding to the bars of the given layout is a convex polygon Π. A vector $\mathbf{V}$ with endpoint V outside this polygon would violate the inequality in Eq. (4.10) for at least one bar. If, on the other hand, V is on a side but not at a vertex of Π, the relation in Eq. (4.9) is satisfied as an equality for only one bar. The corresponding optimal truss consists of only this bar, which is only possible if the load has the direction of this bar.

A vertex of Π is, in general, the intersection of two strip boundaries. At the vertex Q (Fig. 4.1b), for instance, the strip boundaries 1^+ and 2^+ intersect. The corresponding optimal truss consists of bars 1 and 2, which are stressed in tension, and the load is of the form

$$\mathbf{P} = Y_1\mathbf{U}_1 + Y_2\mathbf{U}_2, \tag{4.11}$$

where Y_1 and Y_2 are the nonnegative yield forces of the two bars. According to Eq. (4.11), the load $\mathbf{P}$ is directed from O into the angle $B_1'OB_2'$ (Fig. 4.1b), where B_i' is symmetric to B_i with respect to O.

At the vertex R, on the other hand, the strip boundaries 1^+, 3^-, and 4^- intersect. The corresponding optimal truss may thus consist of bar 1 stressed in tension and bars 3 and 4 stressed in compression. The load then would be of the form

$$\mathbf{P} = Y_1\mathbf{U}_1 - Y_3\mathbf{U}_3 - Y_4\mathbf{U}_4. \tag{4.12}$$

Now $\mathbf{U}_4$ may be represented as a positive linear combination

$$\mathbf{U}_4 = \alpha\mathbf{U}_1 - \beta\mathbf{U}_3, \qquad \alpha > 0, \beta > 0, \tag{4.13}$$

of $\mathbf{U}_1$ and $-\mathbf{U}_3$. Accordingly,

$$\mathbf{P} = Y_1'\mathbf{U}_1 - Y_3'\mathbf{U}_3 \qquad \text{with } Y_1' = Y_1 + \alpha Y_4, \quad Y_3' = Y_3 + \beta Y_4 \tag{4.14}$$

is an alternative form of Eq. (4.12), which shows that we may restrict the present discussion to trusses containing only bars 1 and 3 stressed in tension and compression, respectively. The load of Eq. (4.14) is directed from O into the angle B_3OB_1'.

Continuing to discuss the vertices of the polygon Π in this manner, we find that the optimal truss consists of the two bars 1 and 2, 2 and 3, or 3 and 1, depending on whether the line of action of the load lies inside one of the angles B_1OB_2, B_2OB_3, or B_3OB_1'. If, on the other hand, the line of action of the load coincides with a side of one of these angles, the optimal truss consists of a single bar. Incidentally, if the load acts along OB_4, an optimal truss consisting of only bar 4 is an alternative to the optimal truss consisting of bars 3 and 1.

In concluding the discussion of this example, we note that the strip in Eq. (4.10) for bar 5 is too wide to furnish a side of the polygon Π. Regardless of the direction of the load, the optimal truss will therefore not contain bar 5.

Exercise 4.1. Show that the trusses corresponding to the decompositions in Eqs. (4.12) and (4.14) of the load **P** have the same cost; i.e.,

$$Y_1 l_1 + Y_3 l_3 + Y_4 l_4 = Y_1' l_1 + Y_3' l_3. \tag{4.15}$$

Exercise 4.2. Show that the findings in the next to last paragraph of this section remain valid when the tensile and compressive yield stresses have different absolute values σ_0 and σ_0'.

4.2. Optimal Layout of a Truss in Plastic Design for a Single Loading

A typical problem for optimizing the layout of a truss is shown in Fig. 4.2. The given load P acting at the point A is to be transmitted to the rigid *foundation arc f-f* by means of a truss whose joints are to remain in a given *available domain* Δ. In Fig. 4.2, *f-f* is a circular arc with center O and radius R, and, in the polar coordinates r, ϕ shown in the figure, Δ will be taken as the domain $r \geq R$.

The bars of the truss are to consist of a rigid, perfectly plastic material with tensile and compressive yield stresses $\pm\sigma_0$, and the bounds on the yield force of the typical bar are to be $Y_i^- = 0$ and $Y_i^+ = \infty$. The truss is to

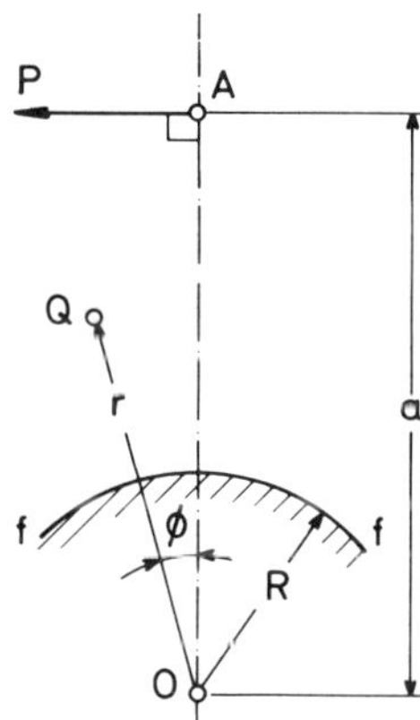

Figure 4.2. Applied load and given rigid foundation arc.

be on the verge of plastic collapse under the given load P, while the total volume of its bars is to be as small as possible.

For a given layout—that is, for given positions of the potential joints of a truss—a kinematically admissible rate of deformation is specified by velocities of these joints that satisfy the kinematic conditions at the supports. When the layout is to be optimized, any point of the domain Δ is a potential joint, and a kinematically admissible rate of deformation is specified by a velocity field in Δ. For the problem considered here, the velocity must vanish along the rigid foundation arc. In view of the bounds $Y_i^- = 0$, $Y_i^+ = \infty$, the optimality condition in Eq. (4.18) requires the strain rates of this velocity field to have constant absolute value c^2 for the line elements of the axes of the bars of the optimal truss, and absolute values not exceeding c^2 for all other line elements.

To discuss consequences of this optimality condition, we denote the principal strain rates at a typical point of the velocity field by ε_1 and ε_2, numbering them in such a manner that $|\varepsilon_1| \geq |\varepsilon_2|$. The direction in which the strain rate has the value ε_1 (or ε_2) will be called the first (or second) principal direction at the point considered; a line that has, at each of its points, the first (or second) principal direction will be called a first (or second) principal line. Since the strain rate in an arbitrary direction through the point has a value in the closed interval bounded by ε_1 and ε_2, the optimality condition above is fulfilled if the axes of the bars of the truss follow principal lines along which the strain rate has the absolute value c^2. In principle, the domain Δ may thus be divided into regions of the following types:

Type R, for which $|\varepsilon_1| = c^2 > |\varepsilon_2|$. The bars follow the first principal lines. Depending on the sign of ε_1, we distinguish types R^+ and R^-.

Type S, for which both ε_1 and ε_2 have either the value c^2 (type S^+) or the value $-c^2$ (type S^-). Any direction may be regarded as principal direction, and the optimal layout is not unique.

Type T, for which both ε_1 and ε_2 have absolute value c^2, but have opposite signs. There are two orthogonal families of bars.

For the problem in Fig. 4.2, for instance, the velocity field with radial and circumferential components

$$v_r = 0, \qquad v_\phi = 2c^2 r \ln(r/R) \tag{4.16}$$

satisfies the conditions $v_r = v_\phi = 0$ along the foundation arc and has strain rates

$$\varepsilon_r = \partial_r v_r = 0, \qquad \varepsilon_\phi = (1/r)\partial_\phi v_\phi + v_r/r = 0,$$
$$\varepsilon_{r\phi} = \tfrac{1}{2}\{(1/r)\partial_\phi v_r + \partial_r v_\phi - v_\phi/r\} = c^2, \tag{4.17}$$

where ∂_r and ∂_ϕ indicate partial derivatives with respect to r and ϕ.

The principal strain rates are thus

$$\varepsilon_1 = c^2, \qquad \varepsilon_2 = -c^2, \tag{4.18}$$

and the field in Eq. (4.1) is of type T. The principal directions at a point Q (Fig. 4.2) form angles $\pm 45°$ with the ray OQ, and the principal lines are the logarithmic spirals

$$r/R = \exp[\pm(\phi - \phi_0)], \tag{4.19}$$

where the upper and lower signs on the right correspond to first and second principal lines, and R, ϕ_0 are the polar coordinates of the intersection of the considered spiral with the foundation arc. The spirals in Eq. (4.19), only some of which are shown in Fig. 4.3, form a dense net. The optimal structure is thus not a truss in the usual sense of this term, but a *truss-like continuum* consisting of two orthogonal families of bars of infinitesimal lengths. Although a structure of this kind is not practical, it uses the smallest possible volume V of structural material for the considered behavioral constraint and thus furnishes a useful basis for comparing the efficiencies of practical structures.

The volume V may be evaluated by applying the principle of virtual power to the collapse load $\mathbf{P}$ and the velocity field in Eq. (4.16). Since all bars of the optimal truss-like continuum experience axial strain rates of absolute value c^2, the internal power of dissipation is $\sigma_0 c^2 V$. If the point of application A of the load has polar coordinates a, O, the external power of dissipation is $2Pc^2 a \ln(a/R)$. The volume V is thus

$$V = (2P/\sigma_0)a\ln(a/R). \tag{4.20}$$

For $a/R = 3$, for instance, $V = 6.592PR/\sigma_0$, and the logarithmic spirals through A intersect the foundation arc at the points B and B' with polar angles $\pm 62.95°$. The truss consisting of the straight bars AB and AB' is found to have volume $V^* = 8.164PR/\sigma_0$; hence, the *efficiency* $100\ V/V^* = 80.7\%$.

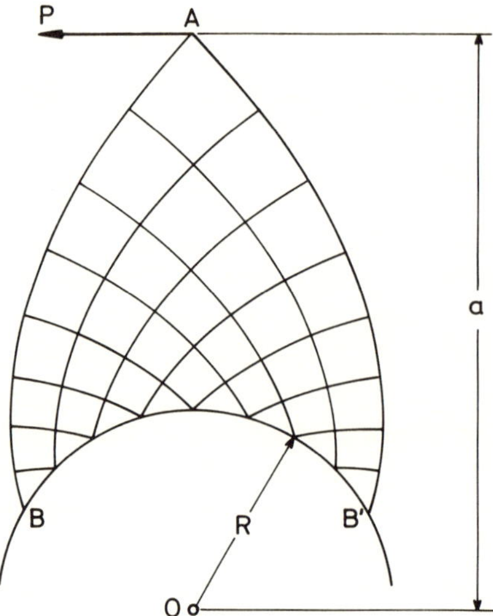

Figure 4.3. Optimal trusslike continuum.

Exercise 4.3. In addition to the horizontal load of magnitude P in Fig. 4.3, let a vertical load, whose magnitude does not exceed P, be applied at the point A. Show that the optimal trusslike continuum that is on the verge of plastic collapse under the joint action of the two loads uses the same amount of structural material as the optimal trusslike continuum that is on the verge of plastic collapse under the sole action of the horizontal load P.

4.3. Michell's Problem

Optimal trusslike continua of the kind considered in Section 4.2 were first discussed by Michell (1904); therefore they are called *Michell continua*. Even before Michell, Maxwell (1890) had used fields of type S in connection with certain problems of optimal layout of trusses. Both authors were concerned with the optimal layout of an *elastic* truss, subject to a single loading, when the axial stresses in the bars of the truss are restricted to a given *allowable range* $\{-\sigma_0, \sigma_0\}$. It will now be shown that the optimality condition for this behavioral constraint is essentially the condition used in Section 4.2, except that, instead of a velocity field and the corresponding strain rates, we now use a *displacement field* and the corresponding *strains*.

A displacement field defined throughout the domain Δ available for the trusses is called *kinematically admissible* if it is continuous, has first derivatives continuous in subregions of Δ, and satisfies the kinematic conditions at the supports. A field $\mathbf{v}(x, y)$ of this kind will be called a *Michell field* if the principal strains derived from it satisfy the conditions

$$|\varepsilon_1| = \varepsilon_0, \qquad |\varepsilon_2| \leq \varepsilon_0, \tag{4.21}$$

where $\varepsilon_0 = \sigma_0/E$ and E is Young's modulus.

Let us suppose that there exists a Michell field $\mathbf{v}(x, y)$ for the problem considered; denote by V the total volume of material in the Michell continuum, whose bars follow principal lines of this field along which the principal strain has absolute value ε_0. Suppose further that the field $\mathbf{v}(x, y)$ is compatible with the given loads; that is, these loads produce axial forces in the bars of the Michell continuum, based on the field $\mathbf{v}(x, y)$, whose signs agree with those of the strains of this field for the directions of these bars. Applying the principle of virtual work to the given loads and this displacement field yields the following expression for the work W of the loads on the displacements of their points of application:

$$W = \sigma_0 \varepsilon_0 V. \tag{4.22}$$

Consider now an elastic truss in which the given loads do not produce axial stresses outside the given allowable range. Apply the principle of virtual work to these loads and the corresponding bar forces S_i^*, on the one hand, and to the displacements $\mathbf{v}(x, y)$ and the corresponding strain ε_i^* of the typical element ds_i^* of bar i^* of the truss on the other hand. Thus,

$$W = \sum_{i^*} S_i^* \int \varepsilon_i^* \, ds_i^* \leq \sum_{i^*} |S_i^*| \int |\varepsilon_i^*| \, ds_i^*, \tag{4.23}$$

where the integration is extended over the bar i^*, and the summation includes all bars of the truss. Because this truss does not violate the behavioral constraint, $|S_i^*| \leq \sigma_0 A_i^*$, where A_i^* is the cross-sectional area of bar i^*. Moreover, because ε_i^* is a value in the closed interval bounded by the principal strains ε_1 and ε_2 at any point of bar i^*, we have $|\varepsilon_i^*| < \varepsilon_0$, by Eq. (4.21). Accordingly, Eq. (4.23) may be rewritten as

$$W < \sigma_0 \varepsilon_0 \sum A_i^* \, ds_i^* = \sigma_0 \varepsilon_0 V^*, \tag{4.24}$$

where V^* is the volume of structural material in the truss. Comparison of Eqs. (4.22) and (4.24) establishes the global optimality of the Michell continuum.

Exercise 4.4. Consider a trusslike continuum and let the allowable range of stress be $[-\sigma_0', \sigma_0]$. Setting $\varepsilon_0 = \sigma_0/E$ and $\varepsilon_0' = \varepsilon_0 \sigma_0/\sigma_0'$, define generalized Michell field as a kinematically admissible displacement field with principal strains ε_1, ε_2 satisfying the following conditions: the values ε_1, ε_2 are in the closed interval $[-\varepsilon_0', \varepsilon_0]$, and at least one of them is at an endpoint of this interval. Show that a trusslike continuum for the given loads requires the smallest possible volume of structural material if its bars follow principal lines of a generalized Michell field, along which the strains have the values ε_0 or ε_0', and if this field is compatible with the given loads.

4.4. Michell Fields

Returning to the optimal *plastic* design of trusses, we shall henceforth use the term *Michell field* for velocity fields of types *R*, *S*, and *T* discussed in Section 4.2. In this section the use of these fields in optimal plastic design of trusses will be explored more systematically.

4.4.1. Fields of Type *R*

Since the axis of only one bar passes through a typical point *P* of a field of this kind, the equilibrium conditions could not be fulfilled at *P* if this axis had a finite radius of curvature there. To a field of type *R* thus corresponds a one-parameter family of straight bars.

For example, if one chooses as x-axis a straight foundation arc to which a load in the positive y-direction, acting at a point *A* of the upper half-plane, is to be transmitted, the velocity field $v_x = 0$, $v_y = c^2 y$ is kinematically admissible. Since its principal strain rates are $\varepsilon_x = 0$ and $\varepsilon_y = c^2$, the field is of type R^+. Its first principal lines are parallel to the y-axis, and the optimal transmission of the given load to the foundation arc is by a bar in the y-direction.

4.4.2. Fields of Type *S*

These fields have the same strain rate c^2 (or $-c^2$) for all directions and thus represent a uniform rate of expansion (or contraction) in the plane of the truss. Since the center of expansion (or contraction) is the only point with vanishing velocity, a field of type *S* cannot be kinematically admissible for a truss with more than one fixed joint. There are, however, problems for which a field of type *S* is appropriate.

Figure 4.4 shows a problem of this kind: The vertices of a square of side $2a$ are the points of application of four loads with common magnitude *P* that have the diagonals as lines of action. A truss that includes the points of application of these loads among its joints is to have a given load factor Λ at plastic collapse and minimum total volume of its bars, which are to

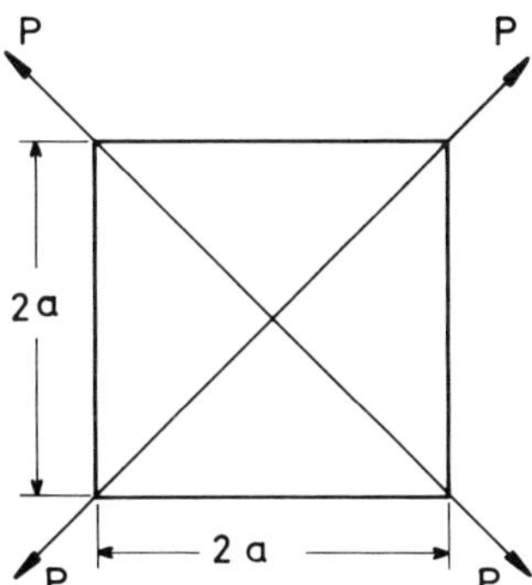

Figure 4.4. Loads acting diagonally at the vertices of a square.

consist of a given, rigid, perfectly plastic material with tensile and compressive yield limits $\pm\sigma_0$.

A plane velocity field of type S^+ with center of expansion at the center of the square is obviously kinematically admissible for this problem. Because any direction is a principal direction of this field, the optimal layout is not unique. For the truss in Fig. 4.4 we may, for instance, let the bars along the sides of the square carry the loads $\alpha\Lambda P$, where $0 \leq \alpha \leq 1$; and the bars along the diagonals, the loads $(1-\alpha)\Lambda P$. The tensile forces in these two groups of bars then have magnitudes $\alpha\Lambda P/\sqrt{2}$ and $(1-\alpha)\Lambda P$. Since all bars are stressed to the yield limit σ_0, the cross-sectional areas of the two kinds of bars are obtained by dividing these forces by σ_0. Because the corresponding lengths are $2a$ and $a\sqrt{2}$, the total volume of material is

$$V = (4\Lambda Pa/\sigma_0)\{2\alpha/\sqrt{2} + (1-\alpha)\sqrt{2}\} = 4\Lambda Pa\sqrt{2}/\sigma_0, \tag{4.25}$$

independently of the way in which the loads have been distributed to the two groups of bars. Maxwell (1890) had already noted this lack of uniqueness of the optimal design when a field of type S is kinematically admissible.

4.4.3. Fields of Type *T*

To derive important relations for a field of this type, consider a point P of this field and the angular velocities Ω and rates of extension ε of line elements at this point. In Fig. 4.5a the line elements marked x and y have the coordinate directions; those marked 1 and 2, the principal directions. The angle between the negative y-axis and the first principal direction will be denoted by θ.

Let the circle c in Fig. 4.5b be the circle of relative velocities for the point P (see the Appendix to this chapter). Because $\varepsilon_1 = -\varepsilon_2 = c^2$ for a field of type T, this circle has radius c^2 and center on the Ω-axis, which is

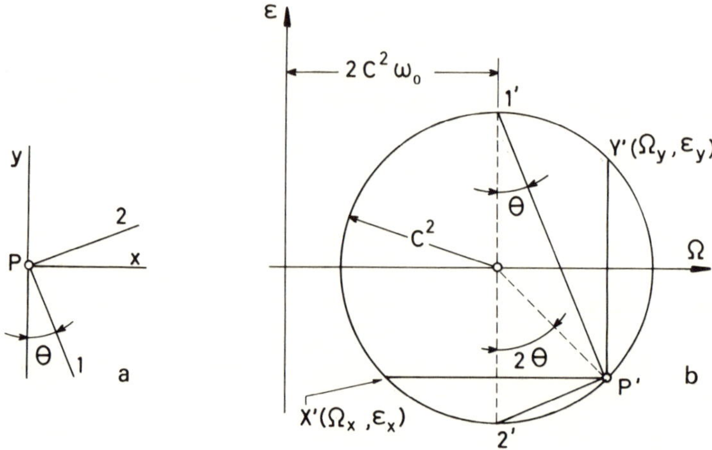

Figure 4.5. (a) Coordinate directions x and y and principal directions 1 and 2 at point P. (b) Circle of relative velocities.

parallel to the x-axis. The abscissa of the center represents the common angular velocity of line elements 1 and 2. It will be convenient to denote this abscissa by $2c^2\omega_0$. Given the circle c and its pole P' (see the Appendix), the angular velocity Ω and the rate of extension ε of a line element through P are found by drawing, through P', a line parallel to this element; the second intersection of this line with the circle has coordinates Ω, ε. In particular, the intersections X' and Y' of c with lines through P' that have the x- and y-directions have coordinates Ω_x, ε_x and Ω_y, ε_y. The pole P' thus has coordinates Ω_y, ε_x.

The lines joining P' to the highest point $1'$ and lowest point $2'$ have the first and second principal directions. The angle $2'1'P'$ is thus θ, and the central angle of the arc $2'P'$ is 2θ. As we see from Fig. 4.5b that

$$\Omega_y = c^2(2\omega + \sin 2\theta), \qquad \varepsilon_x = -c^2 \cos 2\theta. \tag{4.26}$$

Since $\Omega_y = -\partial_y v_x$ and $\varepsilon_x = \partial_x v_x$, eliminating v_x between these relations furnishes $\partial_x \Omega_y + \partial_y \varepsilon_x = 0$ or, using (4.26),

$$\partial_x \omega + \partial_x \theta \cos 2\theta + \partial_y \theta \sin 2\theta = 0. \tag{4.27}$$

When the first (or second) principal direction is chosen as the x-direction, the angle θ has the value $\pi/2$ (or 0). Equation (4.27) thus yields the relations

$$d_1(\omega - \theta) = 0, \qquad d_2(\omega + \theta) = 0, \tag{4.28}$$

where d_1 and d_2 indicate differentiation in the principal directions. These relations show that the difference $\omega - \theta$ is constant along a first principal line, while the sum $\omega + \theta$ is constant along a second principal line. These facts suggest introducing

$$\xi = \frac{\omega + \theta}{2}, \qquad \eta = \frac{\omega - \theta}{2} \tag{4.29}$$

as curvilinear coordinates. The lines $\xi = \text{const}$ and $\eta = \text{const}$ are then second and first principal lines, and

$$\omega = \xi + \eta, \qquad \theta = \xi - \eta. \tag{4.30}$$

In Fig. 4.6 let AB, BC, CD, and DA be finite arcs of the principal lines $\eta = \eta_1$, $\xi = \xi_2$, $\eta = \eta_2$, and $\xi = \xi_1$. Using Eq. (4.30) to evaluate ω and θ at the corners of the curvilinear rectangle $ABCD$, we obtain

$$\omega_B - \omega_A = \omega_C - \omega_D = \theta_B - \theta_A = \theta_C - \theta_D = \xi_2 - \xi_1. \tag{4.31}$$

Thus, if we select two second principal lines—e.g., $\xi - \xi_1$ and $\xi - \xi_2$—and progress from the first to the second along a first principal line, ω and θ change by an amount $\xi_2 - \xi_1$, which is independent of the choice of the first principal line. It can be shown in a similar way that ω and θ, respectively, change by $\eta_2 - \eta_1$ and $-(\eta_2 - \eta_1)$ if we progress along any second principal line from the first principal line $\eta = \eta_1$ to the first principal line $\eta = \eta_2$.

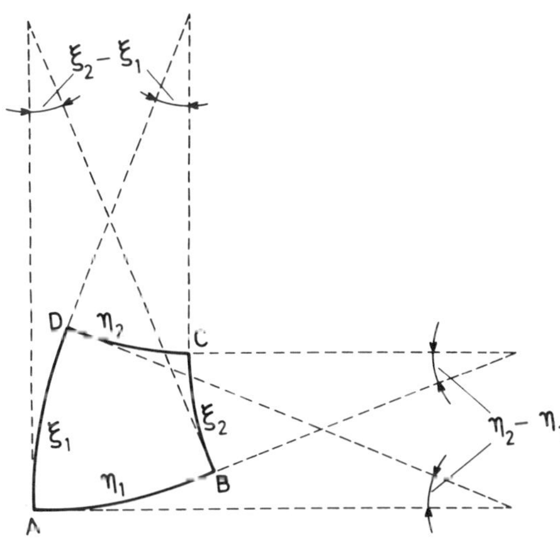

Figure 4.6. Element determined by the arcs of principal lines.

As far as they concern the angle θ, these theorems characterize the geometry of the net of principal lines in a velocity field of type *T*. Hencky (1923) established analogous theorems for the slip line net in plane plastic flow, and Prandtl (1923) discussed their geometric implications. Nets of this geometry are therefore known as Hencky–Prandtl nets. The above investigation has shown that the principal lines of a velocity field of type *T* form a Hencky–Prandtl net [Hegemier and Prager (1969)].

Along a straight principal line, θ is constant. Moreover, the difference of the θ-values at the points of intersection of any principal line of the other family with the considered straight principal line and one of its neighbors is constant. Accordingly, θ is constant along this neighboring line, which is therefore straight. A straight principal line is thus always a member of a family of straight principal lines. The lines of the other family are then, in general, parallel curves that have the straight principal lines as common normals.

Along a straight *first* principal line, not only θ but also $\theta - \omega$ and, hence, ω are constant. Since $\omega + \theta$ is constant along any second principal line, $\omega + \theta$ is constant throughout any field of type *T* whose first principal lines are straight. Similarly, $\omega - \theta$ is constant throughout a field of type *T* with straight second principal lines. Finally, both ω and θ are constant throughout a field of type *T* that has two families of straight principal lines.

The Hencky–Prandtl theorems enable us to construct the net of principal lines approximately when it is known how ω and θ vary along a curve *c* that is not a principal line. In Fig. 4.7, *A* and *B* are neighboring points of *c*, and *C'* is the approximate position of the intersection *C* of the first principal line through *A* with the second principal line through *B*. Since

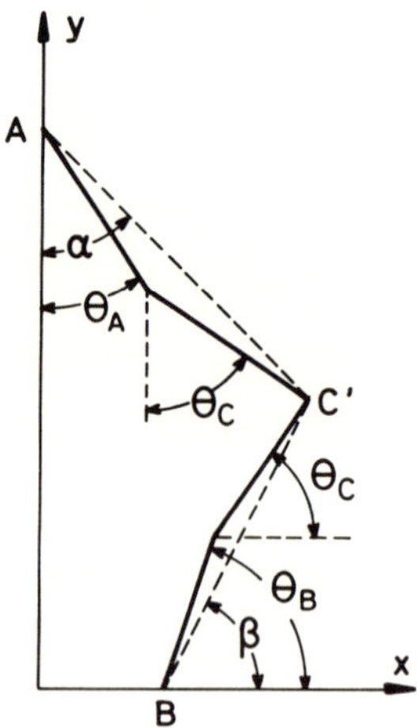

Figure 4.7. Approximate construction of principal lines at points *A* and *B*.

$\omega - \theta$ is constant along AC, and $\omega + \theta$ is constant along BC,

$$\omega_C = \frac{\omega_A + \omega_B - \theta_A + \theta_B}{2},$$
$$\theta_C = \frac{-\omega_A + \omega_B + \theta_A + \theta_B}{2}. \tag{4.32}$$

To find C' we approximate the arc AB by a parabolic arc AC' whose tangents at A and C' form the angles θ_A and θ_C with the y-axis. The chord AC', which forms the angle

$$\alpha = \frac{\theta_A + \theta_C}{2} = \frac{\omega_B - \omega_A + 3\theta_A + \theta_B}{4} \tag{4.33a}$$

with the y-axis, is a first locus for C'. Similarly, the chord BC', which forms the angle

$$\beta = \frac{\theta_B + \theta_C}{2} = \frac{\omega_B - \omega_A + \theta_A + 3\theta_B}{4} \tag{4.33b}$$

with the x-axis, is a second locus for C'.

Repeating the construction for neighboring points on c, we find, in addition to C', other points of a curve c' at which the values of ω and θ are known and which enable us to continue the construction. From a finite arc of c one thus obtains an approximation to the net of principal lines in the *influence zone* of this arc, which is bounded by principal lines through its endpoints.

Another way of specifying a Hencky–Prandtl net is by giving two intersecting principal lines. Let the points A and B in Fig. 4.8a be neighboring points of a first principal line, let A and C be neighboring points of a second principal line, and let the values of θ at the points A, B, C be given. The approximate position D' of the intersection D of the first principal line through C with the second principal line through B is then found as follows. According to the Hencky–Prandtl theorems,

$$\theta_D = \theta_B + \theta_C - \theta_A \tag{4.34}$$

We approximate the arc CD of the first principal line through C by a parabolic arc CD' whose tangents at C and D' form the angles θ_C and θ_D with the y-axis. The chord CD' then forms the angle

$$\gamma = \frac{\theta_C + \theta_D}{2} = \frac{-\theta_A + \theta_B + 2\theta_C}{2} \tag{4.35a}$$

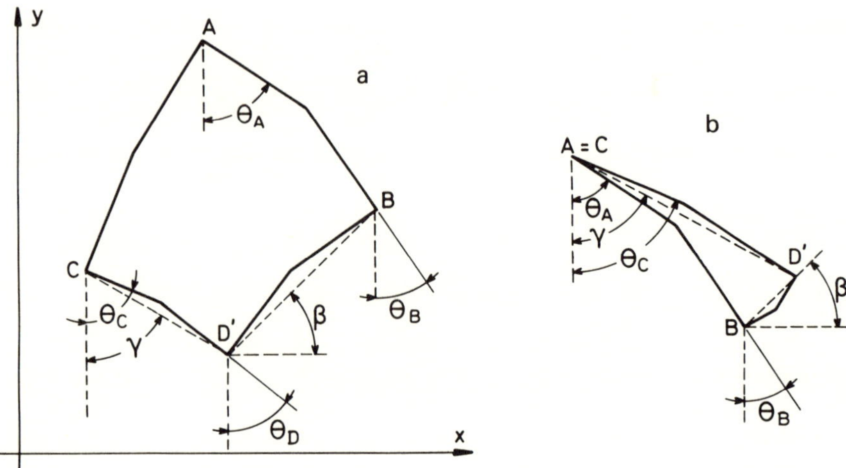

Figure 4.8. (a) Element of approximate Hencky–Prandtl net. (b) Special case where points A and C coincide.

with the y-axis. In a similar way one finds that the chord BD' forms the angle

$$\beta = \frac{\theta_B + \theta_D}{2} = \frac{-\theta_A + 2\theta_B + \theta_C}{2} \tag{4.35b}$$

with the x-axis.

Figure 4.8b illustrates the special case in which the first principal lines intersect each other at A. The point C then coincides with A, and D' is found as shown in the figure, the angles β and γ being given by Eqs. (4.35).

In Fig. 4.9a, P is a point on the foundation arc, whose tangent at P is the line PA. The solid-line circle in Fig. 4.9b is a corresponding circle of

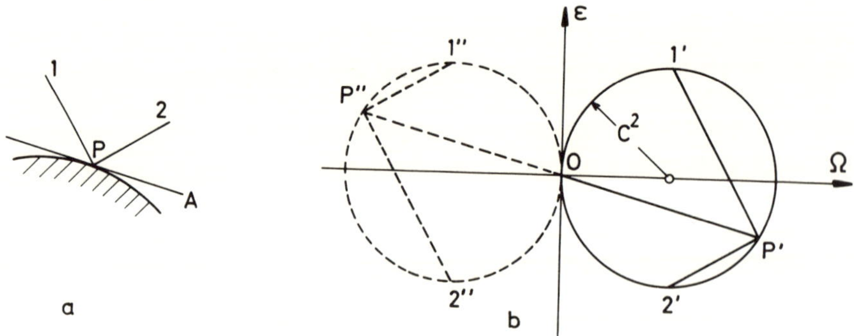

Figure 4.9. (a) Rigid foundation arc. (b) Circles of relative velocities.

relative velocities. Because $\Omega = \varepsilon = 0$ for the elements of the foundation arc, this circle passes through the origin O of the (Ω, ε)-plane, and the line joining O to the pole P' of the circle is parallel to PA. The lines from P' to the highest point $1'$ and the lowest point $2'$ of the circle have the first and second principal directions at P, which form angles of 45° with PA. The abscissa $2\omega c^2$ of the center of the circle has the value c^2. Accordingly, $\omega = \frac{1}{2}$. The dashed circle in Fig. 4.9b indicates an alternative, for which $\omega = -\frac{1}{2}$, while the first principal direction coincides with the former second principal direction. We shall see in Section 4.5 how the appropriate circle is chosen.

The approximate construction of the net of principal lines is greatly simplified if the nodes on the foundation arc are chosen in such a manner that the values of θ at neighboring nodes have a constant difference. In Fig. 4.10 this difference is denoted by 4δ. Since ω is constant on the foundation arc, it follows from Eq. (4.28) that the value of θ at a node equals the arithmetic mean of the values of θ at the intersections of the foundation arc with the principal lines through the considered node. Figure 4.10 shows nodal values of θ that are computed in this way.

If the arc between neighboring nodes on a first principal line is replaced, as above, by a parabolic arc, the angle that the chord of this arc forms with the y-axis is given by the arithmetic mean of the values of θ at the given nodes. For neighboring nodes on a second principal line, the arithmetic mean of the θ-values at these nodes gives the angle between the chord and the x-axis. The numbers at the mesh sides in Fig. 4.10 indicate these angles. It is seen that the typical quadrilateral mesh has right angles at two opposing vertices and angles of $\pi/4 \pm 2\delta$ at the other vertices.

For an analytic treatment of Michell fields, the reader is referred to a book by Hemp (1973).

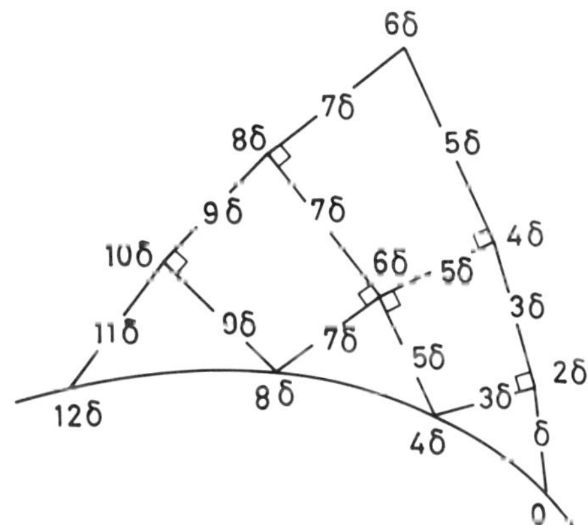

Figure 4.10. Approximate construction of a net of principal lines.

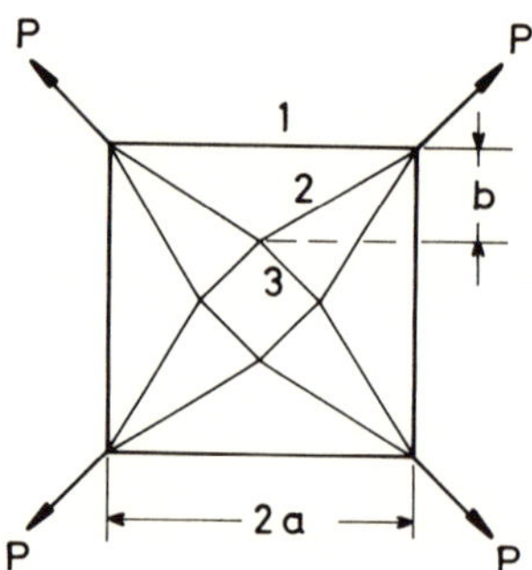

Figure 4.11. Optimal truss (Exercise 4.5).

Exercise 4.5. For the problem in Fig. 4.4, verify that the bars of the truss in Fig. 4.11 have the total volume of Eq. (4.25) independent of the choices of b and the ratio in which the loads are assigned to the bars along the sides of the square and to the remaining bars.

4.5. Use of Michell Fields

All problems of this section concern the optimal plastic design of a truss or trusslike continuum for a given load factor Λ. The bars of the structure are to consist of a given, rigid, perfectly plastic material with tensile and compressive yield limits $\pm\sigma_0$, and the total volume of the bars is to be minimized.

(a) The given load **P** is applied to point A with coordinates 0, h (Fig. 4.12a); it is to be optimally transmitted to the foundation arc, which coincides with the x-axis.

The circle of relative velocities in Fig. 4.12b passes through the origin and has pole P' with coordinates $-2c^2$, 0. It represents the angular velocities Ω and the rates of extension ε of a Michell field of type T with straight principal lines that may be appropriate for the problem on hand. Because the first and second principal directions are given by $P'1'$ and $P'2'$, the truss in Fig. 4.12c corresponds to this field. If the angle α between the positive x-direction and the direction of the load satisfies $-\pi/4 < \alpha < \pi/4$, the bar AB is stressed in tension, which means that the assumed Michell field that has AB as a first principal line is compatible with the load. The layout in Fig. 4.12c is thus optimal for this range of α. As in the derivation of Eq. (4.20), the volume V of the bars of this truss may be found by equating the external and internal powers of dissipation. Because the line elements of the y-axis have angular velocity $-2c^2$ and vanishing rate of extension,

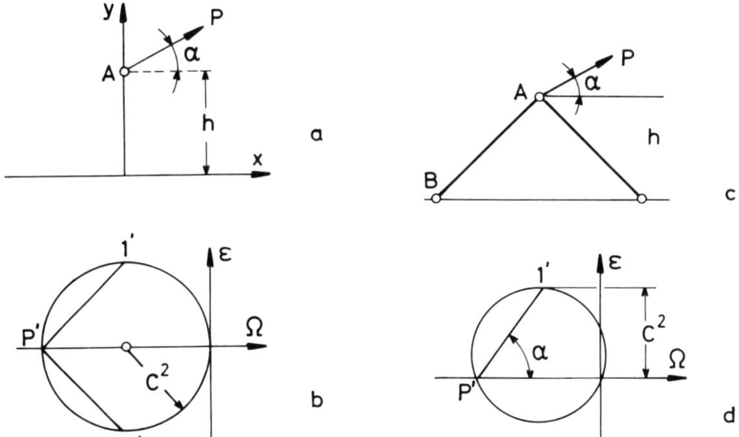

Figure 4.12. (a) Load P applied at point A, and foundation arc coinciding with the x axis. (b) Circle of relative velocities for $-\pi/4 < \alpha < \pi/4$. (c) Optimal two-bar truss. (d) Circle of relative velocities for $\pi/4 < \alpha < 3\pi/4$.

the velocity of point A has positive x-direction and magnitude $2c^2h$. Accordingly,

$$V = (2\Lambda Ph/\sigma_0)\cos\alpha. \tag{4.36}$$

When α satisfies $3\pi/4 < \alpha < 5\pi/4$, the layout in Fig. 4.12c is also optimal, but the circle of relative velocities is obtained from that in Fig. 4.12b by symmetry with respect to the ε-axis.

For α values in the range $\pi/4 < \alpha < 3\pi/4$, however, both bars of the truss in Fig. 4.12c would be stressed in tension, whereas the Michell field furnishing this truss is of type T and, hence, not compatible with the considered loads. For this range of α a field of type R^+ is appropriate. Figure 4.12d shows the corresponding circle of relative velocities. The only direction with rate of extension c^2 is given by $P'1'$, and the optimal transmission of the load to the foundation arc is by a single bar along the line of action of the load. This remark also applies when α has the values $\pi/4$ or $3\pi/4$, and when it satisfies $5\pi/4 \leq \alpha \leq 7\pi/4$.

(b) Let us reconsider the problem of Fig. 4.2 when the load P, instead of being normal to the ray OA, forms an angle α, $0 < \alpha < \pi/4$, with it. The layout in Fig. 4.3 is then no longer appropriate, because the bars at joint A would both be stressed in tension, whereas the Michell field on which this layout is based has strain rates of opposite signs for the directions of these bars.

The optimal layout for $\pi/4 > \alpha > \arcsin(R/a\sqrt{2})$ is due to Hu and Shield (1961). A bar along the line of action of the load transmits this to joint D (Fig. 4.13) of the Michell continuum BDC, whose bars follow logarithmic spirals forming angles of ±45° with the rays from O. The bar AD is tangent to the spiral CD at D. The polar coordinates of D are thus

$$r_D = a\sqrt{2}\sin\alpha, \qquad \phi_D = \alpha - \pi/4. \tag{4.37}$$

Within the annulus $R \le r \le r_D$ the velocity field is given by Eq. (4.17). In the region bounded by the circle $r = r_D$ and its tangent at D, the velocity components are

$$v_r = 0, \qquad v_\phi = 2c^2 r \ln(r_D/R); \tag{4.38}$$

since they correspond to a rigid-body rotation, the optimal layout has no bars within this region. Finally, above this tangent, the velocity field is obtained by superposition of the field of Eq. (4.38) with the simple shear motion specified by the velocity components

$$v_x = 0, \qquad v_y = 2c^2 x \tag{4.39}$$

with respect to the rectangular axes x, y indicated in Fig. 4.13. Since the principal lines of this velocity field form angles of ±45° with the x-axis, the bar AD follows a first principal line. Note that the above velocities are continuous across regional boundaries.

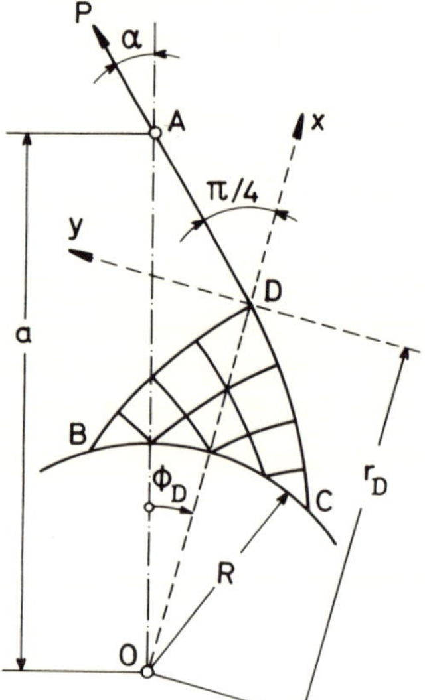

Figure 4.13. Optimal truss for $\pi/4 > \alpha > \arcsin(R/a\sqrt{2})$.

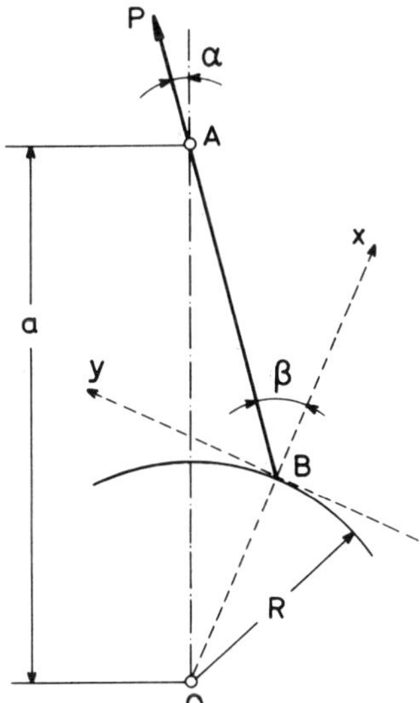

Figure 4.14. Optimal truss for $\alpha < \arcsin(R/a\ 2)$.

If $\alpha < \arcsin(R/a\sqrt{2})$, the optimal structure consists of a single bar AB along the line of action of the load (Fig. 4.14). The corresponding Michell field vanishes identically between the foundation arc and its tangent at B, and has velocity components

$$v_x = c^2 x(1 - \tan^2 \beta), \qquad v_y = 2c^2 x \tan \beta \tag{4.40}$$

with respect to the rectangular axes x, y in Fig. 4.14 and, hence, the principal rates of extension c^2 and $-c^2 \tan^2 \beta$ in the direction AB and the direction normal to AB.

(c) In treating the problem in Fig. 4.12, we have tacitly assumed that the entire x-axis is available for the foundation arc. Figure 4.15 shows the optimal layout if the foundation arc is the segment $-b \leq x \leq b$ of this axis. In the triangle BCD the first and second principal directions are given by CD and BC. In the circular sector CFD the first principal lines are the rays through C, whereas in the sector BDF, the rays through B are the second principal lines. The circular arcs DF and DE determine the layout in $DFAE$ in the manner discussed in connection with Fig. 4.8a.

An important point is that the considered Michell field can be continued beyond the contour $CFAEB$ of the Michell continuum based on this field, because only then can we exclude the possibility that a truss requiring an even smaller amount of structural material could be constructed by using bars lying outside $CFAEB$.

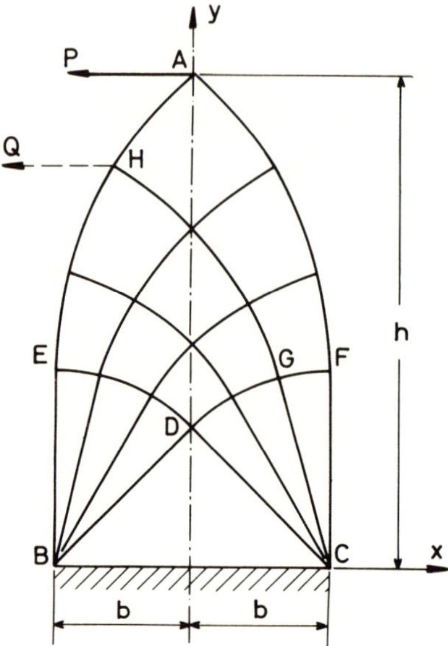

Figure 4.15. Optimal truss for straight foundation arc of limited length $2b$.

If the point of application of the load does not lie on the y-axis, the optimal layout is no longer symmetric with respect to this axis. For the load **Q** in Fig. 4.15, for instance, the optimal layout covers the domain $BDCFGHE$.

(d) In the preceding examples we developed appropriate Michell fields from the given loads and foundation arcs. Alternatively, one may start from a Michell field and try to find conditions of loading and support for which this field is statically and kinematically admissible.

The field in Fig. 4.16a, for instance, is symmetric with respect to the vertical through the point O. In the angular regions AOB (Region 1) and COD (Region 2) the velocity components with respect to the polar coordinate systems indicated in Fig. 4.16a are

$$v_r' = c^2 r', \qquad v_\theta' = -2c^2 r'\theta', \tag{4.41}$$

and

$$v_r'' = -c^2 r'', \qquad v_\theta'' = 2c^2 r''\theta''; \tag{4.42}$$

in the angular region DOA (Region 3) the velocity components with respect to the rectangular axes x, y are

$$v_x = -c^2(x + \pi y/2), \qquad v_y = c^2(y + \pi x/2). \tag{4.43}$$

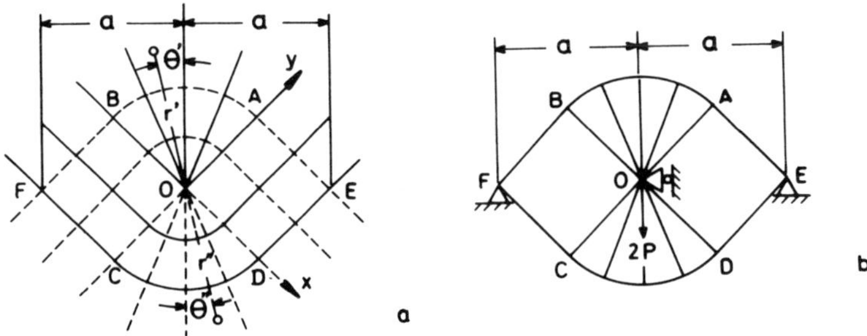

Figure 4.16. (a) Michell field. (b) Optimal truss deduced from the given Michell field.

As one may readily verify, Eqs. (4.41)-(4.43) define a continuous Michell field of type *T*, whose first principal lines are rays from *O* in Region 1, circles with center *O* in Region 2, and parallels to *OA* in Region 3.

According to Eq. (4.43), the velocity of *E* is given by $v_x = -v_y = -c^2a(1 + \pi/2)/\sqrt{2}$; it is thus vertical and directed upwards and has magnitude $v = c^2a(1 + \pi/2)$. By symmetry the point *F* has the same velocity. Superimposing a vertical translation with a downward velocity of this magnitude, we may bring the points *E* and *F* to rest, so that they may serve as fixed supports. The point *O* thereby gains a vertical downward velocity of magnitude *v*.

The structure in Fig. 4.16b, which has fixed supports at *E* and *F*, is based on the resulting velocity field. Even if all its bars were rigid, however, the joint *O* could still have a nonvanishing velocity along *EF*. To exclude this we must place a movable support at *O* that allows only a vertical motion of this joint. The resulting structure (Fig. 4.16b) is optimal for the transmission of a vertical load $2P$ acting at *O* to fixed supports at *E* and *F*. The vertical reaction at *E* has magnitude *P* and produces compressive forces of magnitude $P/\sqrt{2}$ in the bar *EA*, the circular bar *AB*, and the bar *BF*; and tensile forces of the same magnitude in the bars *ED*, *DC*, and *CF*. The volume of material required for these bars is

$$V_1 = (P/2)(a/\sqrt{2})\frac{(4 + \pi)}{\sigma_0}. \tag{4.44a}$$

To the central angle $d\theta'$ of the arc *AB*, there corresponds a radial tensile force of magnitude $Pd\theta/\sqrt{2}$; a similar statement applies to the arc *CD*. The volume of material required for the radial bars is thus

$$V_2 = \frac{\pi(P/\sqrt{2})(a/\sqrt{2})}{\sigma_0}, \tag{4.44b}$$

and the total volume of the structure is

$$V = V_1 + V_2 = \frac{Pa(2+\pi)}{\sigma_0}. \tag{4.45}$$

This volume may also be obtained by equating the external power of the load $2P$—that is, $2Pv = Pc^2a(2+\pi)$—to the internal power $\sigma_0 c^2 V$.

As this example indicates, the inverse procedure that starts from an arbitrary Michell field does not always lead to a practical structure.

Exercise 4.6. As an approximation to the optimal truss-like continuum for the problem in Fig. 4.2, construct a truss with six joints in the manner indicated in Fig. 4.10, using as starting joints the points on the circle of radius R that correspond to the polar angles $-40°$, $0°$, $40°$. The bars BD and CD are to have the same length, and BDA is to be a right angle. Determine the lengths l_i of the bars labeled 1, 2, 3 in Fig. 4.17, the axial forces S_i in these bars, and the volume V^* of structural material required for the six-joint truss when the load factor at plastic collapse is unity and the uniaxial yield stress is σ_0. Comparing V^* to the minimal volume V given by Eq. (4.20), find the efficiency, $\eta = 100\, V/V^*$, of the six-joint truss.

Exercise 4.7. Show that the efficiency decreases when joint D of the truss in Exercise 4.6 is moved along DA to a position D^* between D and A such that the angle DBD^* equals 5°, while the left half of the truss is modified in an analogous manner.

Exercise 4.8. Show that the efficiency also decreases when joint D of the truss in Fig. 4.17 is moved along DB to a position D^{**} between D and

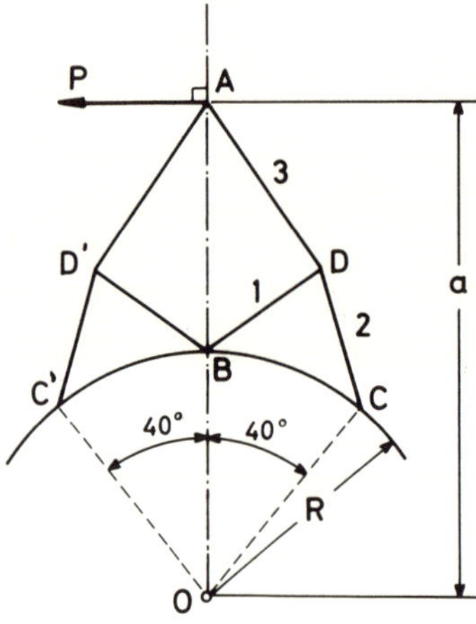

Figure 4.17. Approximate optimal truss with six joints.

B such that the angle DAD^{**} equals 5°, while the left half of the truss is modified in an analogous manner.

4.6. Optimal Design of an Elastic Truss for Given Compliance

Hegemier and Prager (1969) have shown that a given load factor at plastic collapse or a given allowable range of axial stress are not the only behavioral constraints for which the Michell layout is optimal. Only optimal design for given elastic compliance will be discussed here.

Consider again a truss that is to transmit a given load $\mathbf{P}$ from its point of application A to a given rigid foundation arc f-f (Fig. 4.2). The bars of the truss are to consist of a given elastic material with Young's modulus E. The total volume V of the bars is to be as small as possible and subject to the behavioral constraint that the compliance of the truss to the given load is to have the given value $\mathscr{E}$.

Let T be a trusslike continuum based on a Michell field whose principal strains satisfy Eq. (4.21), where ε_0 is chosen to satisfy the behavioral constraint. Thus,

$$\mathscr{E} = E\varepsilon_0^2 V \tag{4.46}$$

because the compliance equals twice the strain energy, and all members of T experience axial strains of absolute value ε_0.

Next, let T^* be any truss transmitting the given load to the foundation arc and satisfying the behavioral constraint. If the given load produces a strain ε_i^* in bar i^* of this truss, and if V_i^* is the volume of i^*, the behavioral constraint requires that

$$\mathscr{E} = E \sum_{i^*} \varepsilon_i^{*2} V_i^*. \tag{4.47}$$

If $\mathbf{v}$ and $\mathbf{v}^*$ are the displacements of joints A of the two structures, it follows from the behavioral constraint that $\mathbf{P} \cdot \mathbf{v} = \mathbf{P} \cdot \mathbf{v}^*$. Since the displacements of the Michell field for T are kinematically admissible for T^*, the principle of minimum potential energy applied to the truss T^* furnishes

$$\mathscr{E} = E \sum_{i^*} \varepsilon_i^{*2} V_i^* \leq E \sum_{i^*} A_i^* \int \bar{\varepsilon}_i^{*2} \, ds_i^*, \tag{4.48}$$

where A_i^* is the cross-sectional area of bar i^*, $\bar{\varepsilon}_i^*$ is the axial strain to which the Michell field for T subjects the typical element ds_i^* of this bar, the integration is extended along the axis of i^*, and the summation includes all bars of T^*.

According to Eq. (4.21),

$$|\bar{\varepsilon}_i^*| \leq \varepsilon_0. \tag{4.49}$$

Using this inequality in Eq. (4.48) furnishes

$$\mathscr{E} \leq E\varepsilon_0^2 \sum_{i^*} A_i^* l_i^* = E\varepsilon_0^2 V^*, \tag{4.50}$$

where l_i^* is the length of bar i^*, and V^* is the total volume of the bars of T^*. Comparison of Eqs. (4.46) and (4.50) yields the inequality

$$V^* - V \geq 0, \tag{4.51}$$

which establishes the global optimality of the Michell continuum.

Another way of stating this result is as follows. The Michell continuum cannot have a greater compliance to the given load than any other truss using the same amount of material. Observing that all members of the Michell continuum experience axial stresses of the same magnitude under the given load, and combining the statement just made with the result obtained in Section 4.3, we have the following theorem: When an allowable range of axial stress is imposed as a behavorial constraint, the Michell continuum cannot have a greater compliance to the given load than any other truss using the same amount of structural material.

Exercise 4.9. The given horizontal load P is to produce a horizontal displacement v of joint J of the truss in Fig. 4.18. Determine the necessary volume V of the elastic bars of the truss (Young's modulus E) as a function of the angle α, and show that the minimum of V corresponds to $\alpha = 45°$ in accordance with the Michell field discussed in connection with Fig. 4.12b.

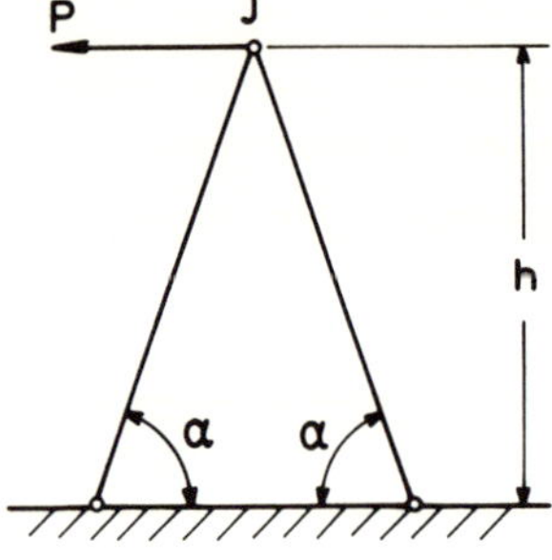

Figure 4.18. Two-bar truss to be optimized for assigned compliance.

4.7. Nearly Optimal Layout of Trusses with a Finite Number of Joints

In Sections 4.2 to 4.6 we have not worried about determining the cross-sectional areas of the trusslike continuum corresponding to the Michell fields. Due to the continuous nature of the structure, there are areas t_1 and t_2 per unit of arc in the directions orthogonal to the local axes of the bars. They can be obtained from the equilibrium equations of an element of the trusslike continuum. A similar problem will be discussed in more detail in Section 6.6 concerned with disks. Here, because Michell structures are impractical, only two features are of direct help in the design of nearly optimal practical trusses: their volumes serve as reference absolute minima, and their general shapes serve as guides for the layout of trusses with a finite number of joints.

If we consider a Michell trusslike continuum to be formed of an infinity of bars, most of which have infinitesimal length and cross section and are connected by an infinite number of joints, it suffices to introduce the cost of the joints in the cost function to reduce the number of joints to a finite value. Consider, for example [Prager (1977)], the problem of Fig. 4.15, in which the single load P must be transferred to hinges at B and C. The cost of the truss is defined by

$$\Gamma = \sum_i A_i l_i + \frac{cn}{\sigma_0}, \tag{4.52}$$

where n is the number of joints, and c/σ_0 is the specific cost of a joint. From the solution of Fig. 4.15, we tentatively construct the three trusses of Fig. 4.19(a)–(c) with 3, 6, and 11 joints, respectively, and minimize the cost, Eq. (4.52), with respect to the geometric parameters.

It turns out that, in each family, the angles marked with a circular arc must have the same value 2α, while the adjacent angles must be right angles. The trusses of Fig. 4.19 are drawn for the particular case $h = 5b$, corresponding to $\alpha = 27.885°$ and $\alpha = 11.310°$ for the six-joint and eleven-joint trusses, respectively. The cost of the three-joint truss is less for values of c/Pb greater than 1.814, whereas the six-joint truss is preferable for $1.814 > c/Pb > 0.471$, and the eleven-joint truss for smaller values of c/Pb. As c tends to zero, the number of joints increases, and the truss approaches the Michell structure of Fig. 4.15.

The procedure just outlined may turn out to be quite difficult when the space available for the structure is bounded and the cost of the joints is relatively low. An alternative can be found in designing a truss for which the cost [defined by Eq. (4.52)] exceeds, by at most a known amount, that

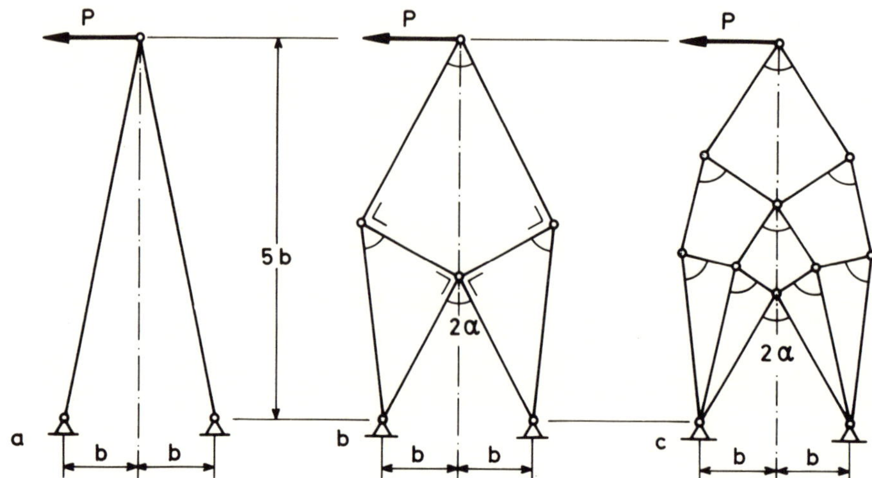

Figure 4.19. (a) Three-joints optimal truss. (b) Six-joints optimal truss. (c) Eleven-joints optimal truss.

of the minimum-cost truss *with the same number of joints*. Such a design has been described by Prager (1978) in the frame of elastic design for assigned limiting stress σ_0, as a truss with *meshwise constant strain* of a magnitude not exceeding the limiting value σ_0/E reached by the principal strains. One can perform a systematic generation of the truss using the circle of relative velocities (see the Appendix to this chapter). The reader is referred to the original paper for more details.

4.8. Some Remarks on Optimal Trusses

The optimal trusses considered so far suffer, from the point of view of practical application, from neglecting the possibility of instability of the bars in compression.† Hence, they are mostly to be regarded, together with the Michell trusslike continua, as forming bases for more practical designs for which their costs can serve as reference minima. For this reason we shall quote briefly only a few more results obtained in this field.

The typical properties of minimum-volume elastic trusses with limited stress intensity have been studied by Chern and Prager (1972), as already discussed in Section 2.4. The three-bar truss considered (Fig. 4.20) is subjected to alternative loads L' and L''. The load L' is always positive, whereas the components P and Q of L'' vary in some nonnegative range. Minimum-volume designs are found to be either two-bar trusses (fully

† More practical engineering approaches will be developed in Volumes 2 and 3 of this treatise.

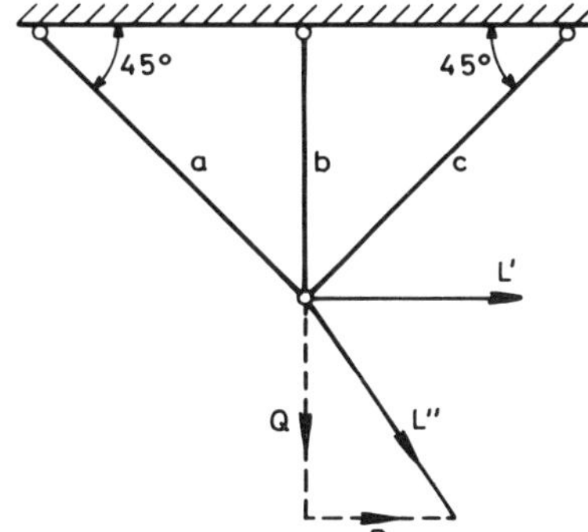

Figure 4.20. Three-bar truss subjected to alternative loads L' and L''.

stressed) or three-bar trusses (fully stressed or not). On the same example, but with the single load Q, it has been pointed out by Prager (1981), after Sved and Ginos (1968), that the minimum-volume design may be singular, in the sense that reducing the volume of a three-bar truss does not give the true minimum volume (which belongs to a one- or two-bar truss), however small the positive cross-sectional area of some bars may be. This singularity comes basically from the fact that the bars that do not tend to vanish in the three-bar truss are understressed because of the compatibility condition in the elastic range. Deleting at least one bar eliminates this compatibility condition and enables jumping down to a smaller volume by stressing the remaining bars at the allowable limit. This phenomenon was shown by Prager (1981) by taking the limit stress σ_{01} of the oblique bars equal to one third the value σ_{02} of the vertical bar. This discussion was extended by Save (1983) to the entire interval $0 \leq \sigma_{01}/\sigma_{02} < \infty$, including comparison with corresponding minimum-volume designs for assigned load factor at plastic collapse and for assigned maximum elastic compliance, as well as discussion of the influence of lower bounds set on cross-sectional areas.

The minimum-volume truss, including unknown layout, for alternative plastic collapse loads has been studied by Nagtegaal and Prager (1973). Whereas the Michell truss is of minimum volume for assigned elastic limit stress, assigned elastic compliance, and assigned plastic collapse load when it is subjected to a one-parameter loading, this is no longer true when alternative states of loading have to be considered. The optimality criterion obtained is completely similar to Section 3.7.2: namely, that a family of collapse modes must be found such that the sum of their strain rates in each bar is equal to a positive constant if the bar exists, and is smaller then or equal to that constant when the bar vanishes. Solutions for the transmission to a rigid foundation arc of two alternative forces concentrated at a given point were obtained by applying the Nagtegaal–Prager superposition method already used in Section 3.8.1.

Minimum-volume elastic trusses subject to multiple loadings and to stress and displacement constraints have been studied analytically by Chern and Prager (1971). Despite the restriction to statically determinate trusses and the neglect of any instability, the solution rapidly becomes intricate. Hence, in a more practical context, and especially for large trusses, a numerical approach becomes necessary; this will be developed in Volume 2 of this treatise.

4.9. Appendix

Circle of Relative Velocities [Prager (1961, Chapter 3, Section 2)]. Figure 4.21 shows a variant of Mohr's circle. The physical plane (Fig. 4.21a) is referred to rectangular axes x, y. The mean rates of rotation and extension of line elements issuing from the generic point P are denoted by Ω and ε. Counterclockwise rotation and extension of the generic line element PA are denoted by Ω_a and ε_a, respectively. The plane of relative velocities (Fig. 4.21b) is referred to rectangular axes Ω_a, ε_a that are parallel to the coordinate axes of the physical plane. The line element PA of the physical plane has as its image in the plane of relative velocities the point A' with coordinates Ω_a and ε_a.

It is well known that the images of all line elements issuing from P fill a circle called the *circle of relative velocities*, which has center Ω, ε. Indeed,

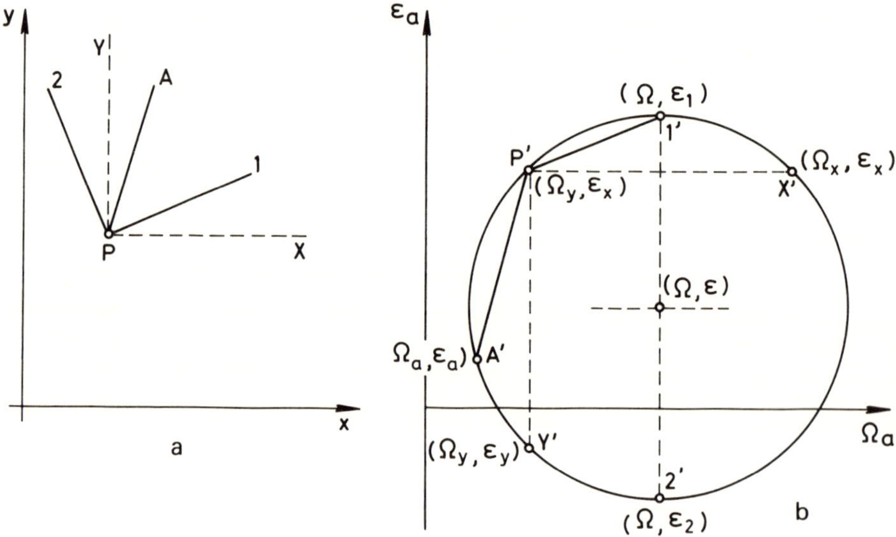

Figure 4.21. (a) Reference directions x and y, principal directions 1 and 2, and typical segment PA at point P. (b) Circle of relative velocities.

rates of rotation and extension differ from strain rates only by superposition of the rigid-body rate of rotation Ω, which is the rotation rate of the principal directions. On this circle there exists a point P', the pole, such that $P'A'$ is parallel to the corresponding line element PA.

As we see from the coordinates of the points X' and Y' that correspond to the line elements PX and PY in the coordinate directions, the pole has coordinates Ω_y, ε_x. The lines joining the pole to the highest point $1'$ and the lowest point $2'$ of the circle of relative velocities indicate the principal directions $P1$ and $P2$ of the strain rate.

5

Optimal Design of Grillages

5.1. Optimal Plastic Design of Rectangular Grillages of Orthogonal Prismatic Beams

We consider a rectangular grillage of horizontal prismatic beams that are parallel to the sides of the rectangle. The beams are to consist of a given, rigid, perfectly plastic material with tensile and compressive yield limits $\pm\sigma_0$. The total volume of the beams is to be minimized subject to the behavioral constraint that the grillage is to be on the verge of plastic collapse under the given uniformly distributed vertical load p.

In treating this problem we shall assume that each beam extends from one side of the grillage to the other and has constant rectangular cross section, whose breadth may vary from beam to beam, but whose height h has a given value that is the same for all beams. Accordingly, the yield moment of the typical beam is $Y_i = \sigma_0 h A_i/4$, where A_i is the cross-sectional area of the beam. Minimization of the total volume of the beams is thus equivalent to minimization of the "cost"

$$\Gamma = \sum_i Y_i l_i, \tag{5.1}$$

where l_i is the length of the typical beam. No explicit bounds will be prescribed for the yield moments of the beams.

Our assumptions obviously fail to require the optimum design to consist of only a finite number of beams. In analogy with the treatment of truss-like continua in Chapter 4, most of the present chapter will therefore be concerned with grillage-like continua—that is, with dense arrangements of beams of infinitesimal breadths. A collapse mechanism of a structure of this kind is specified by a rate of deflection $v(x, y)$, where x and y are rectangular coordinates in the median plane of the grillage. For the rectangular grillages considered in the following, we take the coordinate axes along two adjacent edges in such a manner that the grillage lies in the first quadrant. In the figures of this chapter, built-in, simply supported, and free edges will be respectively indicated by heavy, light double, and light single lines.

Because the typical beam is prismatic, its collapse mechanism corresponds to the formation of plastic hinges that connect rigid segments of the beam. The rate of deflection $v(x, y)$ is thus regionwise linear. For the grillage in Fig. 5.1a, for instance, we tentatively assume $v(x, y)$ to be linear in each of the four regions bounded by the edges and the segments AF, BG, CG, DF, and FG. Since loading and supports are symmetric with respect to the line $y = a$, the points F and G lie on this line. The distance from F to the y-axis will be denoted by βa; the distance from G to the line $x = 2\alpha a$, by γa.

The assumed rate of deflection is completely determined by the magnitude ω_x of the angular velocity of the rotation of $ABGF$ about the x-axis. The angular velocity of $CDFG$ about the line $y = 2a$ also has magnitude ω_x, and the magnitudes of the angular velocities of AFD and BCG about

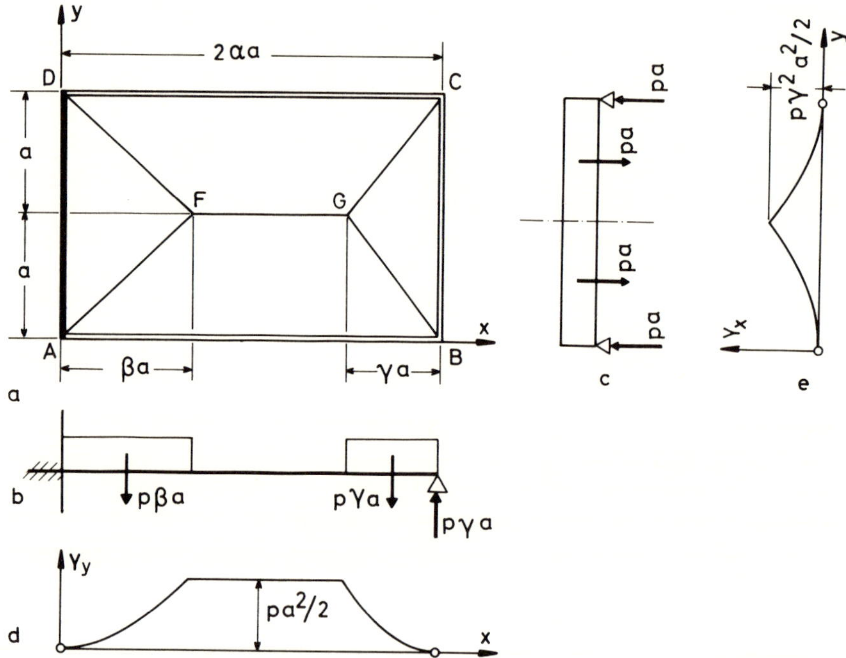

Figure 5.1. (a) Rectangular grillage with piecewise linear velocity field: one edge built in, three edges supported. (b) Beam along $y = a$ and its loading. (c) Beam along $x = \alpha a$ and its loading. (d) Function $Y_y(x)$. (e) Function $Y_x(y)$.

the lines $x = 0$ and $x = 2\alpha a$ will be respectively denoted by ω_y and ω'_y. Because the points F and G have the same rates of deflection, we have

$$\beta\omega_x = \beta\omega_y = \beta\omega'_y\gamma. \tag{5.2}$$

According to the mechanical interpretation of the optimality condition in Eq. (3.9) given at the end of Section 3.1, the optimal grillage-like continuum admits a collapse mechanism for which the power of dissipation per unit yield moment in each prismatic beam numerically equals the length of this beam. Now, the power of dissipation per unit yield moment of a prismatic beam equals the sum of the absolute values of the relative angular velocities in all yield hinges of the beam. For the collapse mechanism considered above, a beam in the x-direction (for short "an x-beam") with a y-value smaller than a has yield hinges with relative angular velocities $-\omega_y$ and ω_y at its left end and its intersection with AF, and a yield hinge with relative angular velocity ω'_y at its intersection with BG. The power of dissipation per unit yield moment is thus $2\omega_y + \omega'_y$, independent of the y-value of the x-beam considered. A beam in the y-direction (for short "a y-beam") with an x-value smaller than βa has yield hinges with relative angular velocity ω_x at its intersection with AF and FD. Its power of dissipation per unit yield moment is thus $2\omega_x$, and the same value is obtained for all other y-beams. Optimality therefore requires that

$$\omega_x = a, \qquad 2\omega_y + \omega'_y = 2\alpha a. \tag{5.3}$$

Eliminating a between these equations and using Eq. (5.2), we find

$$\beta = \frac{2\gamma}{2\alpha\gamma - 1}. \tag{5.4}$$

A further relation between β and γ is obtained from static considerations. An x-beam with a y-value smaller than a has bending moments equal to its yield moment at its points of intersection with AF and BG. Its bending moment, which cannot exceed the yield moment, is therefore constant between these points, and this segment of constant bending moment is unable to carry any load. Consequently, the full load on the region $ABGF$ is transmitted by y-beams to the edge AB. Similarly, the full load on the region $CDFG$ is transmitted by y-beams to the edge CD, whereas the loads on AFD and BCG are transmitted by x-beams to the edges AD and BC, respectively.

Figure 5.1b shows the beam along $y = a$ and the loads it carries. The condition that the clamping moment, which equals the negative yield moment, be equal and opposite to the bending moments at F and G

furnishes the yield moment per unit width in the y-direction,

$$Y_x(a) = \frac{p\gamma^2 a^2}{2}, \tag{5.5}$$

and the relation

$$\beta = \gamma\sqrt{2}. \tag{5.6}$$

Using Eq. (5.6) in Eq. (5.4), we obtain

$$\gamma = \frac{1+\sqrt{2}}{2\alpha}. \tag{5.7}$$

Figure 5.1c shows the beam along $x = \alpha a$ and its load. The yield moment per unit width in the x-direction is

$$Y_y(\alpha a) = pa^2/2. \tag{5.8}$$

For y-beams to the left of F, the common length $l(x)$ of the load-carrying segments decreases linearly from a at $x = \beta a$ to zero at $x = 0$, while the yield moment per unit width in the x-direction is $Y_y(x) = pl^2(x)/2$. A similar remark applies to y-beams to the right of G. Figure 5.1d shows this variation of $Y_y(x)$, and Fig. 5.1e shows the variation of $Y_x(y)$. The curves in these figures are parabolas whose vertices have been circled. The cost, Eq. (5.76), is obtained by multiplying the areas of the diagrams in Fig. 5.1(d)–(e) by $2a$ and $2\alpha a$, respectively, and adding the products. Accordingly,

$$\Gamma = pa^4\left[2\alpha - \beta - \gamma + \frac{\beta+\gamma}{3} + \frac{2\alpha\gamma^2}{3}\right]. \tag{5.9}$$

Substituting Eqs. (5.6) and (5.7) into Eq. (5.9), one finds

$$\Gamma = pa^4\left[2\alpha - \frac{(1+\sqrt{2})^2}{6\alpha}\right]. \tag{5.10}$$

The reader is urged to verify this result by evaluating Γ as the product of p by the integral of $v(x, y)$ over the area of the grillage.

For the preceding discussion to be valid, the point G must be to the right of F or coincide with F; that is, $\beta + \gamma$ must not exceed 2α. This condition yields

$$\alpha \geq \frac{1+\sqrt{2}}{2}. \tag{5.11}$$

The investigation of the case in which this condition is not fulfilled is proposed in Exercise 5.1.

As a second example, we consider the boundary conditions in Fig. 5.2a, which differ from those in Fig. 5.1a by the fact that the edge BC is free. We tentatively assume that the rate of deflection is linear in each of the regions $ABHF$, $HCDF$, and AFD, and that the x-beams are simply supported at their right ends by a beam with finite yield moment Y along BC to which they transmit the loads in the triangle BCG.

For the angular velocities ω_x and ω_y of the regions $ABHF$ and AFD about the coordinate axes, we again have

$$\omega_x = \omega_y \beta, \tag{5.12}$$

and optimality requires that

$$\omega_x = a, \qquad \omega_y = \alpha a. \tag{5.13}$$

It follows from Eqs. (5.12) and (5.13) that

$$\beta = 1/\alpha. \tag{5.14}$$

Except for the fact that the relations between α, β, and γ will differ from those obtained above, the loads carried by the beam along $y = a$ and by any y-beam between F and G are as shown in Fig. 5.1(b)–(c), which means that the functions $Y_y(x)$ and $Y_x(y)$ are given by Fig. 5.1(d)–(e). The

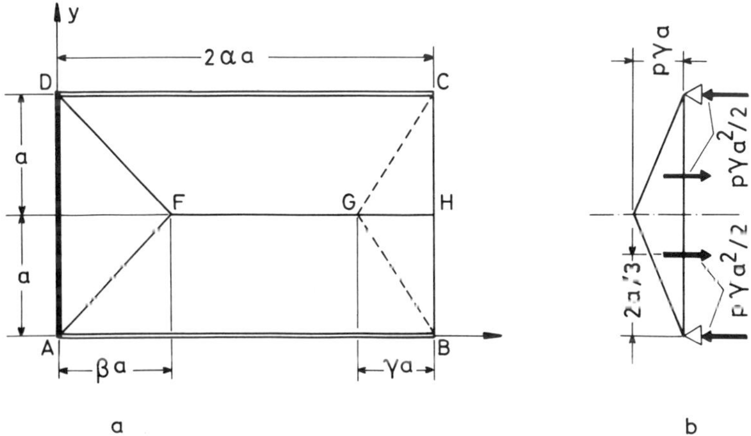

Figure 5.2. (a) Rectangular grillage with piecewise linear velocity field: one edge built in, two edges supported, one edge free. (b) Loads carried by the edge beam BC.

cost Γ of the grillage-like continuum is again given by Eq. (5.9), but to this must now be added the cost Γ' of the edge beam along BC. Before we can evaluate Eq. (5.9), however, we must express γ in terms of α. The fact that the bending moments of the beam along $y = a$ must equal $Y_x(a) = p\gamma^2a^2/2$ at F and G, and $-Y_x(a)$ at the built-in end, again yields Eq. (5.6). From this and Eq. (5.14) we obtain

$$\gamma = 1/\alpha\sqrt{2}. \tag{5.15}$$

Using Eqs. (5.14) and (5.15) in Eq. (5.9), we find the cost of the grillage-like continuum:

$$\Gamma = pa^4\left[2\alpha - \frac{1+\sqrt{2}}{3\alpha}\right]. \tag{5.16}$$

Figure 5.2b shows the loads carried by the edge beam BC, whose yield moment and cost are $Y = p\gamma a^2/3$ and

$$\Gamma' = \frac{2p\gamma a^4}{3} = \frac{pa^4\sqrt{2}}{3\alpha}. \tag{5.17}$$

The total cost of the structure is thus

$$\Gamma + \Gamma' = pa^4\left[\frac{2\alpha - 1}{3\alpha}\right]. \tag{5.18}$$

Note that this total cost may also be found by using Eqs. (5.13) and (5.14) in

$$\Gamma + \Gamma' = p\iint v(x, y)\, dx\, dy$$

$$= pa\omega_x\{\tfrac{1}{3}\cdot 2a\cdot \beta a + \tfrac{1}{2}\cdot 2a(2\alpha - \beta)a\}. \tag{5.19}$$

Use of Eq. (5.14) in Eq. (5.19) again yields Eq. (5.18).

It is interesting to compare the costs of the structures in Figs. 5.1a and 5.2a. For $\alpha = 1.5$, for instance, the second structure requries 18.1% more material than the first. For $\alpha = 2.0$, however, only 9.1% more material is needed.

Exercise 5.1. Determine the cost of the grillage-like continuum with the boundary conditions in Fig. 5.1a when the condition of Eq. (5.11) is not fulfilled.

Exercise 5.2. Find the cost of the grillage-like continuum in Fig. 5.1a when the continuum is built-in along CD.

5.2. Optimal Layout of a Grillage in Plastic Design for a Single Loading

The grillages of the preceding section were required to consist of *prismatic* beams of *given layout*. For the remainder of this chapter these two requirements will be dropped. The general problem to be treated concerns a grillage whose horizontal beams are to have rectangular cross sections of given height h that is to be the same for all beams, and breadths that vary along the beams. Moreover, the layout of the beams is not prescribed, but is the choice of the designer.

The grillage is to have a given plan form [in the (x, y)-plane] and to be supported in a given manner along its edge or an adequate part of its edge. Its beams are to consist of a given, rigid, perfectly plastic material with tensile and compressive yield stresses $\pm\sigma_0$ and to utilize a minimal amount of this material subject to the behavioral constraint that the grillage is on the verge of plastic collapse under the given distributed vertical *downward* load $p(x, y)$.

Since the yield moment of the typical beam is now a function $Y_i(s)$ of the distance s measured along the axis of this beam, Eq. (5.1) for the "cost" must now be replaced by

$$\Gamma = \sum_i \int Y_i(s)\, ds, \tag{5.20a}$$

where the integration is extended along the typical beam, and the summation includes all beams. No explicit bounds will be prescribed for the yield moment $Y_i(s)$, which thus equals the absolute value of the bending moment $M_i(s)$. The cost Γ may therefore be written as

$$\Gamma = \sum_i \int |M_i(s)|\, ds. \tag{5.20b}$$

The integral in this expression is called the *moment area* of the beam because it is the total area of the bending moment diagram of this beam when the sign of the bending moment is disregarded.

Because our formulation of the problem does not require the use of only a finite number of beams, the desired optimal structure will be a *grillage like continuum* rather than a grillage in the usual sense of the term. A collapse mechanism of this kind of structure is specified by a rate of deflection $v(x, y)$ that is *kinematically admissible for optimality* in the sense that it and its first derivatives are continuous and satisfy the kinematic conditions of the support. Thus v must vanish along a simply supported edge, and v and its normal derivative must vanish along a built-in edge.

For a given direction at a typical point of the (x, y)-plane, the rate of curvature corresponding to the rate of deflection $v(x, y)$ is defined as the negative second derivative of v in the direction considered. Since there are no explicit bounds on the yield moments of the beams, the optimality condition in Eq. (3.17) requires that the rate of curvature have a constant absolute value c^2 for the line elements of the axes of the beams of the optimal grillage-like continuum, and absolute values not exceeding c^2 for all other line elements of the (x, y)-plane. Because a collapse mechanism is only defined to within an arbitrary constant positive factor, the value of the constant c^2 is not relevant, and we shall henceforth use $c^2 = 1$.

In discussing this optimality condition, we denote the principal rates of curvature by κ_1 and κ_2, numbering them in such a manner that $|\kappa_1| \geq |\kappa_2|$. The direction in which the rate of curvature has the value κ_1 (or κ_2) will be called the first (or second) principal direction, and a line that has, at each of its points, the first (or second) principal direction will be called a first (or second) principal line.

Because the rate of curvature in an arbitrary direction at the typical point of the (x, y)-plane has a value in the closed interval bounded by the principal curvatures at this point, the optimality condition above is satisfied if the axes of the beams of the grillage-like continuum follow principal lines along which the absolute value of the rate of curvature is one. In principle, the domain of the optimal grillage-like continuum may therefore be divided into regions of the following types.

Type R, for which $|\kappa_1| = 1 > |\kappa_2|$. The beams follow the first principal lines. Depending on the sign of κ_1, we distinguish the types R^+ and R^-. In an R-region the axis of only one beam passes through the typical point of the (x, y)-plane.

Type S, for which κ_1 and κ_2 both have either the value 1 (type S^+) or the value -1 (type S^-). Since any direction may be regarded as a principal direction, the optimal layout of beams is not unique in an S-region.

Type T, for which κ_1 and κ_2 both have absolute value 1, but opposite signs. A T-region admits two orthogonal families of beams, but the designer in many cases may decide to use only one of them.

Figure 5.3(a)–(c) show optimal grillage-like continua, each of which is based on a rate of deflection field of one of these types. The edges $y = \pm x \tan \alpha$ of the grillage in Fig. 5.3a are simply supported, and the edge

$x = b$ is free. The rate of deflection

$$v(x, y) = \frac{x^2 \tan^2 \alpha - y^2}{2} \tag{5.21}$$

is kinematically admissible. Since the mixed derivative $\partial_{xy}v$ vanishes, the coordinate directions are principal directions, and the principal rates of curvature are

$$\kappa_1 = -\partial_{yy}v = 1, \qquad \kappa_2 = -\partial_{xx}v = -\tan^2 \alpha. \tag{5.22}$$

The rate of deflection field, Eq. (5.21), thus satisfies the optimality condition, provided that $\alpha \leq 45°$. For $\alpha < 45°$ the field is of type R^+, and the unique optimal layout consists of beams in the y-direction.

Because each beam has a unit rate of curvature, the equality of internal and external powers of dissipation shows that the cost, Eq. (5.20b), is given by

$$\Gamma = \iint p(x, y)v(x, y)\, dx\, dy, \tag{5.23}$$

where the integration is extended over the plan form of the structure. Note that Eq. (5.23) shows the rate of deflection to be the *influence function* for the cost Γ: the cost caused by a concentrated downward unit load at the point x, y is given by the value of $v(x, y)$.

For the simply supported circular grillage of radius a in Fig. 5.3b, the rate of deflection field

$$v(x, y) = \frac{a^2 - x^2 - y^2}{2} \tag{5.24}$$

is kinematically admissible. Because $\partial_{xy}v$ vanishes, the coordinate directions are principal directions. The corresponding rates of curvature are $\kappa_1 = \kappa_2 = 1$. The field is thus of type S^+, and the optimal layout is not unique. To show this directly, consider a concentrated downward load P at the center A of the grillage. To transmit this to the edge we may use a beam along any diameter. Its moment area equals one-half of the product of the span $2a$ and the bending moment $Pa/2$ at the point of application of the load. Accordingly,

$$\Gamma = Pa^2/2. \tag{5.25}$$

Alternatively, the considered load may be transmitted by the beam BB' (Fig. 5.3b) to the beams CD and $C'D'$, and by these to the edge. The beam

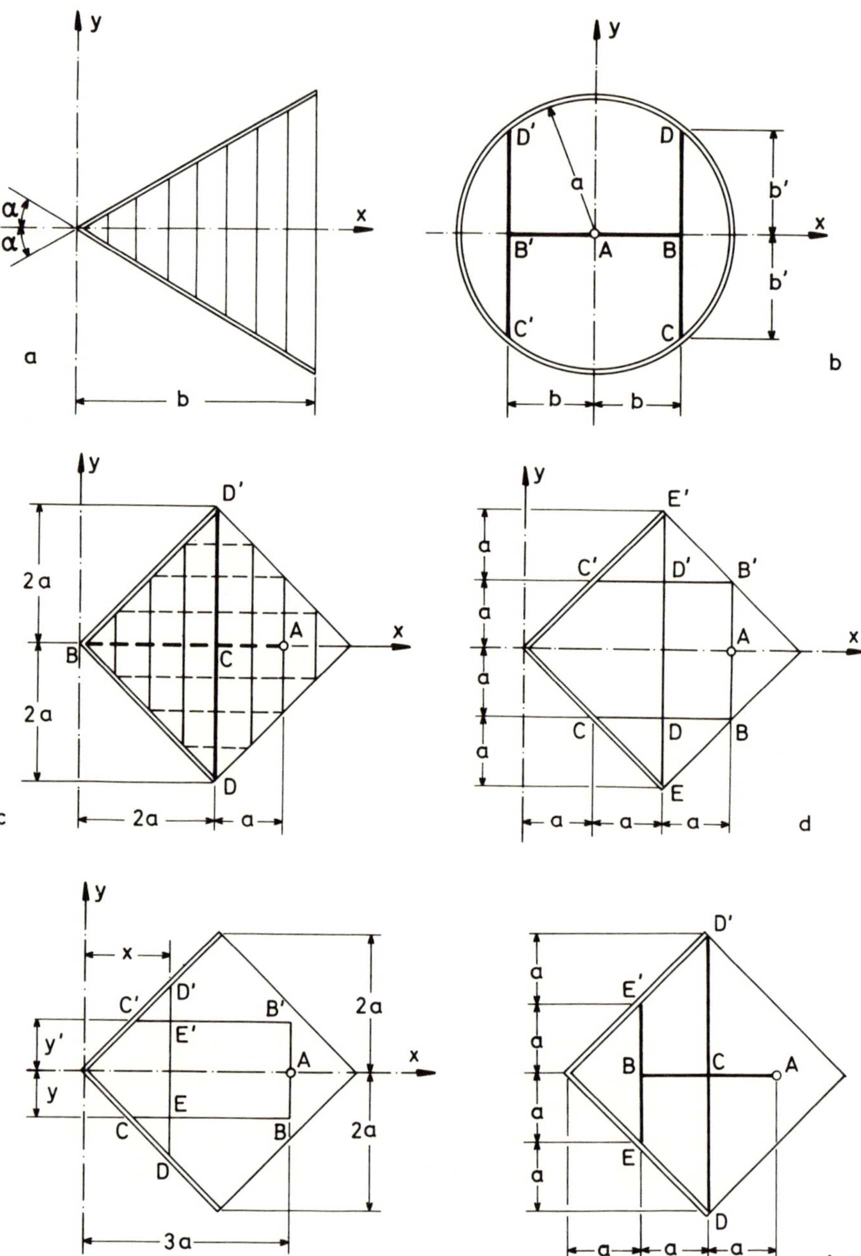

Figure 5.3. (a) Optimal grillagelike continuum; two edges supported, one edge free. (b) Simply supported circular optimal grillage for central load. (c) Possible optimal grillages. (d), (e), (f) Examples of optimal grillage for concentrated load at point A.

BB' has bending moment $Pb/2$ at A and, hence, moment area $Pb^2/2$. The beam CD carries the load $P/2$ at B; it has bending moment $Pb'/4$ at B and, hence, moment area $Pb'2/4 = P(a^2 - b^2)/4$. The moment area of the structure thus again has the value in Eq. (5.25). Note that this value equals the product of P by the rate of deflection, Eq. (5.24), at the point of application of this load.

The edges $y = \pm x$ of the square grillage in Fig. 5.3c are simply supported, while the other two edges are free. The rate of deflection

$$v(x, y) = \frac{x^2 - y^2}{2} \tag{5.26}$$

is kinematically admissible. The corresponding principal rates of curvature are $\kappa_1 = 1$ and $\kappa_2 = -1$ in the y- and x-directions. The field is thus of type T.

In Fig. 5.3c the beams with positive rates of curvature are shown by full lines; those with negative rates of curvature, by dashed lines. Although this layout of potential beams is unique, we may choose different groups of beams to transmit a given concentrated downward load P to the supported edges. If, for instance, P is applied at the point A (Fig. 5.3c), we may use the beam AB that is supported at C by the beam DD'. The moment area of the beam AB is $3Pa^2/2$. Since the load transmitted at C to the beam DD' is $3P/2$, the moment area of this beam is $3Pa^2$. Accordingly, $\Gamma = 9Pa^2/2$, which agrees with the product of P by the rate of deflection, Eq. (5.26), at point A, where $x = 3a$ and $y = 0$.

Alternatively, we may, for instance, use the beam BB' (Fig. 5.3d) to transmit the load P to the beams BC and $B'C'$, which are supported at D and D' by the beam EE'. We leave it to the reader to verify that the moment areas are $Pa^2/2$ for each of the beams BB', BC, and $B'C'$, and $3Pa^2$ for the beam EE'. The sum of these moment areas again gives $\Gamma = 9Pa^2/2$.

The structures considered by no means provide the only optimal transmission of the load P from A to the supported edges. Figure 5.3e shows another optimal arrangement of beams. We may, moreover, simultaneously use several of these optimal arrangements, dividing the load P between them. Note, however, that the structure in Fig. 5.3f is not optimal. Although this structure uses beams in the principal directions of the field in Eq. (5.26), an upward load would be transmitted at B by beam AB to beam EE'. This beam would therefore have a negative rate of curvature rather than the positive unit rate of curvature associated with the field of Eq. (5.26).

Exercise 5.3. Show that the grillage in Fig. 5.3e has cost $\Gamma = 9Pa^2/2$ independent of the choice of positive variables x, y, y', provided that $x > y$, $x > y'$, $y < a$, $y' < a$.

Exercise 5.4. Determine the cost Γ^* and the efficiency η^* of the grillage in Fig. 5.3f, which carries the load P at point A.

5.3. Morley Fields

In the preceding section an optimality condition has been established that characterizes the rate-of-deflection field of an optimal grillage-like continuum. In a different context, rate-of-deflection fields satisfying this condition were first used by Morley (1966), who discussed the minimum reinforcement of concrete slabs of constant thickness. In the following, kinematically admissible rate-of-deflection fields whose principal rates of curvature satisfy the conditions

$$|\kappa_1| = 1, \qquad |\kappa_2| \leq 1, \tag{5.27}$$

will therefore be called *Morley fields.*

In the discussion of the optimal layout of the beams of a grillage-like continuum, the rate of curvature and twist associated with the Morley field $v(x, y)$ of the collapse mechanism play an important role. At a typical point P of the (x, y)-plane, a direction will be specified by the angle α that it forms with the positive x-direction. For brevity this will be called direction α. The rates of curvature and twist, κ_α and τ_α, for direction α are defined as the negative second derivative of $v(x, y)$ in the direction α and the negative mixed second derivative of $v(x, y)$ in the directions $\alpha - \pi/2$ and α. Thus,

$$\begin{aligned}\kappa_\alpha &= -(\partial_x \cos\alpha + \partial_y \sin\alpha)^2 v \\ &= -(\partial_{xx} v \cos^2\alpha + \partial_{yy} v \sin^2\alpha) - 2\partial_{xy} v \cos\alpha \sin\alpha \\ &= \tfrac{1}{2}(\kappa_0 + \kappa_{\pi/2}) + \tfrac{1}{2}(\kappa_0 - \kappa_{\pi/2}) \cos 2\alpha - \tau_0 \sin 2\alpha,\end{aligned} \tag{5.28a}$$

$$\begin{aligned}\tau_\alpha &= -(\partial_x \cos\alpha + \partial_y \sin\alpha)(\partial_x \sin\alpha - \partial_y \cos\alpha) v \\ &= -(\partial_{xx} v - \partial_{yy} v) \cos\alpha \sin\alpha + \partial_{xy} v(\cos^2\alpha - \sin^2\alpha) \\ &= \tfrac{1}{2}(\kappa_0 - \kappa_{\pi/2}) \sin 2\alpha + \tau_0 \cos 2\alpha.\end{aligned} \tag{5.28b}$$

According to these equations, the locus of points with coordinates κ_α and τ_α, with respect to axes parallel to those of x and y, is a Mohr circle of radius

$$r = [\tfrac{1}{4}(\kappa_0 - \kappa_{\pi/2})^2 + \tau_0^2]^{1/2} \tag{5.29}$$

whose center lies on the κ_α-axis and has abscissa $(\kappa_0 + \kappa_{\pi/2})/2$. The point

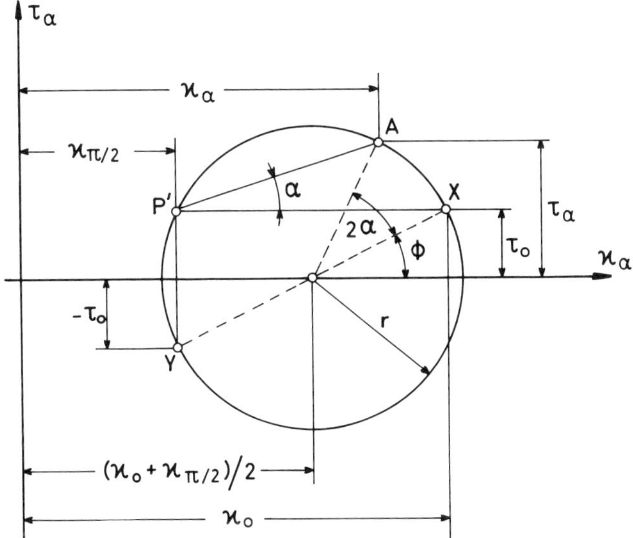

Figure 5.4. Mohr circle for curvatures and twist.

P' with coordinates $\kappa_{\pi/2}$, τ_0 is the *pole* of this circle (Fig. 5.4). The line with direction α through this pole intersects the circle at the point with coordinates κ_α, τ_α. In preparation for a verification of this statement, we note that the parallels to the axes of κ_α and τ_α through P' should intersect the circle at points X and Y with coordinates κ_0, τ_0 and $\kappa_{\pi/2}$, $-\tau_0$. If the angle between the diameter YX and the κ_α-axis is denoted by ϕ, Fig. 5.4 shows that

$$r \cos \phi = \frac{\kappa_0 - \kappa_{\pi/2}}{2}, \qquad r \sin \phi = \tau_0. \tag{5.30}$$

Since the radii of points X and A form the angle 2α, it follows from Fig. 5.4 that the coordinates of point A are $(\kappa_0 + \kappa_{\pi/2})/2 + r\cos(2\alpha + \phi)$ and $r \sin(2\alpha + \phi)$. Using Eq. (5.30) in these expressions shows that the coordinates of A are indeed κ_α and τ_α, as given by Eqs. (5.28a) and (5.28b).

Having discussed the variation of κ_α and τ_α with direction α at a fixed point, we now investigate how the principal directions of a Morley field vary with position. The rates of curvature and twist with respect to the axes x, y are defined as

$$\kappa_0 = -\partial_{xx} v, \quad \kappa_{\pi/2} = -\partial_{yy} v, \quad \tau_0 = \partial_{xy} v, \tag{5.31}$$

and hence satisfy the compatibility conditions

$$\partial_x \kappa_{\pi/2} + \partial_y \tau_0 = 0, \qquad \partial_y \kappa_0 + \partial_x \tau_0 = 0. \tag{5.32}$$

With the substitutions

$$\kappa_0 \to \sigma_y, \qquad \kappa_{\pi/2} \to \sigma_x, \qquad \tau_0 \to \tau_{xy}, \tag{5.33}$$

these compatibility conditions take the form of the equilibrium conditions for the stress components σ_x, σ_y, τ_{xy} of a field of plane stress in the absence of body forces. It is well known that, with respect to the principal lines of stress, the equilibrium conditions are

$$d_1\sigma_1 + \frac{\sigma_1 - \sigma_2}{\rho_2} = 0, \qquad d_2\sigma_2 + \frac{\sigma_2 - \sigma_1}{\rho_1} = 0, \tag{5.34}$$

where σ_1, σ_2 are the principal stresses, ρ_1 and ρ_2 are the radii of curvature of the principal lines, and d_1, d_2 indicate differentiation along these lines. The radius of curvature ρ_1 at a point P of a first principal line is regarded as positive if the vector from the center of curvature C_1 to P has positive direction along the second principal line at P (Fig. 5.5). An analogous statement applies to the sign of ρ_2.

It follows from Eqs. (5.33) and (5.34) that, referred to the principal lines, the compatibility conditions in Eq. (5.32) have the form

$$d_1\kappa_2 + \frac{\kappa_2 - \kappa_1}{\rho_2} = 0, \qquad d_2\kappa_1 + \frac{\kappa_1 - \kappa_2}{\rho_1} = 0 \tag{5.35}$$

[Shield (1960a)]. The second equation shows that the first principal lines are straight throughout a region of a Morley field in which κ_1 and κ_2 are distinct and κ_1 does not change sign. Accordingly, the beams in regions of types R^+, R^-, and T are straight. Although curved beams could be used in regions of types S^+ and S^-, straight beams will not cause an increase of structural weight. Therefore only straight beams will be used in our examples.

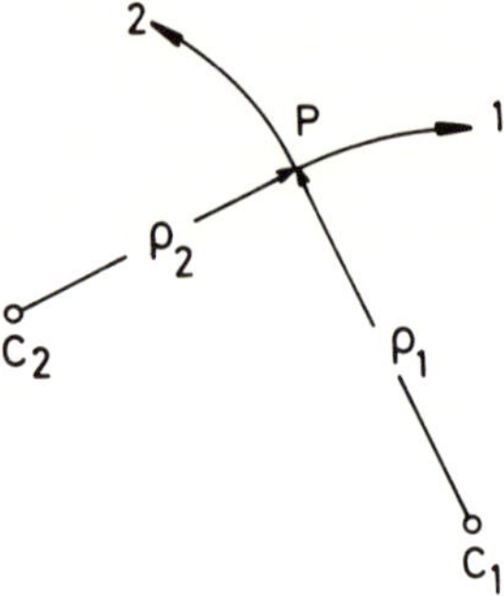

Figure 5.5. Radii of curvature of principal lines at point *P*.

In a region of type T the beams of each of the two orthogonal families are straight. Each of these families thus consists of parallel beams. The single family of straight beams in a region of types R^+ or R^-, however, need not consist of parallel beams. The remainder of this section will be concerned with the general description of fields of type R^+. The reader will readily perceive how this description must be modified to adapt it to regions of type R^-.

For the sake of brevity we shall use the word "beam" to also designate the axis of a beam. Taking the typical beam b of a region of type R^+ as the x-axis and choosing the origin O at the location of the greatest rate of deflection v_0 of the beam, we propose to study the variation of the rate of deflection $v(x, y)$ over a strip Σ of infinitesimal width with centerline b. Because b is a principal line, the rate of twist $\tau_0 = \partial_{xy}v(x, 0)$ vanishes along b, which means that all line elements of the strip Σ that are normal to b experience the same angular velocity:

$$\omega = \partial_y v(x, 0). \tag{5.36}$$

Moreover, the rate of curvature of the beam has constant value $\kappa = 1$. In the equations of the remainder of this section, we shall use the symbol κ, rather than its numerical value 1, in order to more clearly exhibit the dimensions of the various terms. Note that the dimension of κ is $L^{-1}T^{-1}$, while v_0 and ω have dimensions LT^{-1} and T^{-1}. Along the strip Σ the rate of deflection therefore varies in accordance with

$$v(x, y) = v_0 + \omega y - \kappa x^2/2. \tag{5.37}$$

Since the rate of deflection field

$$\bar{v}(x, y) = v_0 + \omega y - \frac{\kappa(x^2 + y^2)}{2}, \tag{5.38}$$

which is of type S^+, varies in precisely the same fashion over the strip Σ, we shall say that it is tangential to the field $v(x, y)$ along this strip.

Each beam in a region of type R^+ thus has a tangential field of type S^+, and the rate of deflection in the region may be described as the *envelope* of these tangential fields, which form a one-parameter family, as do the beams of the region.

The point O, ω/κ, at which the tangential rate of deflection $\bar{v}(x, y)$ in Eq. (5.38) assumes its greatest value, will be called the *extremal point* for the beam. (The neutral term "extremal" is used because, in a region of type R^-, the tangential rate of deflection at the extremal point for a beam is not a maximum but a minimum.) The rate of deflection

$$\bar{v}(O, \omega/\kappa) = v_0 + \omega^2/2\kappa \tag{5.39}$$

at the extremal point for the beam will be called *the extremal rate of deflection* for this beam. Finally, the locus of extremal points for beams of a region of type R will be called the *extremal line* of this region.

To define a region of type R^+ we may specify the coordinates $x = \xi(s)$, $y = \eta(s)$ of its extremal points and its extremal rates of deflection $\zeta(s)$ as functions of the arc length s of the extremal line. The tangential fields are then

$$\bar{v}(x, y; s) = \zeta(s) - \{[x - \xi(s)]^2 + [y - \eta(s)]^2\}/2. \tag{5.40}$$

In Fig. 5.6 let E, b, and F be the extremal points for a given value of s, the corresponding beam, and the foot of the perpendicular dropped from E onto b. If differentiation with respect to s is denoted by a prime, it follows from Eq. (5.40) that b has the equation

$$(x - \xi)\xi' + (y - \eta)\eta' = -\zeta'/\kappa. \tag{5.41}$$

Since ξ', η' are the components of the unit tangent vector $\mathbf{t}$ of the extremal line, Eq. (5.41) is the equation of a line normal to $\mathbf{t}$, the vector FE being $(\zeta'/\kappa)\mathbf{t}$. Comparison with the special situation discussed in connection with Eqs. (5.37) and (5.38) shows that any line element AB emanating from a point A of b and having the direction of $\mathbf{t}$ experiences the angular velocity

$$\omega = \zeta', \tag{5.42}$$

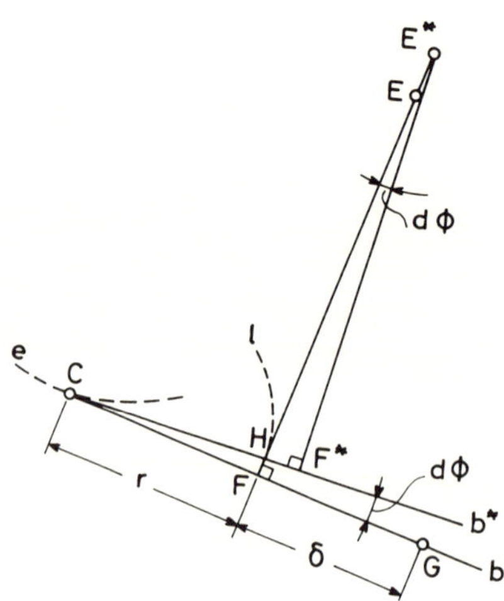

Figure 5.6. Extremal point E, beam b, and variations.

where a positive value of ω indicates that B has a greater rate of deflection than A.

We now determine the principal rate of curvature κ_2 in the direction normal to b at the typical point G of this beam (Fig. 5.6), denoting the distances FG by δ and regarding it as positive if F lies between G and the intersection C of b and the neighboring beam b^* whose extremal point E^* has coordinates $\xi + \xi'\, ds$, $\eta + \eta'\, ds$. The foot of the perpendicular dropped from E^* onto b^* will be denoted by F^*; the intersection of EF and b^*, by H. Since $FE = \zeta'/\kappa$, $EE^* = ds$, and, to within higher-order quantities, $F^*E^* = HE^* = (\zeta' + \zeta''\, ds)/\kappa$, we have $FH = (1 - \zeta''/\kappa)\, ds$. Moreover, according to Eq. (5.42), the angular velocities of line elements along FH and F and H are ζ' and, to within higher-order quantities, $\zeta' + \zeta''\, ds$. At F, that is, for $\delta = 0$, the rate of curvature in the direction normal to b is thus

$$\kappa_2(0) = -\frac{\kappa\zeta''}{\kappa - \zeta''}. \tag{5.43}$$

Noting that the segment CF in Fig. 5.6 is the radius of curvature r of the second principal line through F, and that the second principal line through G has radius of curvature $\delta + r$, we may write the first equation in Eq. (5.35) in the form

$$\frac{d\kappa_2}{d\delta} = \frac{\kappa - \kappa_2}{\delta + r}. \tag{5.44}$$

Integrating this differential equation, with initial condition $\kappa_2 = \kappa_2(0)$ for $\delta = 0$, we find

$$\kappa_2(\delta) = \frac{\kappa\delta + \kappa_2(0)r}{\delta + r}. \tag{5.45}$$

In view of Eqs. (5.43) and (5.45), the second relation in Eq. (5.27) restricts ζ'' and δ to the ranges

$$\kappa/2 \leq \zeta'' \leq \infty, \qquad -\frac{\kappa r}{2(\kappa - \zeta'')} \leq \delta \leq \infty. \tag{5.46}$$

The tangential rate of deflection, Eq. (5.40), vanishes along a circle of radius $\sqrt{2\zeta(s)/\kappa}$, which is centered at the extremal point for the given value of s and intersects the corresponding beam at the points specified by

$$\delta = \pm 1/\kappa[2\kappa\zeta(s) - \zeta'^2(s)]^{1/2}. \tag{5.47}$$

This circle will be called the *null circle* for the given value of s.

Let us assume that, for $s_1 \leq s \leq s_2$, the values of δ in Eq. (5.47) are in the range specified by the second relation in Eq. (5.46). The beams corresponding to $s_1 \leq s \leq s_2$ then represent the optimal layout for a grillage-like continuum that is simply supported along the envelopes of the null circles and free along the beams for $s = s_1$ and $s = s_2$. Note that, at the endpoints of any beam, the tangents to the simply supported edges form equal angles with the beam.

An alternative description of a region of type R^+ is due to Rozvany (1973c). Let the curve e in Fig. 5.6 be the envelope of the beams, specify a particular beam by the angle ϕ it forms with a fixed direction, and denote by $\omega(\phi)$ the common angular velocity of line elements normal to this beam. Moreover, denote by ϕ_0 the direction of the beam b in Fig. 5.6, denote by $v_0(\phi_0)$ its greatest rate of deflection, and denote by $r(\phi_0)$ the distance of the location F of this greatest rate of deflection from the point of contact C of the beam with the envelope e. Finally, let l be the evolute of e described by F, and specify the location of the greatest rate of deflection of the beam with the direction ϕ by its distance $d(\phi)$ from the intersection of the beam with l. To within higher-order terms the distance E^*F^* in Fig. 5.6 is ω/κ. The distance HF is thus $\omega\, d\phi/\kappa$ and

$$d(\phi) = \int_{\phi_0}^{\phi} \omega(\phi)\, d\phi. \tag{5.48}$$

Since the locus of points with the greatest rates of deflection are thus known from Eq. (5.48), the variation of their distances $r(\phi)$ from e is also known. Since the rate of deflection of the beam b^* in Fig. 5.6 has a maximum at F^*, it is, to within higher-order quantities, equal to that at H, which is $v_0 + \omega r\, d\phi$. Accordingly,

$$v_0(\phi) = v_0(\phi_0) + \int_{\phi_0}^{\phi} \omega(\phi) r(\phi)\, d\phi. \tag{5.49}$$

Finally, the principal rate of curvature in the direction normal to a beam at its point of maximum rate of deflection is

$$\kappa_2(0) = \frac{d\omega/d\phi}{r(\phi)}, \tag{5.50}$$

and the variation of κ_2 along the beam is given by Eq. (5.45).

Exercise 5.5. In Fig. 5.6 introduce the point I of the beam b at which $\kappa_2 = -\kappa$, and show that, in terms of the distances $CI = \rho_0$ and $IG = \rho$, the rate of curvature κ_2 at G is

$$\kappa_2 = \frac{\kappa(\rho - \rho_0)}{\rho + \rho_0}.$$

5.4. Matching of Morley Fields

For each of the optimal grillages in Fig. 5.3(a)-(c), the rate of deflection was given by a single expression [Eqs. (5.21), (5.24), and (5.26)]. Accordingly, each of these examples uses a single Morley field (of types R^+ in Fig. 5.3a, S^+ in Fig. 5.3b, and T in Fig. 5.3c). Situations of this kind, however, are exceptions. As a rule the plan form of an optimal grillage is divided into regions of various types. Across the common boundary of two neighboring regions, the rate of deflection as well as its first derivatives must be continuous, which means that the rates of curvature and twist in the direction of the common boundary must be continuous.

To illustrate a way of matching regions of different types to satisfy these continuity conditions, we shall discuss the optimal layout of beams in the neighborhood of a corner formed by two supported straight edges. In the figures of this section the diagram showing the Mohr circles for the various regions will be superimposed on the plan form of the grillage.

Along a built-in edge the rate of deflection and its normal derivative vanish. The rates of curvature and twist for the direction of the edge are thus zero. Because the rate of twist vanishes, the edge is a principal line. The corresponding rate of curvature is $\kappa_2 = 0$. The other principal direction is normal to the edge and, for positive rate of deflection, the corresponding rate of curvature is $\kappa_1 = -1$. The Mohr circle for the region adjacent to a built-in edge therefore intersects the κ-axis at the origin O and the point with the abscissa -1 (Fig. 5.7). The region adjacent to a built-in edge is thus of type R^- with first principal lines normal to the edge.

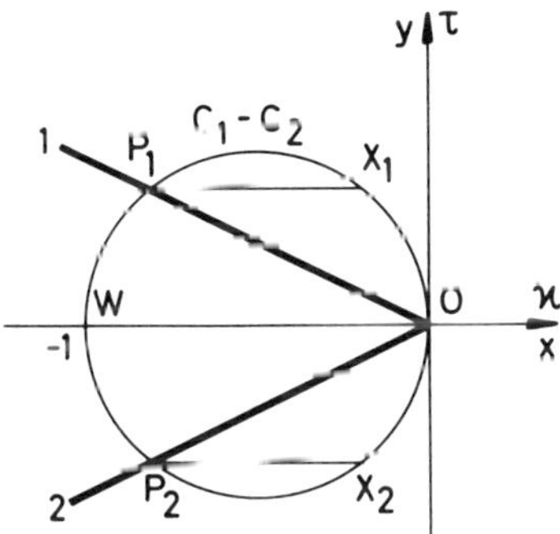

Figure 5.7. Mohr circle for region adjacent to a built-in edge.

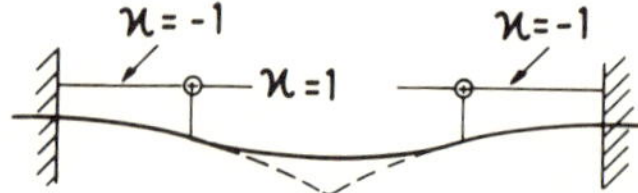

Figure 5.8. Optimal rate of deflection of a doubly built-in beam.

Let the lines $O1$ and $O2$ in Fig. 5.7 represent built-in edges forming the corner O. The fact that they are symmetric with respect to the x-axis, which coincides with the κ-axis, does not involve a loss of generality, because we are free to choose the bisector of the angle formed by the edges as x-axis. The regions adjacent to these edges will be called Region 1 and Region 2. The circle in Fig. 5.7 is the Mohr circle for both these regions of type R^- and will be labelled $C_1 = C_2$. The pole of the circle C_1 for Region 1 is the point of intersection P_1 of the edge $O1$ with the circle. Indeed, since the second point of intersection of this edge with the circle is the origin O, the rates of curvature and twist along the edge are zero. Similarly, the pole of the circle C_2 for Region 2 is the intersection P_2 of $O2$ with the circle.

If each of the two regions extended up to the negative x-axis, the rates of curvature and twist along the two sides of this axis could be represented by the coordinates of the points X_1 and X_2 in Fig. 5.7; that is, the rate of curvature would be continuous across this axis, but not the rate of twist, contrary to the basic features of a boundary. This behavior resembles that of the optimal rate of deflection of a doubly built-in beam (Fig. 5.8): The segments of negative unit rate of curvature cannot be extended to the center of the span, but must be matched with a central segment of positive unit rate of curvature. Similarly, the regions of type R^- near the built-in edges of our grillage-like continuum must be matched with a region of type R^+ near the negative x-axis (Region 3).

Instead of giving the built-in edges and determining the interfaces between regions of types R^+ and R^-, let us choose the circle C_3 in Fig. 5.9 as the Mohr circle for the region of type R^+ and determine the corresponding directions of edges and interfaces, retaining the symmetry with respect to the x-axis, which does not imply any loss of generality. Note that this circle passes through the point 1, 0, and it must have two distinct intersections, I_{13} and I_{23}, with the left-hand circle in Fig. 5.9, which is again labelled $C_1 = C_2$ because it is the Mohr circle for Regions 1 and 2.

On account of the symmetry with respect to the x-axis, the beams of Region 3 will be normal to this axis. The pole of circle C_3 is thus the rightmost point P_3 of this circle. Because the rates of curvature and twist along the interface OQ_1 of Regions 1 and 3 must be continuous across this interface, they are represented by the coordinates of I_{13}, and the interface is parallel to the line P_3I_{13}. The intersection of this line with the circle C_1 is the pole P_1 of this circle, and OP_1 is the built-in edge of Region 1. Note

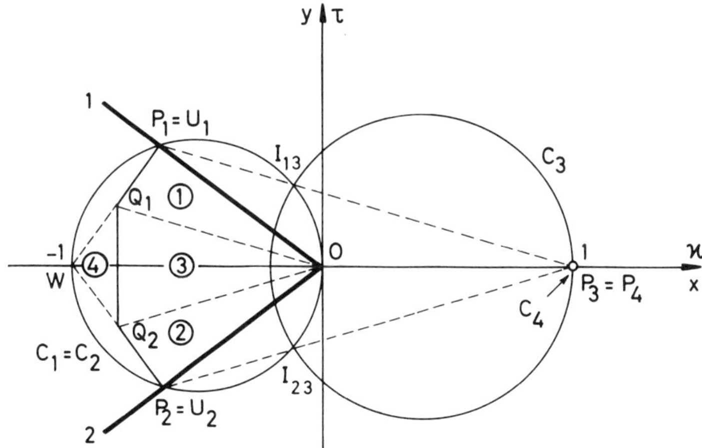

Figure 5.9. Mohr circles for curvature and twist.

that $Q_1P_1 = Q_1W$, because OP_3 and OW have unit length and the lines OQ_1 and P_3P_1 are parallel.

When the built-in edges are given, the interfaces of Regions 1 and 2 with Region 3 are therefore found as follows: From a point W that has the same distance from the built-in edges, drop perpendiculars WU_1 and WU_2 onto these edges. The desired interfaces are then the lines joining the corner O to the centers Q_1 and Q_2 of these perpendiculars. The beams in Regions 1 and 2 are normal to the built-in edges of these regions, while the beams of Region 3 are parallel to Q_1Q_2 and simply supported on cantilevers from the edges.

The point W in Fig. 5.9 is the extremal point for the beam Q_1Q_2 (see Section 5.3), and the rate of deflection in Region 3, which is of type R^+, may be continued by a rate of deflection of type S^+ in the region Q_1WQ_2 (Region 4). The Mohr circle for this region shrinks to the point $P_4 = P_3$.

The circle C_3, which has its center on the κ-axis and intersects it at $\kappa = 1$, may be specified by its second point of intersection with the κ-axis. As this intersection moves from O to W, the angle formed by the built-in edges varies from 0 to π. When this angle exceeds π, the transition from Region 1 to Region 2 is achieved by a region of type S^- (Region 3) as shown in Fig. 5.10. The boundaries $O1'$ and $O2'$ of this region are the normals to the edges $O1$ and $O2$ at O, and its Mohr circle shrinks to the point P_3 with abscissa -1. In Region 3 the optimal layout of beams is not unique; we may, for instance, use beams in the y-direction that are simply supported on the cantilevers $O1'$ and $O2'$.

We next discuss the optimal layout of beams near a corner formed by a built in and a simply supported edge. The Mohr circle for Region 1 near the built-in edge is the circle C_1 in Fig. 5.11. Let us assume that the circle C_2 in this figure is the Mohr circle for Region 2 near the simply supported edge, and that the rightmost point P_2 of this circle is its pole. This means that the beams in Region 2 have y-direction. This assumption does not imply a loss of generality because we are free to choose the direction of the y-axis.

Except for the facts that the right-hand circle, its rightmost point, and the upper point of intersection of the two circles are now labelled C_2, P_2, and I_{12} instead of C_3, P_3, and I_{13}, the pole P_1 of circle C_1 is found as in Fig. 5.9. The interface OQ_1 of Regions 1 and 2 is parallel to P_2P_1. As in Fig. 5.9, the point Q_1 in Fig. 5.11 is the center of the segment WP_1, and the line OP_1 is the built-in edge.

Because the rate of curvature along the simply supported edge is zero, the direction of this edge is given by the line joining the pole P_2 of the circle C_2 to the intersection A of this circle with the τ-axis. The intersection of the simply supported edge (i.e., the parallel to P_2A through O) with the circle C_1 will be labelled V_2. The beam Q_1V_2, which has the y-direction, is simply supported on the edge at V_2 and on the cantilever P_1Q_1 at Q_1; its extremal point is W. The rate of deflection in Region 2, which is of type R^+, may be continued by a rate of deflection of type S^+ in the region Q_1V_2W (Region 3). The Mohr circle for this region shrinks to the point $P_3 = P_2$. If the feet of the perpendiculars from P_1 and Q_1 onto the x-axis are labelled B and B', we have $(WV_2)^2 = (OW)(WB')$ and $(WP_1)^2 = (OW)(WB) = 2(WV_2)^2$ because $(WB) = 2(WB')$. When the two edges are given, the interface of the two regions is therefore formed as follows: from a point W, whose distances from the built-in and simply supported edges have ratio $\sqrt{2}:1$, drop the perpendiculars WU_1 and WV_2 onto these edges. The desired interface then is the line joining the corner O to the center Q_1 of WU_1. The beams in the region adjacent to the built-in edge are normal to it, and the beams in the region adjacent to the simply supported edge are parallel to Q_1V_2.

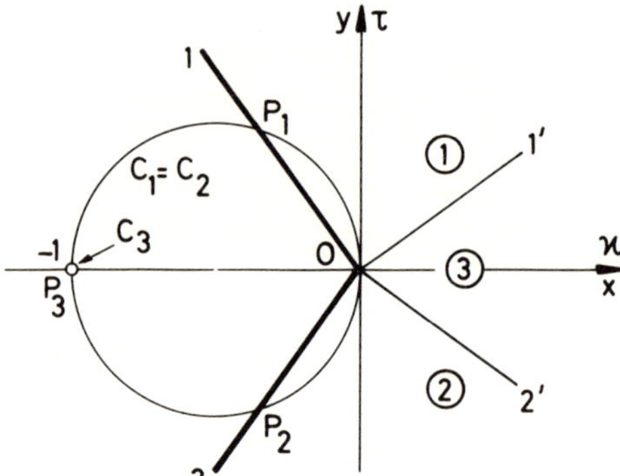

Figure 5.10. Mohr circle for curvature and twist in region of type S^-.

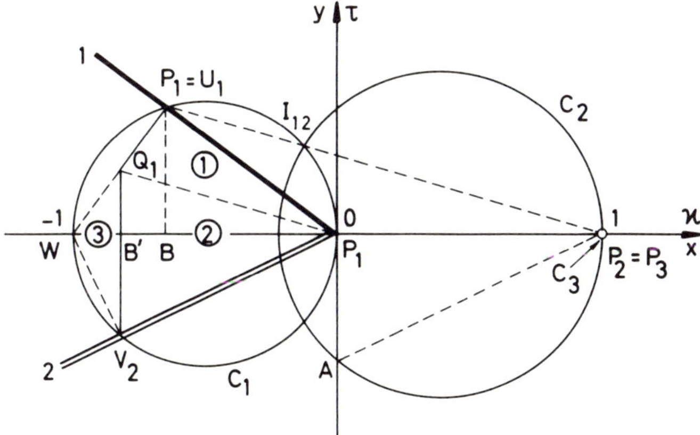

Figure 5.11. Mohr circles for curvature and twist.

The point W is the extremal point of the beam Q_1V_2. The feet U_1 and V_2 of the perpendiculars dropped from it onto the edges will be called the *base points* associated with this extremal point. In the following, the letters U or V will denote base points on built-in or simply supported edges, and the subscripts given to these letters will refer to the edges on which these base points lie.

The circle V_2 in Fig. 5.11, which has its center on the κ-axis at $\kappa = 1$, may be specified by its second point of intersection with the κ-axis. As this point moves from O to W, the angle formed by the edges varies from 0 to $3\pi/4$. For the latter value of this angle the region near the simply supported edge is of type T. Figure 5.12 shows the optimal arrangement of beams for this case. A concentrated load at the point A may be transmitted in various ways to the edges. We may, for instance, use the beam BC, which is simply supported on the edge at B and on the cantilever OC at C. Alternatively, we may use the beam AD, which is simply supported on the edge at D and on any one of the beams in the y-direction—for instance on the beam $B'C'$, which is in turn simply supported on the edge at B' and on the cantilever OC' at C'. Figure 5.12b shows the Mohr circles and their poles for the beam arrangement in Fig. 5.12a.

Figure 5.13 concerns the case where the angle formed by the two edges exceeds $3\pi/4$. At the corner O the derivative of the rate of deflection vanishes in the direction of each edge, hence, in any direction. The grillage-like continuum must therefore be regarded as clamped at O, and the distances of the point W in Fig. 5.13 from this clamped corner and the simply supported edge have the ratio $\sqrt{2}:1$. The beams of the region adjacent to the built-in edge (Region 1) are normal to this edge, while the beams of the region adjacent to the simply supported edge (Region 2), which is of

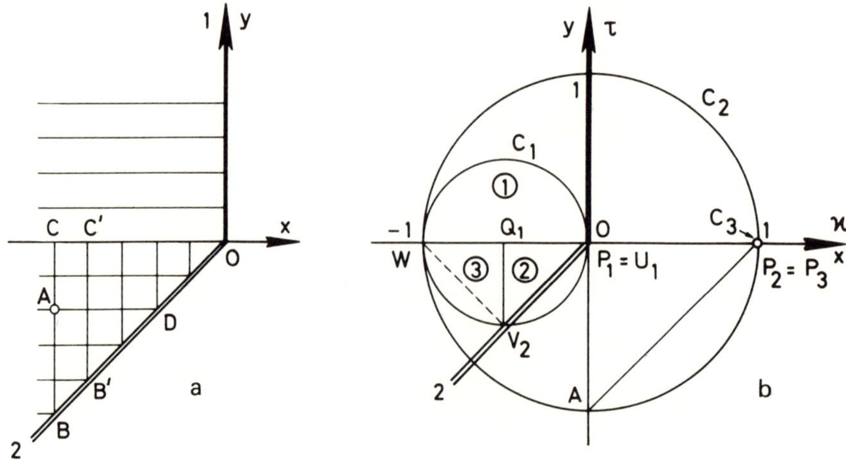

Figure 5.12. (a) Optimal beam arrangement. (b) Corresponding Mohr circles.

type T, form angles of 45° with this edge, as in the lower section of Fig. 5.12a. The transition between Regions 1 and 2 is achieved by a region of type S^- (Region 3), whose boundaries are the negative x-axis and the normal to the built-in edge at O. The beams of Region 3 are simply supported on cantilevers from the clamped point along the boundaries of this region. The direction of these beams is not unique.

Figure 5.14(a)-(c) concern the optimal beam arrangements near a corner formed by two simply supported edges. When the edges form an acute angle, we have a single region of type R^+ (Region 1), whose beams

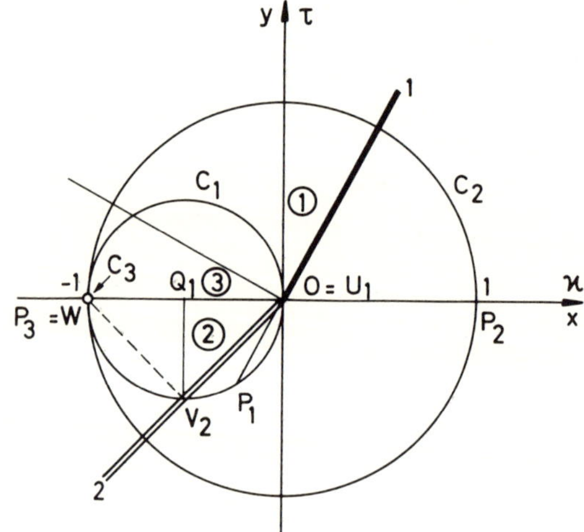

Figure 5.13. Mohr circle, angle formed by the edges exceeding $3\pi/4$.

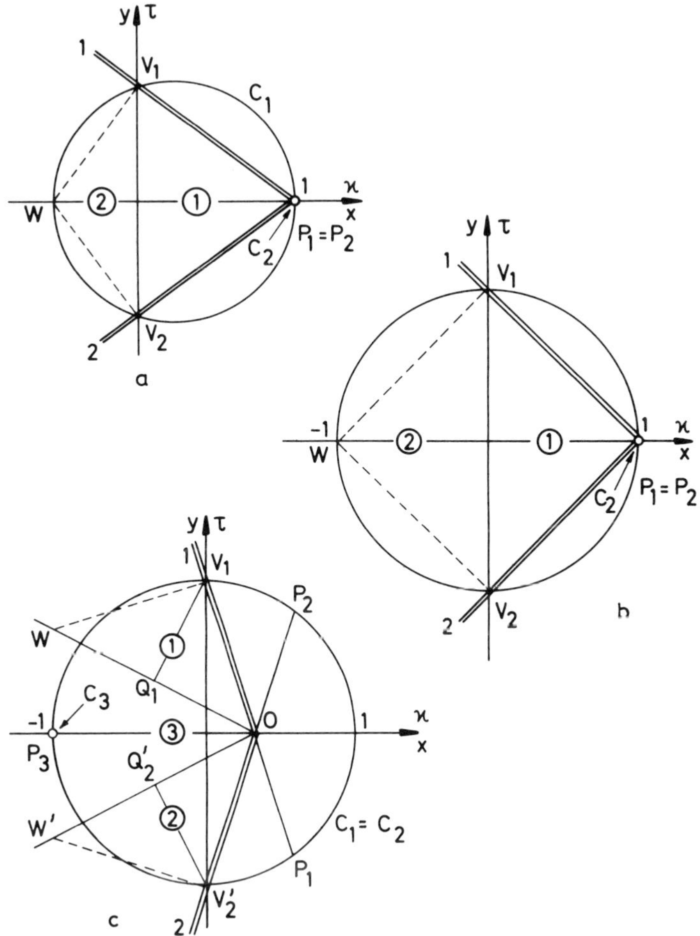

Figure 5.14. (a), (b), (c) Variation of optimal beam arrangements near a simply supported corner.

are normal to the bisector of this angle (Fig. 5.14a). When the edges form a right angle, the type of this region changes to T (Fig. 5.14b). In Fig. 5.14(a)–(b) the beam V_1V_2 has the extremal point W, and the rate of deflection field of Region 1 may be continued by a field of type S^+ in the region V_1V_2W (Region 2). The Mohr circle for this region shrinks to the point P_2, which coincides with P_1. When the angle between the edges is obtuse, we have a region of type T adjacent to each edge (Regions 1 and 2). The beams of these regions form angles of 45° with the edges. Since the corner O must be regarded as clamped, the beams OW and OW' (Fig. 5.14c) are cantilevers. The transition between Regions 1 and 2 is achieved by a region of type S^- (Region 3). Its beams are simply supported on the cantilevers OW and OW'; their direction is not unique. Note that the distances WO and WV_1 of the point W from the clamped corner O and the simply supported edge OV_1 have the ratio $\sqrt{2}:1$. The segment Q_1V_1, where Q_1 is the center of WO, indicates the direction of one beam family in Region 1, the other family being orthogonal to this. The beam arrangement in Region 2 is obtained from that in Region 1 by symmetry with respect to the x-axis.

The remainder of this section is concerned with the matching of the Morley field near a built-in or simply supported edge along the x-axis (Figs. 5.15 and 5.16) with the Morley field near a point U with coordinates $(0, h)$ at which the grillage is clamped. The point U may, for instance, be conceived as a point at which the grillage is supported by an inflexible column. It will be helpful to think of the point U as the limiting case of a small circle along which the grillage-like continuum is clamped. Since near a clamped edge the beams of optimal grillage-like continuum are normal to this edge, we have a region of type S^- near U (Region 1) with cantilevers emanating from U.

In Fig. 5.15 the edge along the x-axis is supposed to be built in; the distances WU and WU' of the point W from U and from this edge are equal, and Q and Q' are the centers of the segments WU and $W'U'$. As the discussion in connection with Fig. 5.9 shows, the beam QQ' and the cantilevers UQ and UQ' on which it is simply supported form an optimal layout for an element of clamped edge at U that is normal to UQ. Since the small circle of clamping, of which the point U is the limiting case, has edge elements in all directions, we must repeat the construction in Fig. 5.15

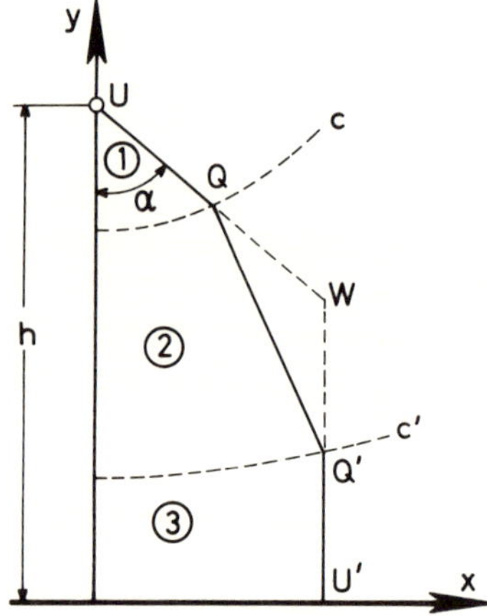

Figure 5.15. Matching Morley fields between a built-in edge along x and a clamping point U.

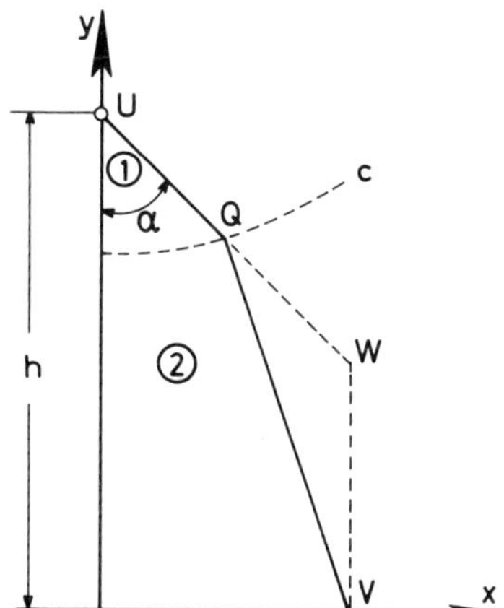

Figure 5.16. Matching Morley fields between a simply supported edge along x and a clamping point U.

for other values of the angle α between UW and the negative y-direction. We leave it to the reader to derive the equations of loci c and c' of points Q and Q' (Exercise 5.6). The region between c and c' (Region 2) is of type R^+, and the region between c' and the x-axis (Region 3) is of type R^-.

In Fig. 5.16 the edge along the x-axis is supposed to be simply supported; the distances WU and WV of the point W from U and this edge have the ratio $\sqrt{2}:1$, and Q is the center of WU. Using the information derived from Fig. 5.11 and proceeding in a similar manner as above, we conclude that the cantilever UQ and the beam QV, which is simply supported on this cantilever at Q and on the edge at the foot V of the perpendicular dropped from W onto the edge, are typical beams of the optimal layout, which is obtained by repeating the construction in Fig. 5.16 for other values of the angle α. We leave it to the reader to derive the equation of the locus c of the points Q (Exercise 5.7). The region between c and the x-axis is of type R^+.

The constructions in Figs. 5.15 and 5.16 were first given in Rozvany (1972a) and Rozvany *et al.* (1973). Prager and Rozvany (1977) first used Mohr circles to determine the optimal layouts near corners formed by two supported edges after Rozvany (1973c, d) and Rozvany *et al.* (1973) had used other methods.

Exercise 5.6. Derive the equations of loci c and c' of points Q and Q' in Fig. 5.15.

Exercise 5.7. Derive the equation of the locus c of the point Q in Fig 5.16.

5.5. Optimal Polygonal Grillages

5.5.1. Polygonal Grillages Without Free Edges

This part of Section 5.5 is concerned with the optimal layout of beams of polygonal, grillage-like continua without free edges. The optimal layouts near corners discussed in the preceding section are readily combined to furnish the optimal layout for the entire grillage. Consider, for instance, the square grillage in Fig. 5.17, which is built in along the edges AB and BC and simply supported along the edges CD and DA. The extremal point W is uniquely determined by the fact that its equal distances WU_1 and WU_2 from the built-in edges are $\sqrt{2}$ times as great as its equal distances WV_3 and WV_4 from the simply supported edges. The points Q_1 and Q_2 are the centers of segments WU_1 and WU_2. The regions AU_1Q_1 and BQ_1U_1, which are both of type R^- with beams normal to AB, form a single region of this type. The region AQ_1V_4, which is of type R^+ with beams parallel to Q_1V_4, can be matched across Q_1V_4 with a region of type S^+ in Q_1WV_4, and the region BQ_2Q_1, which is of type R^+ with beams parallel to Q_2Q_1, can be matched across Q_2Q_1 with a region of type S^+ in Q_2WQ_1. These two regions of type S^+ form a single region of this type in $Q_1Q_2V_4$. The corners C and D can be treated in a similar manner, furnishing a region of type S^+ in $Q_2V_3V_4$, which may be merged with that in $Q_1Q_2V_4$ into a single region of type S^+ in $Q_1Q_2V_3V_4$.

Except for loads acting at points of this region and region DV_4V_3, which is of type T, the way in which a load is transmitted to the edges is unique. A load at point E, for instance, is transmitted by beam FH to cantilevers GF and IH and then, by these, to edges AB and BC.

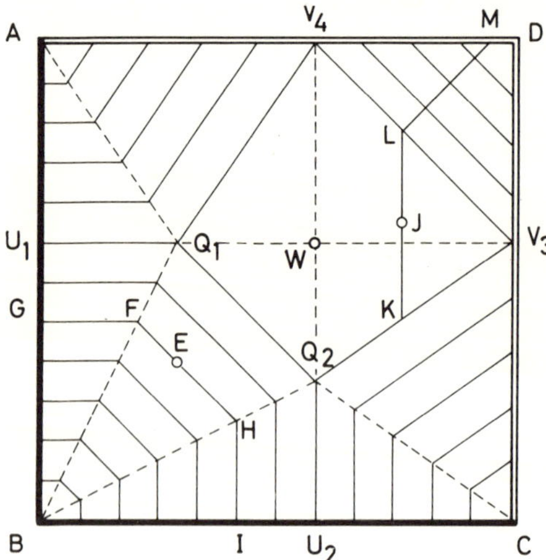

Figure 5.17. Square grillage, edges AB and BC built in, edges AD and DC supported.

In region $Q_1Q_2V_3V_4$, however, beams may be given any direction. A load at point J, for instance, may be transmitted by beam KL to points K and L. The load component at K is transmitted in unique fashion to the edges: namely, by beam Q_2V_3 to edge CD at V_3 and to cantilever U_2Q_2 and by this to edge BC at U_2. The load component at L, however, may be transmitted either by beam V_3V_4 directly to edges CD and DA, or by beam LM to edge DA at M and to any one of the beams with direction V_3V_4 to edges CD and DA. This lack of uniqueness of load transmission in a region of type T has already been pointed out in connection with Fig. 5.12a.

Note that the bending moments of the beams participating in the transmission of a concentrated load to the edges are statically determinate, and that the yield moment everywhere equals the absolute value of the bending moment.

Figure 5.18 shows a rectangular grillage that is supported in the same manner as the square grillage in Fig. 5.17. There are now two extremal points: W and W'. The region to the left of U_2V_4 and that to the right of $U_2'V_4'$ correspond to the regions to the left and right of U_2V_4 in Fig. 5.17. Region $Q_2Q_2'V_4'V_4$ is of type R^+, and region $U_2U_2'Q_2'Q_2$ is of type R^-.

Figure 5.19 shows a trapezoidal grillage with built-in edges AB and BC and simply supported edges CD and DA, which form an acute angle. There are two extremal points: W and W'. The segments WU_1 and WU_2 are equal and $\sqrt{2}$ times as long as segment WV_4. Similarly, segment $W'U_2'$ is $\sqrt{2}$ times as long as the equal segments $W'V_3'$ and $W'V_4'$. Note that beams Q_2V_4 and $Q_2'V_4'$ are parallel on account of the similarity of the figures formed by points W, Q_2, U_2, V_4 and W', Q_2', U_2', V_4'. The region boundary Q_2Q_2' is straight. Regions ABQ_1 and $BCQ_2'Q_2$ are of type R^-, and regions $Q_1Q_2V_4$ and $Q_2'V_3'V_4'$ are of type S^+. All other regions are of type R^+.

In the trapezoidal grillage of Fig. 5.20, the two simply supported edges form an obtuse angle at corner U, which must be regarded as clamped. There are now three extremal points: W, W', and W''. The distances of W'

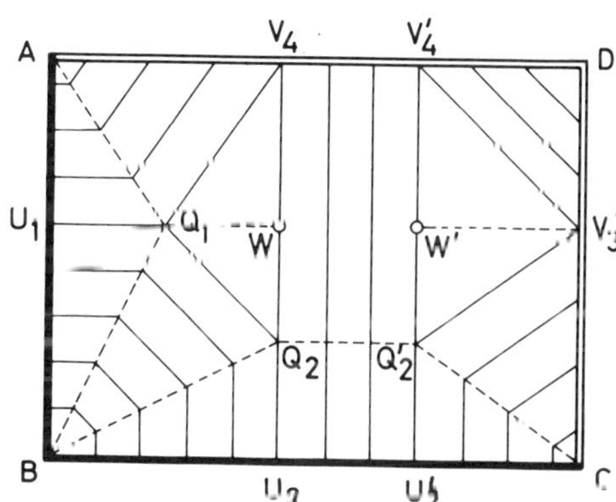

Figure 5.18. Rectangular grillage; edges AB and BC built in, edges AD and DC supported.

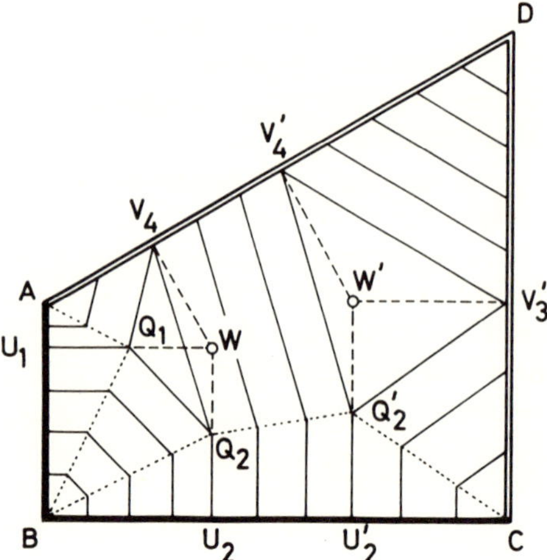

Figure 5.19. Trapezoidal grillage, acute angle between supported edges.

from the clamped corner U and from the built-in edge BC are equal and $\sqrt{2}$ times as great as the distance from the simply supported edge UA. Similar remarks apply to the distances of W'' from U, U_2'', and V_3''. The points Q_1, Q_2, Q_2', Q_2'', Q', and Q'', respectively, are the centers of WU_1, WU_2, $W'U_2'$, $W''U_2''$, $W'U$, and $W''U$. The beams Q_2V_4 and $Q_2'V_4'$ are parallel on account of the similarity of the figures formed by points W, Q_2, U_2, V_4 and points W', Q_2', U_2', V_4', but beams $Q'Q_2'$ and $Q''Q_2''$ are not parallel. The parabolic boundaries $Q'Q''$ and $Q_2'Q_2''$ of region $Q'Q_2'Q_2''Q''$, which is of type R^+, and the directions of its beams are found as in Fig. 5.15. Regions ABQ_1 and $BCQ_2''Q_2'Q_2$ are of type R^-; regions $Q_1Q_2V_4$,

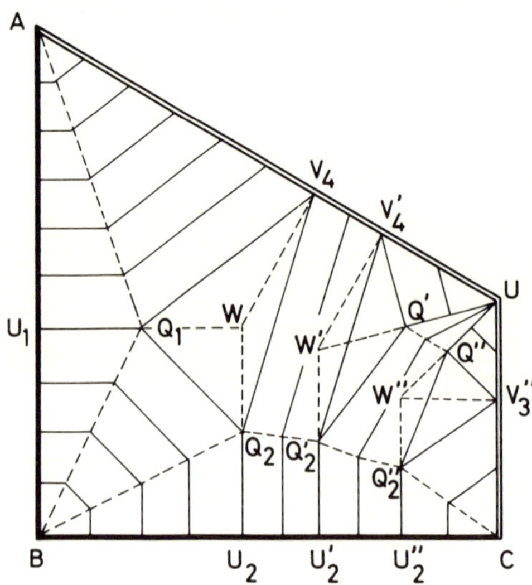

Figure 5.20. Trapezoidal grillage; obtuse angle between supported edges.

$Q_2'Q'V_4'$, and $Q''Q_2''V_3''$ are of type S^+, and regions $Q'UV_4'$ and $Q_2''V_3''U$ are of type T; region $UQ'Q''$ is of type S^-, and all other regions are of type R^+.

The grillage in the main body of Fig. 5.21 has an obtuse corner at U and a reentrant corner at U_4, both of which must be regarded as clamped. The reader will readily verify that the distances of the extremal points W, W', and W'' from the associated edges have the required ratios. The pentagonal region of type S^+ that is associated with the extremal point W' splits into a triangular region associated with the extremal point W_1' (inset of Fig. 5.21) and a quadrilateral region associated with the extremal point W_2' when the ratio AU/AB, which is $\frac{1}{2}$ in the main part of Fig. 5.21, is increased as in the inset of this figure. The parabolic arcs QQ', Q_1Q_1', $Q_2'Q_2''$, and $Q_4'Q_4''$ are constructed as in Fig. 5.15.

The simply supported pentagonal grillage in Fig. 5. 22 must be regarded as clamped at the obtuse corner U. It is symmetric with respect to the line AV_3'. Symmetrically situated points will be given the same label except for a bar added to labels of points below the axis of symmetry. Of the four extremal points, W and W' are on the axis of symmetry, while W'' and $\bar{W}''$ are not. The regions of type S^+ associated with W and W' are $V_1Q\bar{Q}\bar{V}_1$ and $Q'V_3'\bar{Q}'$. Between them lies the region $QQ'\bar{Q}'\bar{Q}$, which is of type T with bars of positive rate of curvature parallel to $Q\bar{Q}$. Regions UQV_1, $UU_2''Q''$, and $U_2''BU_3''$ are also of type T; only their bars with positive rate of curvature are indicated in the figure. Region $UQ''Q'Q$ is of type S^- with cantilevers from the clamped corner U, region $Q''U_2''U_3''$ is of type S^+, and

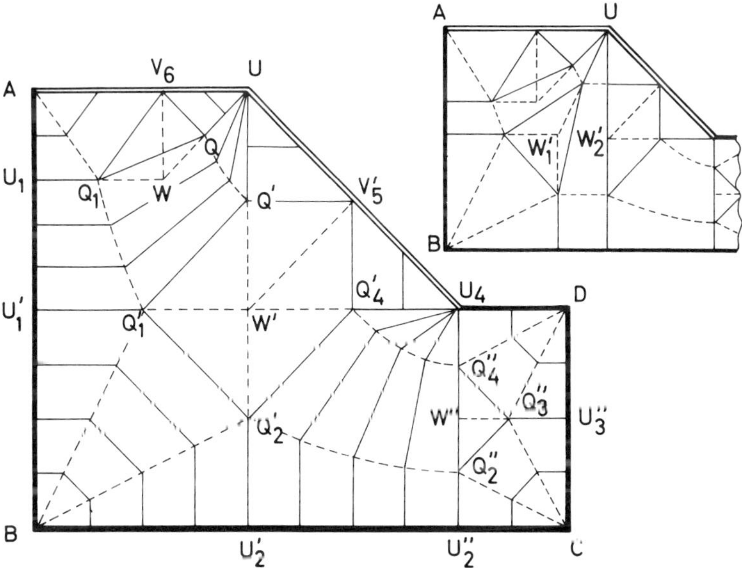

Figure 5.21. Polygonal grillage with reentrant corner.

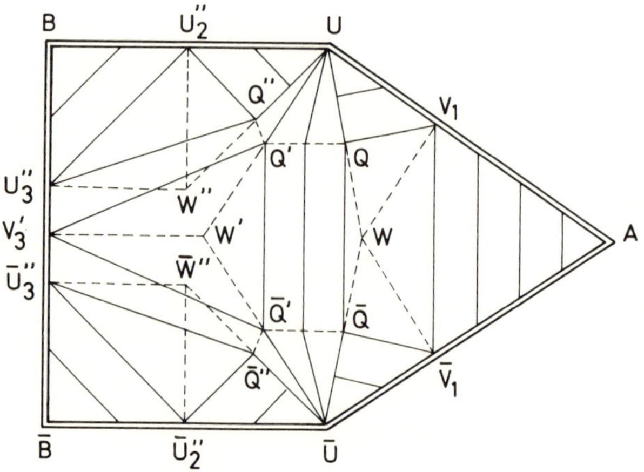

Figure 5.22. Pentagonal supported grillage.

regions $AV_1\bar{V}_1$ and $Q'Q''U_3''V_3'$ are of type R^+. The hyperbolic boundary $Q'Q''$ is constructed as in Fig. 5.16.

For further examples of optimal grillages with supported edges the reader is referred to Rozvany (1972a, b, 1976).

5.5.2. Polygonal Grillages with Free Edges

A particularly simple type of optimal layout of a grillagelike continuum with free and supported edges arises when the optimal layout required by the supported edges happens to have beams along the free edges. For example, the layout shown in rectangle $ABU_2'V_4'$ of Fig. 5.18 is the optimal layout for a grillage that is built in along AB and BU_2', simply supported along $V_4'A$, and free along $U_2'V_4'$. Similarly, the layout shown in $U_1BU_2Q_2Q_1$ of Fig. 5.18 is the optimal layout for a grillage that is built in along U_1B and BU_2 and free along $U_2Q_2Q_1U_1$. Cases of this kind, however, are exceptions. It will be shown in the following that, as a rule, the rate-of-deflection field is of type T along a free edge.

Consider, for instance, the square grillage in Fig. 5.23a that is simply supported along edges AB and AB' and free along edges BC and $B'C$ and carries a concentrated load P at C. The rate-of-deflection field

$$v = \frac{x^2 - y^2}{2} \tag{5.51}$$

is kinematically admissible for optimality. It is of type T and has principal rates of curvature $\kappa_1 = 1$ and $\kappa_2 = -1$ in the y- and x-directions. Figure 5.23a shows the simplest way of optimally transmitting the given load to

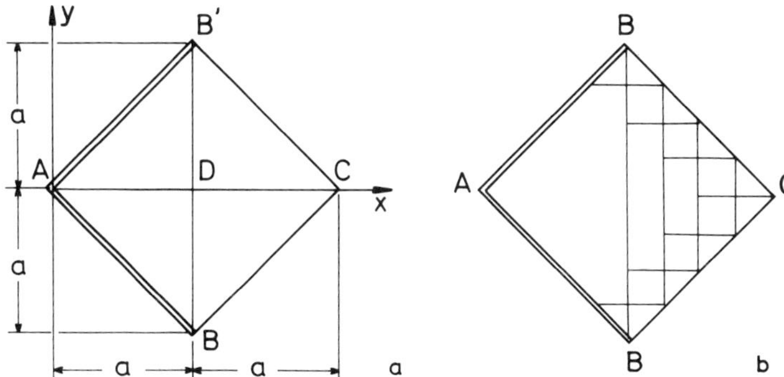

Figure 5.23. (a), (b) Alternative optimal square grillages for concentrated load at point C.

points of support: The beam AC transmits an upward force P to A and a downward force $2P$ to the beam BB' at D. Each of these two beams has moment area $\int |M|\, ds = Pa^2$, and the sum of the moment areas of the two beams—that is, $2Pa^2$—equals the power of the load P on the rate of deflection of its point of application, as it should according to Eq. (5.20b).

As has already been pointed out in connection with Fig. 5.3, there is an infinity of ways in which the given load may be optimally transmitted to points of support. In the present context a limiting case of the one shown in Fig. 5.23b is of interest: The side BC is divided into n equal segments—e.g., into four segments as in Fig. 5.23b; from the left endpoint of each segment a beam starts in the positive y-direction that ends on the line $B'C$, and from the right endpoint of each segment a beam starts in the negative x-direction that has length $2a/n$. We leave it to the reader to work out the moment areas of all beams in the layout of Fig. 5.23b and verify that their sum is again $2Pa^2$ (Exercise 5.12).

Note that, for any value of the integer n, the layout corresponding to that in Fig. 5.23b consists of n uniformly spaced beams in the y-direction in the right half of the grillage and $2n - 1$ beams of common length $2a/n$ in the x-direction. For $n \to \infty$ we thus have what has been called a *beam weave* [Prager and Rozvany (1977)] along the free edges and otherwise only beams in the y-direction, which densely fill the right half of the grillagelike continuum.

Figure 5.24 shows a rhombic grillage with simply supported edges AB and AB' and free edges BC and $B'C$, the angle at A being $2\alpha < \pi/2$. For part ABB' of this grillage, the rate of deflection field

$$v_1 = (x^2 \tan^2 \alpha - y^2)/2 \tag{5.52}$$

is kinematically admissible for optimality. This field is of type R^+ with principal rates of curvature $\kappa_1 = 1$ and $\kappa_2 = -\tan^2 \alpha$ in the y- and x-

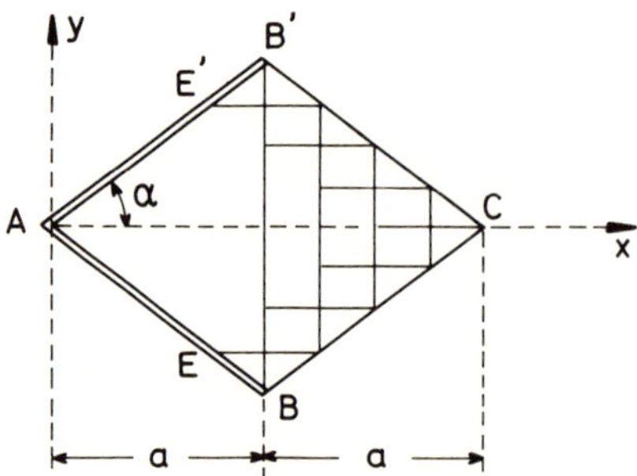

Figure 5.24. Optimal rhombic grillage with concentrated load at corner *e*.

directions. It can be matched across BB' with the field

$$v_2 = [x^2 - (2ax - a^2)(1 - \tan^2 \alpha) - y^2]/2, \tag{5.53}$$

which is of type T with principal rates of curvature $\kappa_1 = 1$ and $\kappa_2 = -1$ in the y- and x-directions.

For a concentrated load at C the layout in Fig. 5.24 corresponds to that in Fig. 5.23b; in contrast to the latter, it is not optimal because the two beam segments to the left of BB' have rate of curvature $-\tan^2 \alpha$, whose absolute value is smaller than unity. This violation of the optimality condition becomes less important if, instead of four beams in the y-direction, we use $n > 4$ beams. The reactions at the points corresponding to E and E' in Fig. 5.24 are then each $nP/2$, and the beam segments to the left of BB' together contribute the amount $(nP/2)(a/n)^2 = Pa^2/2n$ to the moment area, but only $\tan^2 \alpha$ times this amount to the power of dissipation in the grillage, which equals the power $Pv_2(2a, 0)$ of the load. For any other beam segment, moment area and power of dissipation are equal because the rate of curvature has unit magnitude. Accordingly,

$$Pv_2(2a, 0) = \sum \int |M| \, ds - \frac{(1 - \tan^2 \alpha) Pa^2}{2n}, \tag{5.54}$$

where the sum includes *all* beams. Using Eq. (5.53) in Eq. (5.54) furnishes the total moment area

$$\sum \int |M| \, ds = \tfrac{1}{2} Pa^2 \left[1 + 3 \tan^2 \alpha + \frac{1 - \tan^2 \alpha}{n} \right], \tag{5.55}$$

which approaches its smallest value as $n \to \infty$. Whereas the use of beam weaves along the free edges was optional in the optimal grillage for $\alpha = \pi/2$, it becomes mandatory for $\alpha < \pi/2$. Although the optimal layout with its beam weaves along the free edges is not practical, it yields the global

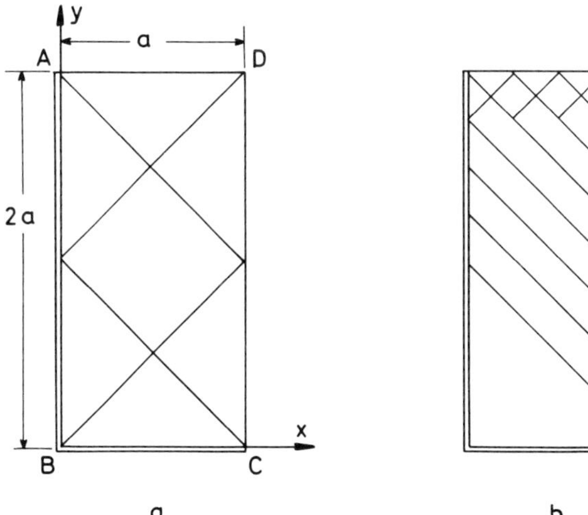

Figure 5.25. (a), (b) Alternative optimal rectangular grillages, concentrated load at point D.

minimum of cost that is needed for the evaluation of the efficiency of practical designs. For $\alpha = 30°$, for instance, it follows from Eq. (5.55) that the efficiency rises from 90.0% for $n = 3$ to 94.7% for $n = 6$.

Figure 5.25a shows a rectangular grillage that is simply supported along edges AB and BC and free along the other edges. The rate of deflection field

$$v = xy \tag{5.56}$$

is kinematically admissible for optimality. It is of type T with principal rates of curvature $\kappa_2 = -1$ in the direction of the bisector of the first quadrant and $\kappa_1 = 1$ in the direction normal to this. The layout shown in Fig. 5.25a affords the simplest way of transmitting a concentrated load P at the corner D to points of support, but the layout in Fig. 5.25b of its limiting case, with beam weaves along the free edges and only beams with positive unit rate of curvature in the rest of the grillage, will have the same cost.

Let us now turn side BC about C so that angle ABC, denoted by 2α becomes acute. We first give a general description of the optimal layout (shown in Fig. 5.26a for $2\alpha = 45°$) before studying it in detail. In the triangle BCE the rate of-deflection field is of type R^+ with beams normal to the bisector of angle ABC. This field is matched across CE with a field in $AECF$ that is also of type R^+, but whose beams are not parallel to each other. This field is in turn matched across AF with a field of type T in AFD. There are beam weaves along the free edges AD and CD, but only a single beam passes through any point not lying on these edges. Figure 5.26a shows some of these beams, all of which have rate of curvature $\kappa_1 = 1$.

The field in triangle BCE is determined in accordance with Fig. 5.14a. Since its beams are normal to the bisector of the angle at B, the boundary CE of this field forms the angle $\beta_0 = \pi/2 - \alpha$ with edge AB (Fig. 5.26b).

The angle BGI, which the typical beam GI in $AECF$ forms with edge AB, will be denoted by β. As we see from the Mohr circle for point G (Fig. 5.26c), the second principal rate of curvature at G is

$$\kappa_2(G) = -\cot^2 \beta. \tag{5.57}$$

If there is to be a beam weave along the free edge segment CF, the second principal rate of curvature at I must be

$$\kappa_2(I) = -1. \tag{5.58}$$

Let $G'I'$ in Fig. 5.26b be a neighboring beam, and denote the intersection of lines GI and $G'I'$ by O, the distance OI by ρ_0, and the distance IG by ρ. According to Exercise 5.5, the second rate of curvature in Eq. (5.57) has value

$$\kappa_2(G) = -\cot^2 \beta = \frac{\rho - \rho_0}{\rho + \rho_0}. \tag{5.59}$$

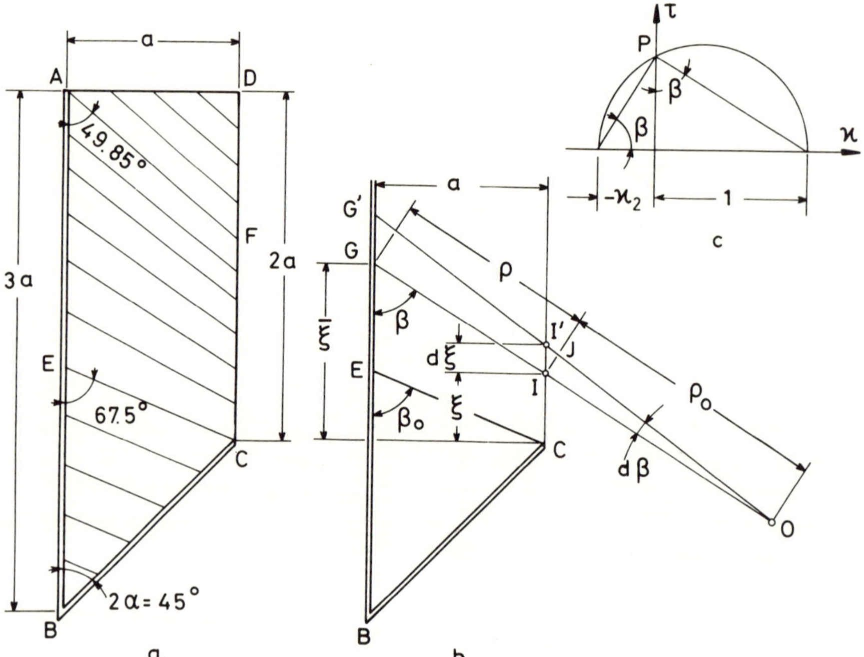

Figure 5.26. (a) Optimal grillage, concentrated load at point D. (b) Morley field in triangle BCE. (c) Mohr circle for point G.

Substituting $\rho = -a/\sin\beta$ into Eq.(5.54) and solving for ρ_0, we find

$$\rho_0 = \frac{-a}{\sin\beta\cos 2\beta}. \tag{5.60}$$

Let the length of segment CI be denoted by ξ, so that the line element II' is $d\xi$. The projection IJ of II' normal to OI then has length $d\xi \sin\beta$; also, since β decreases as we go from I to I', IJ has length $-\rho_0\, d\beta_0$. Setting these two expressions for length equal to each other, we obtain

$$d\xi/d\beta = -\rho_0/\sin\beta. \tag{5.61}$$

Substituting Eq. (5.60) into Eq. (5.61) and integrating the resulting differential equation for $\xi(\beta)$, with initial condition $\xi(\beta_0) = 0$, yields

$$\xi = \ln\left|\frac{\tan(\beta + \pi/4)}{\tan(\beta_0 + \pi/4)}\right| + \cot\beta_0 - \cot\beta. \tag{5.62}$$

The distance marked $\bar{\xi}$ in Fig. 5.26b is

$$\bar{\xi} = \xi + a\cot\beta. \tag{5.63}$$

To find the angle BAF (Fig. 5.20a), we must substitute Eq. (5.62) into Eq. (5.63), set $\bar{\xi} = 2a$, and solve the resulting transcendental equation for β.

The preceding discussion of the optimal rate-of-deflection field in $AECF$ was based on the assumption that there is a beam weave along CF. To show that there is no alternative to this, we treat ρ in Eq. (5.54) as a variable and write this equation as

$$\kappa_2(\rho) = \frac{\rho - \rho_0}{\rho + \rho_0}. \tag{5.64}$$

This equation gives the variation of κ_2 along a beam in a field of type R^+ whose beams are not parallel to each other. In Eq (5.64), ρ_0 stands for the distance from point I, where $\kappa_2 = -1$, to the point of contact O of the axis of this beam with the envelope of the axes of the beams, and ρ stands for the distance of a typical point of the beam from I.

Since the optimality condition restricts κ_2 to the interval $[-1, 1]$, it follows from Eq. (5.64) that negative values of ρ, which correspond to points between I and O, are not admissible. The point I at which $\kappa_2 = -1$ must therefore lie on, or to the right of, line CF. Moreover, Eq. (5.64) shows that $\kappa_2(\rho)$ increases monotonically with ρ. If this point I did lie to

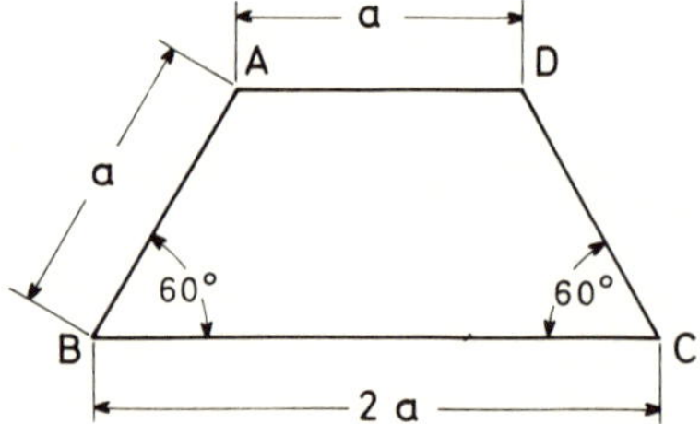

Figure 5.27. Symmetric trapezoidal grillage.

the right of the line CF, the second rate of curvature κ_2 would nowhere have absolute value unity along the considered beam. There could, therefore, be no cross beam to support this beam, for which the simple support at G alone is not sufficient. Point I must therefore lie on CF, as we assumed in the preceding discussion.

Trapezoidal grillages, such as those in Figs. 5.27 and 5.28, with free and supported edges can be treated in a similar manner (Exercises 5.11 and 5.13).

5.5.3. Polygonal Grillages with Edge Beams

The supports considered in Sections 5.5.1 and 5.5.2 were rigid in the sense that they precluded any deflection of the supported edge. Instead of being rigidly supported all along an edge of a polygonal grillage, the beams ending there may be simply supported by an edge beam, which is in turn simply supported at its two ends. As for the beams of the grillage, edge beams are to have rectangular sections of continuously varying width, their common height $\bar{h}$ being greater than the common height of the other beams. Setting

$$\alpha = h/\bar{h}, \tag{5.65}$$

we note that to the yield moment $\bar{Y}$ of an edge beam corresponds the cross sectional area

$$\bar{A} = 4\bar{Y}/\sigma_0\bar{h} = 4\alpha\bar{Y}/\sigma_0 h. \tag{5.66}$$

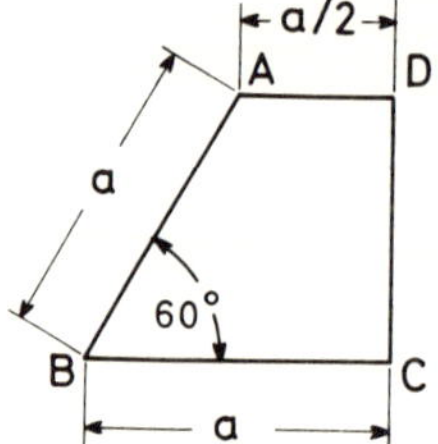

Figure 5.28. Nonsymmetric trapezoidal grillage.

The combined volume of grillage and edge beams is thus proportional to

$$\Gamma = \sum \int Y\,ds + \sum \int \alpha \bar{Y}\,d\bar{s}, \tag{5.67}$$

where the first and second integrations are, respectively, extended over the typical beam or edge beam, while the first and second sums are to include all beams or edge beams.

If there are no explicit bounds on Y or $\bar{Y}$, the type of argument used at the beginning of Section 5.2 shows that the existence of a collapse mechanism is sufficient for global optimality, provided this mechanism has absolute rates of curvature equalling α or unity in the direction of any edge beam or grillage beam and not exceeding unity for any other direction.

We leave it to the reader to prove that the cost defined by Eq. (5.67) may, as in Eq. (5.23), be evaluated as the power of the given loading on the rates of deflection satisfying these optimality conditions.

As a first example, consider a square grillage with four edge beams of length $2a$ that are simply supported at the corners of the square. Figure 5.29 shows the lower left corner of the grillage and the Mohr circle for the vicinity of this corner. The rightmost point I of this circle has unit abscissa. The pole of the circle is its highest point P', which must have abscissa α because parallels through P' to the edges of the grillage must intersect the circle at points X and Y with common abscissa α. The line $P'I$ indicates the direction of the beams of the grillage in the vicinity of the lower left corner. The regime represented by the given circle prevails in the triangle ORS, where R and S are the centers of the edges through O. Along RS this regime is matched by one of type S^+ with unit rate of curvature in any direction.

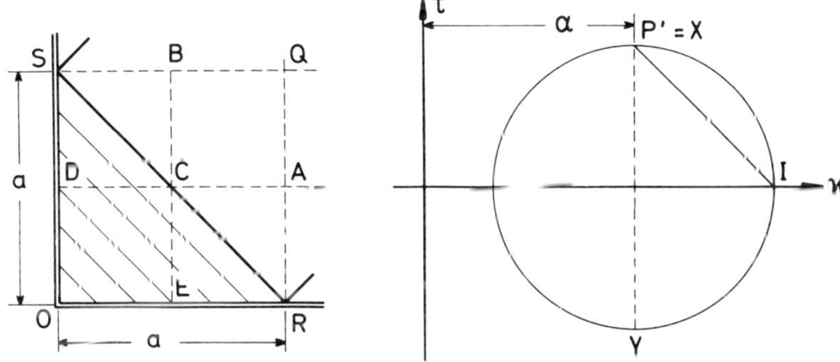

Figure 5.29. Corner of square grillage with edge beams and corresponding Mohr circle.

The rate of deflection v_1 in triangle ORS must satisfy the conditions

$$\kappa_0 = -\partial_{xx}v_1 = \alpha, \quad \kappa_{\pi/2} = -\partial_{yy}v_1 = \alpha, \quad \tau_0 = \partial_{xy}v_1 = 1 - \alpha,$$
$$\partial_x v_1(a, 0) = 0, \quad \partial_y v_1(0, a) = 0, \quad v_1(0, 0) = 0. \tag{5.68}$$

Accordingly,

$$v_1 = -\tfrac{1}{2}\alpha(x^2 + y^2) + (1 - \alpha)xy + \alpha a(x + y). \tag{5.69}$$

In triangle QRS, where Q is the center of the square, we have a regime of type S^+, and the rate of deflection v_2 must satisfy the conditions

$$\kappa_0 = -\partial_{xx}v_2 = 1, \quad \kappa_{\pi/2} = -\partial_{yy}v_2 = 1, \quad \tau_0 = \partial_{xy}v_2 = 0,$$
$$\partial_x v_2(a, y) = 0, \quad \partial_y v_2(x, a) = 0, \quad v_2(a, 0) = v_1(a, 0). \tag{5.70}$$

Thus,

$$v_2 = -\tfrac{1}{2}(x^2 + y^2) + a(x + y) - \tfrac{1}{2}(1 - \alpha)a^2. \tag{5.71}$$

One readily verifies that the rate of deflection as well as its first derivatives are continuous across the line RS.

According to Eq. (5.23), the cost Γ of the optimal grillage for a uniformly distributed loading p per unit area is $4p$ times the sum of the integrals of v_1 and v_2 over triangles ORS and RQS, respectively. To obtain these integrals we use the formula of Exercise 5.14. The rates of deflection at the relevant points in Fig. 5.29 are

$$
\begin{aligned}
v_R &= \frac{\alpha(OR)^2}{2} = \frac{\alpha a^2}{2},\\
v_Q &= v_R + \frac{(RQ)^2}{2} = \frac{(1 + \alpha)a^2}{2},\\
v_A &= v_Q - \frac{(QA)^2}{2} = \frac{(3 + 4\alpha)a^2}{8} = v_B,\\
v_C &= v_R - \frac{(RC)^2}{2} = \frac{(1 + 2\alpha)a^2}{4},\\
v_E &= v_R - \frac{\alpha(RE)^2}{2} = \frac{3\alpha a^2}{8} = v_D.
\end{aligned}
\tag{5.72}
$$

The contributions of triangles ORS and RQS to the cost are thus

$$\Gamma_1 = \frac{pa^2}{6}(v_C + v_D + v_E) = \frac{(1+5\alpha)pa^4}{24},$$
$$\Gamma_2 = \frac{pa^2}{6}(v_A + v_B + v_C) = \frac{(4+6\alpha)pa^4}{24}, \tag{5.73}$$

and the total costs of grillage and edge beams is

$$\Gamma = 4(\Gamma_1 + \Gamma_2) = (5 + 11\alpha)pa^4/6. \tag{5.74}$$

For a fixed height h of the grillage beams, the cost of a grillage with edge beams of height $\bar{h}$ thus exceeds that of the grillage with rigid simple supports along all edges by 220 $h/\bar{h}$ %.

If the grillage is rectangular rather than square, it may be appropriate to give the longer-edge beams a greater height. Figure 5.30 shows the lower left quarter of a rectangular grillage of half sides a and $b > a$, with $\alpha_x > \alpha_y$. The necessary modifications of Fig. 5.29 are readily seen. While the rightmost point I of the Mohr circle for the region near the lower left corner again has unit abscissa, parallels to the edges through pole P' must now intersect the circle at points X and Y with abscissas α_x and α_y and ordinates of equal magnitudes but opposite signs. The abscissa of the center and the radius of the circle thus have the values $(\alpha_x + \alpha_y)/2$ and $1 - (\alpha_x + \alpha_y)/2$. Equating values of JP' obtained from triangles IJP' and KJP' and expressing $\tan 2\theta$ in terms of $\tan\theta$, one finds

$$\tan\theta = \left[\frac{1-\alpha_x}{1-\alpha_y}\right]^{1/2}. \tag{5.75}$$

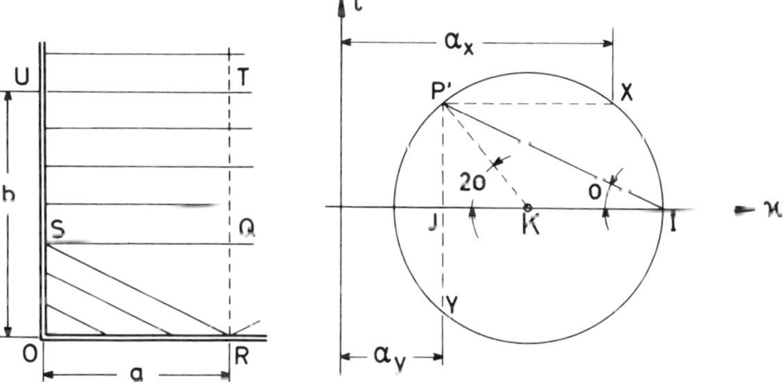

Figure 5.30. Corner of rectangular grillage with edge beams and corresponding Mohr circle.

The regime represented by the Mohr circle prevails in triangle *ORS*, where *R* is the center of the lower edge, and *RS* forms the angle θ with *RO*. Along *RS* this regime is matched by one of type S^+ that has unit rate of curvature in all directions. This regime prevails in triangle *RQS*, where

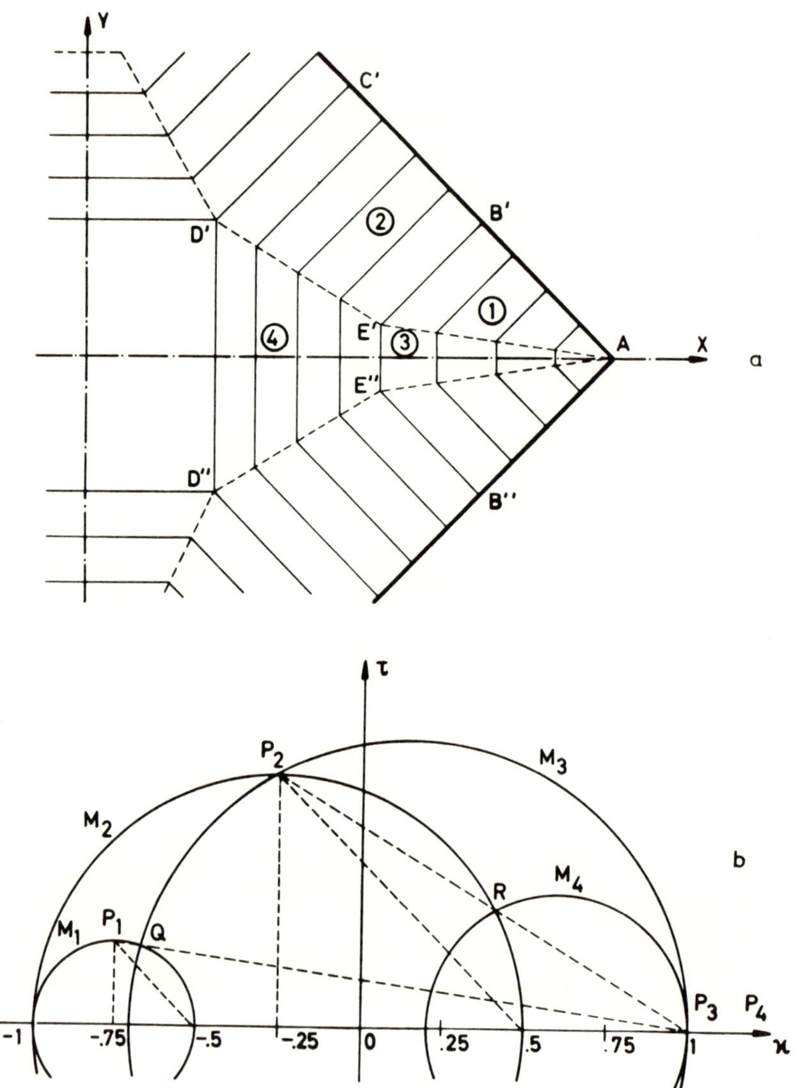

Figure 5.31. (a) Corner of a square grillage with edge beams of large yield strength in torsion. (b) Corresponding Mohr circles.

RQ and SQ are parallel to the edges of the grillage. The Mohr circle of this regime shrinks to point I. In rectangle $SQTU$, where U is the center of the long side through O, we have a regime with principal rates of curvature $\kappa_x = 1$ and $\kappa_y = \alpha_y$, the rate of curvature of the long-edge beams. The Mohr circle of this regime thus has J and I as leftmost and rightmost points and J as pole.

The somewhat unrealistic assumption that the beams of the grillage are simply supported on the edge beams is equivalent to the assumption that they are firmly joined to edge beams with negligible yield torque. At the other extreme, one might consider edge beams whose very large yield torque precludes any twisting of the edge beams.

Figure 5.31a shows the optimal layout in the neighborhood of a corner of a square grillage of this kind with $\alpha = \frac{1}{2}$ for all edge beams. The edge beams are supposed to be clamped at the corners of the grillage. Accordingly, the upper edge beam in Fig. 5.31a has rates of curvature $-\frac{1}{2}$ in the outer quarter AB' of its span and $\frac{1}{2}$ in the inner quarter $B'C'$. The grillage beams that are rigidly connected to this beam are normal to it and have rate of curvature -1. The Mohr circles M_1 and M_2 in Fig. 5.31b correspond to Regions 1 ($A'B'E'$) and 2 ($B'C'D'E'$). The common leftmost point of these circles has negative unit abscissa, while their rightmost points have the abscissas $-\frac{1}{2}$ and $\frac{1}{2}$, and their poles P_1 and P_2 are their highest points.

The hogging beams such as $B'E'$ and their images by symmetry with respect to the x-axis support sagging beams that are normal to this axis and have unit rate of curvature. The Mohr circles M_3 and M_4 correspond to Regions 3 ($AE'E''$) and 4 ($E'D'D''E''$); their poles are at their common rightmost point, which has unit abscissa. The interface AE' of Regions 1 and 3 has the direction of line P_1P_3, which must contain the intersection Q of circles M_1 and M_3, because the rates of curvature and twist along this interface must be independent of whether the interface is regarded as lying in Regions 1 or 3. Similarly, the interface of Regions 2 and 4 is parallel to line P_2P_4, which must contain the intersection R of circles M_2 and M_4. Finally, $D'D''$ is the side of a square central region of type S^+.

The optimal layouts shown in Figs. 5.29 through 5.31 are due to Lowe and Melchers (1974), who did not, however, use Mohr circles in the way they have been used here. The abscissas κ of the centers and the radii ρ of these circles are readily found to be

$$
\begin{aligned}
M_1 &: \kappa = -0.75, & \rho &= 0.25, \\
M_2 &: \kappa = -0.25, & \rho &= 0.75, \\
M_3 &: \kappa = 0.15, & \rho &= 0.85, \\
M_4 &: \kappa = 0.60, & \rho &= 0.40.
\end{aligned}
$$

The coordinates of points Q and R are

$$Q: \kappa = -0.6660, \qquad \tau = 0.2380,$$

$$R: \kappa = 0.4118, \qquad \tau = 0.3529.$$

Exercise 5.8. For a square grillage, each edge of which is either simply supported or built in, determine the optimal layouts for all manners of support other than that in Fig. 5.17.

Exercise 5.9. Find the optimal layout for an equilateral triangular grillage with one built-in edge and two simply supported edges.

Exercise 5.10. Determine the optimal layout for the trapezoidal grillage $ABCD$ of Fig. 5.27 when all edges are simply supported.

Exercise 5.11. Find the optimal layout for the trapezoidal grillage $ABCD$ of Fig. 5.27, when edge AD is free while the other edges are simply supported.

Exercise 5.12. Determine the moment areas of the beams in Fig. 5.23b when the grillage carries a concentrated load P at C, and show that their sum equals $2Pa^2$.

Exercise 5.13. Find the optimal layout for the trapezoidal grillage $ABCD$ of Fig. 5.28 when edges AB and BC are simply supported while the other edges are free, and determine the total cost for a load P at D if $\gamma = Y$.

Exercise 5.14. Verify that, for any quadratic function f of rectangular coordinates x, y, the integral of f over a triangle ABC, with orthogonal sides AB and AC of lengths b and c, equals the product of $bc/6$ and the sum of the values of f at the centers of the three sides.

Exercise 5.15. Using the integration formula of Exercise 5.14, determine the cost, Eq. (5.23), of the optimal square grillages of side $2a$ that are either simply supported or built in along all four sides and carry uniformly distributed loads of intensity p.

Exercise 5.16. Consider an optimal equilateral triangular grillage simply supported on equal edge beams that are, in turn, simply supported at their ends. Draw the Mohr circle for the neighborhood of a corner and find its radius ρ and abscissa κ of its center.

5.6. Partially Discretized Grillages

5.6.1. Introduction

In the preceding sections we discussed the optimal layout and optimal variable cross sections of the beams of *grillage-like continua* consisting of dense arrangements of beams of vanishingly narrow cross sections. We hereafter consider grillages made of *discrete main beams* and dense arrangements of secondary beams of infinitesimal cross sections. The domain occupied by the grillage is divided into subdomains, in each of which the main beams are parallel to one another and the secondary beams are normal to the main beams. Each secondary beam is simply supported on two neighboring main beams (or on a main beam and an edge of the grillage) to which it transmits the loads. The main beams of a subdomain are simply supported on the edges of the grillage or on main beams of adjacent subdomains. We shall first regard the number m of main beams as given and determine the optimal locations of these main beams. For this purpose we shall derive the optimality criterion expressing the stationarity of the total cost Γ with respect to the location of a typical main beam. We shall next determine the value of m that renders Γ minimum. In order to arrive at finite values of the optimal number of main beams, we introduce a constant term in the specific cost of the main beams. Considering only single loading cases and denoting by Y_1, Y_2, γ_1, γ_2 the yield moments and the specific costs of the main and secondary beams, respectively, we have

$$\gamma_1 = a_1 Y_1 + b, \qquad \gamma_2 = a_2 Y_2, \tag{5.76}$$

where a_i and b are nonnegative constants.

5.6.2. Optimality Condition for Main-Beam Locations

In order to derive the optimality condition for the locations of the main beams, let us consider the part of the layout of a grillage shown in Fig. 5.32. The main beams are parallel to the y-axis and simply supported on the edges $y = \pm f(x)$, whereas the secondary beams are parallel to the x-axis and simply supported on the main beams or on the edges. Let the load $p(x, y)$ be directed along the positive z-axis normal to the (x, y)-plane. Consider a generic main beam with abscissa x_i and neighboring beams with abscissas x_{i-1} and x_{i+1} (Fig. 5.32a). A necessary condition for optimality of the abscissa x_i is that the cost Γ remain stationary upon variation of x_i by the infinitesimal amount dx_i. We recall that the moment-dependent part of the cost can be evaluated as the power of the loads in the collapse mechanism (see the end of Section 3.1). Hence, the total cost Γ will exceed the power

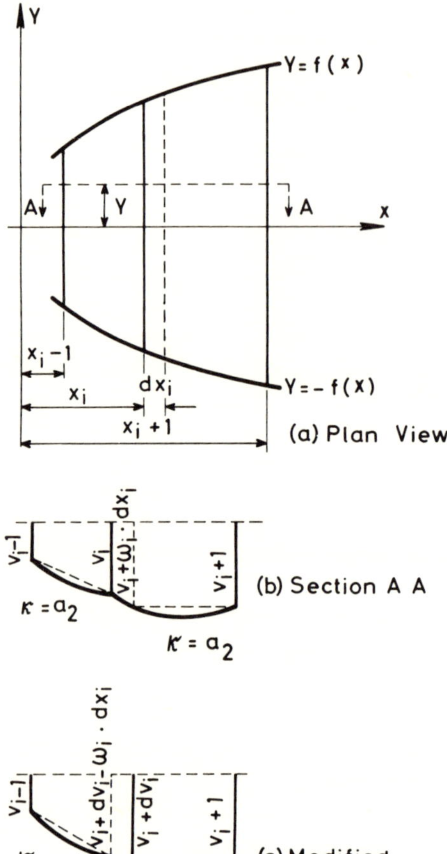

Figure 5.32. (a) Partially discretized grillage: boundary and main beams. (b) Velocity field for section *AA*. (c) Modified velocity field.

of the loads by the product of the constant b with the sum of the lengths of all main beams.

Because main beams have constant rate of curvature $\partial\gamma_1/\partial Y_1 = a_1$, the deflection rate of the beam with abscissa x_i is

$$v_i(x_i, y) = \tfrac{1}{2}a_1[f^2(x_i) - y^2], \qquad -f(x_i) \leq y \leq +f(x_i). \tag{5.77}$$

Figure 5.32b shows the deflection rates of the secondary beams with the ordinate y. Values at the abscissas x_{i-1}, x_i, and x_{i+1} are given by Eq. (5.77), valid for any i. Let us denote by ω_i^- and ω_i^+ the clockwise positive rates of slopes of the secondary beams at $x_i - 0$ and $x_i + 0$, respectively. The right-hand secondary beam in Fig. 5.32b thus has deflection rate $v_i + \omega_i^+ \, dx_i$ at abscissa $x_i + dx_i$. Because the secondary beams have constant rate of curvature a_2, we have

$$2\omega_i^- = \frac{2(v_i - v_{i-1})}{x_i - x_{i-1}} - a_2(x_i - x_{i-1})$$

$$= a_1[f^2(x_i) - f^2(x_{i-1})] - \frac{a_2(x_i - x_{i-1})^2}{x_i - x_{i-1}}. \tag{5.78}$$

Similarly,

$$2\omega_i^+ = a_1[f^2(x_{i+1}) - f^2(x_i)] - \frac{a^2(x_{i+1} - x_i)^2}{x_{i+1} - x_i}. \tag{5.79}$$

Equation (5.78) is valid as long as the secondary beams are supported on two main beams—that is, if $|y(x_{i-1})| \leq f(x_{i-1})$, assuming $df(x)/dx \equiv f' > 0$.

For larger values of $|y(x_{i-1})|$ the left ends of the secondary beams are supported by the edges. If we write the equations of the edges in the form $x = \pm\phi(y)$, these secondary beams will have lengths $x_i - \phi(y)$ and

$$2\omega_i^- = \frac{2v_i - a_2[x_i - \phi(y)]^2}{x_i - \phi(y)}$$

$$= \frac{a_1[f^2(x_i) - y^2] - a_2[x_i - \phi(y)]^2}{x_i - \phi(y)}. \tag{5.80}$$

Figure 5.32c shows the situation after shifting the ith main beam to the abscissa $x_i + dx_i$. Its span is increased by $2f'(x_i)\, dx_i$, and its deflection rate at the ordinate y is increased by

$$dv_i = (\partial v_i/\partial f)(df/dx_i)\, dx_i = a_1 f(x_i) f'(x_i)\, dx_i. \tag{5.81}$$

Note finally that, to within higher-order quantities, the deflection rate at the abscissa x_i in Fig. 5.32c is $v_i + dv_i - \omega_i^- \, dx_i$.

Now, comparing Fig. 5.32(b)–(c), we remark that, due to the fact that all secondary beams have the same constant curvature, a vertical segment between the dotted line and the curve has the same length in Fig. 5.32 (b)–(c) for any abscissa x. Accordingly, the difference between the deflection rates in the two figures varies linearly from zero at x_{i-1} to $dv_i - \omega_i^- \, dx_i$ at x_i, and from $dv_i - \omega_i^+ \, dx_i$ at $x_i + dx_i$ to zero at x_{i+1}. But these linear functions may also be regarded as the influence functions for $V_i^-(dv_i - \omega_i^- \, dx_i)$ and $V_i^+(dv_i - \omega_i^+ \, dx_i)$, where V_i^- and V_i^+ are the shear forces of the secondary beams immediately to the left and to the right of the ith main beam. Hence, the power of the loads at points with ordinates between y and $y+dy$ on the difference of the deflection rates of Fig. 5.32(b)-(c) is given by

$$[(V_i^+ - V_i^-)\,dv_i + (V_i^-\omega_i^- - V_i^+\omega_i^+)\,dx_i]\,dy. \tag{5.82}$$

Integrating Eq. (5.82) between $-f(x_i)$ and $f(x_i)$ and adding the change $2bf'(x_i)\,dx_i$, we obtain the cost variation $d\Gamma$ caused by the shift of the ith main beam by dx_i. Equating this variation to zero, we obtain the optimality condition

$$2bf'(x_i) + \int_{-f(x_i)}^{f(x_i)} [(V_i^+ - V_i^-)a_1 f(x_i)f'(x_i) + V_i^-\omega_i^- - V_i^+\omega_i^+]\,dy = 0. \tag{5.83}$$

A condition of the type in Eq. (5.83) must be satisfied for each main beam.

Equation (5.83) has been derived by Rozvany (1975), using the calculus of variations, and, in the above manner, by Rozvany and Prager (1976), who applied it to a uniformly loaded, simply supported, square grillage with either the main beams parallel to one side of the square (see Exercise 5.17) or with the arrangement of the main beams patterned on that of the beams of an optimal grillage-like continuum for the same loading and support conditions.

Exercise 5.17. Consider a simply supported square grillage of side $2L$ subjected to a uniformly distributed load p. Assume the m main beams to be parallel to one side. Show that, with specific costs in Eq. (5.76), the optimal abscissas of the main beams are such that (a) the inner main beams are equally spaced, and (b) the outer main beams are at distance x_1 from the edges given by

$$x_1/L = 4 - \frac{2n[1-g(4-n^2)]^{1/2}}{4-n^2} \qquad \text{for } n \neq 2,$$

$$x_1/L = \tfrac{1}{2}(1+4g) \qquad \text{for } n = 2,$$

with $n = m-1$ and $g = a_1/6a_2$. Study the variation of the cost with m and $k = b/2a_2pL^3$ for $g = 0$ and $g = 0.05$.

6

Optimal Design of Plates, Shells, and Disks

6.1. Introduction

It was emphasized in Section 3.7.2 that all optimality criteria for beams were derived from a fundamental inequality of the same type, Eq. (3.180), relating two energies or powers. Hence, there is no need for special proofs of the optimality criteria for plates, shells, and disks, provided the proper generalized variables are used, as described in Chapter 1. This is true for multipurpose as well as for single-purpose structures. There is, however, one exceptional case for which the extension of the optimality criterion from beams to plates, shells, and disks is not straightforward—namely, the behavioral constraint of *limited stress intensity.* In contrast to the other behavioral constraints considered in this book, it is *local* and *not* related to a minimum or stationarity principle. Also the definition of "stress intensity" is subject to various generalizations and, hence, must be carefully discussed.

Consider a structure, the behavior of which is described by the generalized variables Q_j, q_j, $\dot{q}_j$, $j = 1, \ldots, n$, as defined in Section 1.1. Designing this structure means giving the design variable R as a function of the coordinate x on the midsurface of the structure.

In order to make sure of elastic behavior under given loads, with a prescribed safety margin with respect to some limiting state (e.g., first yield, or fracture), an upper bound is set on some "stress intensity" defined as a scalar function of the stresses and the design variable:

$$f(Q_j, R) \leq 1. \tag{6.1}$$

The design variable $R(x)$ is the measure of the specific resistance (or strength) of the structure at point x *with respect to the limiting state*

$$f(Q_j, R) - 1. \tag{6.2}$$

We assume that the locus with Eq. (6.2) is convex and *varies isotropically with R.*

Suppose now that, for given loading, it is possible to associate an elastic stress field $Q_j(x)$ and a kinematically admissible *virtual* strain field $q_j(x)$ with a design $R(x)$ in such a way that $Q_j(x)$ is everywhere at the limiting state (Eq. (6.2) holds for all x) and $q_j(x)$ is found by applying the normality law to the locus of Eq. (6.2), exactly as the plastic strain rate is found to be normal to a yield locus. We can then, from the convexity and normality properties, establish the fundamental inequality, Eq. (3.180), in which $D = \Sigma_j Q_j q_j$ and $D^* = \Sigma_j Q_j^* q_j$ are *virtual* specific energies, where Q_j^* is the elastic stress field for a design $R(x)$ that satisfies Eq. (6.1) under the given loads. Hence, the optimality condition is completely analogous to that for assigned limit load; namely,

a. the structure is everywhere at the limiting state;
b. there exists a field of *virtual* strains "associated" with the elastic stress field in the sense defined above, and such that at every point on the midsurface the corresponding (virtual) specific energy per unit value of the design variable is equal† to the derivative of the specific cost with respect to the design variable.

When the specific cost γ is a convex function of R, the optimality condition is a necessary and sufficient condition for absolute optimality, whereas it is only a necessary condition for *local* optimality in the absence of convexity of $\gamma(R)$.

We see that the optimal design must be an elastic uniform strength design (or a so-called fully stressed design), but with a supplementary condition on an associated virtual strain field. This was first shown by Michell (1904) for trusses, Heyman (1959) for beams and frames, and Shield (1960b) for plates, shells, and disks. In the particular case where the stress intensity is identified with the specific elastic energy per unit of strength, the real elastic strain field for the uniform strength design is "associated" with the stress field because of the relations of linear elasticity in Eq. (1.13), and it can be regarded as the desired virtual field.

Because the stress field satisfies Eq. (6.2) all over the structure, if the specific cost is proportional to the strength, the optimality condition is automatically fulfilled, as discussed, for minimum-volume designs, by Save (1968) and by Masur (1970) in a wider context. Hence, the uniform strength design in this case is an optimal elastic design for the assigned limited stress intensity. With the supplementary assumption that the elastic stiffness of the structural element is proportional to its strength, the elastic, uniform strength design has constant specific elastic energy per unit of stiffness and,

† When lower and (or) upper bounds are set on R, equality is replaced by inequalities, as in Section 3.1.1.

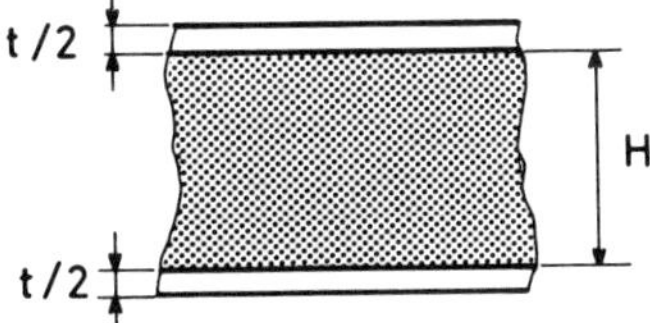

Figure 6.1. Sandwich cross section.

accordingly, is optimal for some assigned upper bound on the elastic compliance (see Section 3.3).

Finally, if the limiting stress condition in Eq. (6.2) is also the condition of full plasticity of the structural element, the elastic uniform strength design is at incipient plastic collapse under the given loads because the elastic strain field can be regarded as the corresponding plastic velocity field, since it satisfies the normality rule. We conclude that, when the assumptions cited above are satisfied, optimal designs of a given structure under given loading for limited stress intensity, for limited elastic compliance, and for assigned load factor at plastic collapse are proportional to one another.

This situation occurs for minimum-volume design of disks or sandwich structures of incompressible elastic material that obeys the von Mises yield condition, as shown by Save (1968). A sandwich structure consists of a central core that resists transversal shear and two face sheets that resist in plane stress only (see Fig. 6.1). Because the thickness $t/2$ of each sheet is small with respect to the height of the core, stresses are assumed uniform through the sheet thickness, and the distance between the midsurfaces of the sheets is approximated by the (given) thickness H of the core. Hence, all elastic stiffnesses, as well as the full plastic moments and forces of such a structural element, are proportional to the sheet thickness, which also measures the specific cost. On the other hand, the condition of limited stress intensity is also the yield condition of the structural element, and its left-hand side is the specific elastic strain energy. All the sufficient conditions are thus satisfied for proportionality of the three types of optimal designs. It is easily verified that disks behave as sandwich structures.

In conclusion, let us point out that proportionality can sometimes occur in other conditions, as specific examples will show (see Section 6.2.6).

6.2. Circular Plates with Axisymmetric Loading

6.2.1. Basic Relations

We adopt the usual Kirchhoff–Love assumptions of the structural theory of thin plates subjected to bending, and we use a system of cylindrical coordinates as shown in Fig. 6.2. The radial and circumferential directions

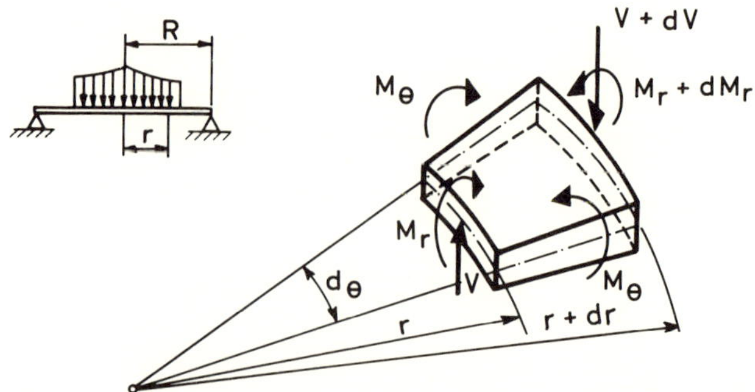

Figure 6.2. Axisymmetrically loaded circular plate and plate element.

are principal directions of moments (M_r, M_θ) and curvatures (κ_r, κ_θ) that are the generalized variables. We list the basic relations [see, for example, Way (1962) and Save and Massonnet (1972)].

6.2.1.1. Kinematics. The curvatures are given by

$$\kappa_r = -d^2w/dr^2, \qquad \kappa_\theta = (-1/r)(dw/dr), \tag{6.3}$$

where $w(r)$ is the deflection field. Obviously, similar relations hold for rates of curvature $\dot{\kappa}_r$, $\dot{\kappa}_\theta$ and velocity $\dot{w}(r)$.

6.2.1.2. Statics. By elimination of the shear among the equilibrium relations, we obtain

$$\frac{d}{dr}(rM_r) - M_\theta + \int_0^r pr\,dr = 0,$$

or, taking derivatives,

$$\frac{d^2}{dr^2}rM_r - \frac{dM_\theta}{dr} + rp = 0, \tag{6.4}$$

where $p(r)$ is the distributed load. The shear force V is given by $V = dM_r/dr + (M_r - M_\theta)/r$.

6.2.1.3. Linear Elasticity. The relations expressing Hooke's law for an isotropic plate are

$$\kappa_r = \frac{M_r - \nu M_\theta}{B}, \qquad \kappa_\theta = \frac{M_\theta - \nu M_r}{B}, \tag{6.5}$$

or, equivalently,

$$M_r = \frac{B(\kappa_r + \nu\kappa_\theta)}{1 - \nu^2}, \qquad M_\theta = \frac{B(\kappa_\theta + \nu\kappa_r)}{1 - \nu^2}, \tag{6.6}$$

where ν is Poisson's ratio.

For solid plates of thickness t,

$$B = Et^3/12, \tag{6.7}$$

whereas for sandwich plates with a core of height H and two equal face sheets each with thickness $t/2$,

$$B = EHt^2/4. \tag{6.8}$$

The specific strain energy is

$$\phi = \frac{B(\kappa_r^2 + \kappa_\theta^2 + 2\nu\kappa_r\kappa_\theta)}{2(1 - \nu^2)}, \tag{6.9}$$

and the specific complementary energy is

$$\psi = \frac{M_r^2 + M_\theta^2 - 2\nu M_r M_\theta}{2B}. \tag{6.10}$$

As we pointed out in Section 1.2.1, $\phi = \psi$.

6.2.1.4. Perfect Plasticity. The plate element is completely defined by its yield locus in the (M_r, M_θ) space, to which the normality flow law must be applied to obtain the plastic rates of curvature $\dot\kappa_r$ and $\dot\kappa_\theta$. If we denote by Y the yield moment in uniaxial pure bending and by σ_0 the yield stress in tension, we have

$$Y = \sigma_0 t^2/4 \tag{6.11}$$

for the solid plate, and

$$Y = \sigma_0 Ht/2 \tag{6.12}$$

for the sandwich plate. The Tresca and von Mises conditions are the two most commonly used yield conditions for metal plates.

The *yield condition of Tresca* is

$$\max(|M_r|, |M_\theta|, |M_r - M_\theta|) = Y. \tag{6.13}$$

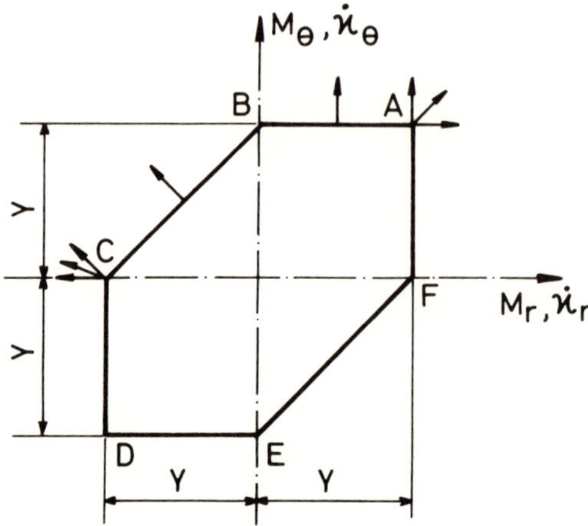

Figure 6.3. Yield condition of Tresca for circular plate element.

The corresponding yield locus is shown in Fig. 6.3. The specific dissipation $D = M_r\dot{\kappa}_r + M_\theta\dot{\kappa}_\theta$ is given by the following:

$$\left.\begin{array}{lll} \text{For sides } AB \text{ or } DE: & D = Y|\dot{\kappa}_\theta|, & \text{(a)} \\ \text{For sides } BC \text{ or } EF: & D = Y|\dot{\kappa}_\theta| = Y|\dot{\kappa}_r|, & \\ & \quad \text{because } \dot{\kappa}_r = -\dot{\kappa}_\theta & \\ & \quad \text{and } |M_r - M_\theta| = Y, & \text{(b)} \\ \text{For sides } AF \text{ and } DC: & D = Y|\dot{\kappa}_r|, & \text{(c)} \\ \text{For corners } A \text{ or } D: & D = Y|\dot{\kappa}_r + \dot{\kappa}_\theta|, & \text{(d)} \\ \text{For corners } B \text{ or } E: & D = Y|\dot{\kappa}_\theta|, & \text{(e)} \\ \text{For corners } C \text{ or } F: & D = Y|\dot{\kappa}_r|. & \text{(f)} \end{array}\right\} \quad (6.14)$$

We recall that at corners the normality law implies that the strain rate vector is a linear combination with positive coefficients of the strain rate vectors for the adjacent sides.

The yield condition of von Mises is

$$M_r^2 + M_\theta^2 - M_rM_\theta = Y^2. \quad (6.15)$$

The corresponding yield locus is shown in Fig. 6.4. The rates of curvature are

$$\dot{\kappa}_r = \alpha(2M_r - M_\theta), \qquad \dot{\kappa}_\theta = \alpha(2M_\theta - M_r), \quad (6.16)$$

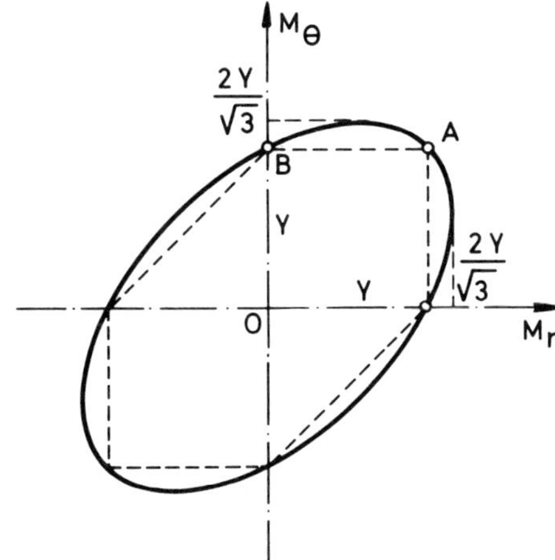

Figure 6.4. Yield condition of von Mises for circular plate element.

from which we obtain

$$M_r = \frac{2\dot{\kappa}_r + \dot{\kappa}_\theta}{3\alpha}, \qquad M_\theta = \frac{2\dot{\kappa}_\theta + \dot{\kappa}_r}{3\alpha}, \tag{6.17}$$

where α is a nonnegative scalar function of r. From Eqs. (6.15) and (6.17) we obtain

$$\alpha = \frac{Y}{3(\dot{\kappa}_r^2 + \dot{\kappa}_\theta^2 + \dot{\kappa}_r\dot{\kappa}_\theta)}. \tag{6.18}$$

The specific dissipation $D = M_r\dot{\kappa}_r + M_\theta\dot{\kappa}_\theta$ is given by

$$D = \frac{2[3(\dot{\kappa}_r^2 + \dot{\kappa}_\theta^2 + \dot{\kappa}_r\dot{\kappa}_\theta)]^{1/2}}{Y}. \tag{6.19}$$

A parametric form of Eq. (6.15) can be given as follows:

$$M_r = \frac{2Y\sin(\beta + \pi/6)}{\sqrt{3}}, \qquad M_\theta = \frac{2Y\sin(\beta - \pi/6)}{\sqrt{3}} \tag{6.20}$$

with $\tan\beta = \sqrt{3}\sin\varphi$, φ being the parameter $(0 \leq \varphi \leq \pi)$.

For reinforced concrete plates *the yield condition of Johansen* is used. If we denote by Y_r, Y_r', Y_θ, Y_θ' the yield moments in positive and negative bendings in the principal directions of a section of unit breadth of the plate,

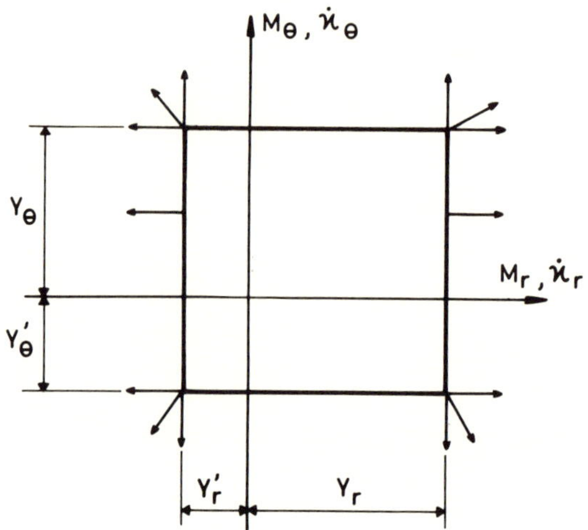

Figure 6.5. Yield condition for reinforced concrete circular plate element.

the rectangular yield locus of Johansen is defined by

$$M_r = Y_r, \quad M_r = -Y'_r, \quad M_\theta = Y_\theta, \quad M_\theta = -Y'_\theta. \tag{6.21}$$

It is shown in Fig. 6.5.

When the reinforcement areas in both the upper and lower layers and in both principal directions can be varied independently, we have four independent design variables. Because the moment arm of the internal forces is fairly constant in reinforced concrete plates [see Wood (1961)], the yield moments are proportional to the reinforcement areas and are thus taken as design variables in the minimization of reinforcement volume. The problem is then identical to that of the grillagelike continuum of Section 5.2. On the other hand, if the four yield moments are equal—that is, if the reinforcement is made of two identical isotropic layers—we recover the sandwich plate.

6.2.2. Assigned Limit Load for a Tresca Sandwich Plate

6.2.2.1. We want to design a sandwich plate for *minimum volume of the face sheets.* As a first step, *no bound is set on the thickness* $t(r)$ *that may vary continuously with* r [see Onat *et al.* (1957)]. Because Y is proportional to t by the relation of Eq. (6.12), we take Y as the design variable.

From the relations of Eq. (6.3) we obtain the compatibility equation

$$\dot{\kappa}_r = r\frac{d\dot{\kappa}_\theta}{dr} + \dot{\kappa}_\theta. \tag{6.22}$$

Since the specific cost is $\gamma = t = 2Y/\sigma_0 H$, the optimality condition is that the ratio of the specific dissipation to the yield moment be a positive constant:

$$D/Y = c^2. \tag{6.23}$$

The relevant plastic regimes to be used with transverse downward loadings are determined as follows. First, we exclude sides *AB*, *DE*, *BC*, *EF*, and corners *B* and *E*. For any of these the combination of Eqs. (6.23) and (6.14) would require $\dot{\kappa}_\theta$ to be constant. Substituting this in Eq. (6.22) would then give $\dot{\kappa}_r = \dot{\kappa}_\theta$ for a finite interval in r, a result that contradicts the existence of nonvanishing curvatures in these regimes. Second, sides *AF* and *DC* are excluded because they imply $\dot{\kappa}_\theta = 0$ and, by Eq. (6.21), $\dot{\kappa}_r = 0$. Hence, the only acceptable regimes are corners *A* (or *D*) and *C* (or *F*), for which one can verify that no contradiction appears.

In regime A, taking into account the relations in Eqs. (6.3) and [6.14(d)] and the fact that $\dot{\kappa}_r$ and $\dot{\kappa}_\theta$ are nonnegative, the condition of Eq. (6.23) gives

$$\frac{d^2\dot{w}}{dr^2} + \frac{1}{r}\frac{d\dot{w}}{dr} = -c^2, \tag{6.24}$$

and hence

$$\dot{w}(r) = -c^2 r^2/4 + c^2 b^2 \ln(b/r) + d, \tag{6.25}$$

where b and d are constants of integration.

In regime C, where $-\dot{\kappa}_r \geq \dot{\kappa}_\theta \geq 0$, we obtain in like manner

$$d^2\dot{w}/dr^2 = -c^2 \tag{6.26}$$

and

$$\dot{w} = c^2 r(r - f) + g, \tag{6.27}$$

where f and g are constants of integration.

We now successively consider two support conditions.

Condition (a): *Simple support.* Regime *A* is acceptable throughout the plate. Because $\dot{w}$ is finite at the center $r = 0$ and vanishes at the boundary $r = R$, Eq. (6.25) becomes

$$\dot{w} = \tfrac{1}{4}c^2 R^2(1 - r^2/R^2). \tag{6.28}$$

The design $Y(r)$ is obtained from the equilibrium equation, Eq. (6.4), where

Y is substituted for M_θ and M_r. Integrating, with boundary conditions $Y = M_r = 0$ for $r = R$ and the shear force vanishes at the center (in the absence of a concentrated central force), we obtain [see Latta (1962)]

$$Y \equiv \sigma_0 Ht/2 = \int_r^R \frac{1}{\xi}\, d\xi \int_0^\xi \rho p(\rho)\, d\rho. \tag{6.29}$$

If there is a central force, a term $(P/2\pi)\ln(R/r)$ must be added to the right-hand side of Eq. (6.29). We obtain t infinite at the center, but the volume of the sheets remains finite.

Condition (b): *Built-in support.* At the center, regime A applies because of axial symmetry; at the boundary, $\dot{\kappa}_r < 0$ and regime C must be used. Hence, there will be a central region in regime A and a ring, adjacent to the support, in regime C. Let r_0 be the radius of the central region.

For $0 \le r \le r_0$ Eq. (6.25) must be used. For $r_0 \le r \le R$ Eq. (6.27) applies. For $r = R$ we must have $\dot{w} = 0$ and $d\dot{w}/dr = 0$ at the built-in support with no hinge allowed in order to obtain $D/Y = c^2$. For $r = r_0$, $\dot{w}$ and $d\dot{w}/dr$ are continuous. We obtain $r_0 = \frac{2}{3}R$ and

$$\dot{w} = \begin{cases} \dfrac{c^2(2R^2 - 3r^2)}{12} & \text{for } 0 \le r \le \frac{2}{3}R, \\[2ex] \dfrac{c^2(r-R)^2}{2} & \text{for } \frac{2}{3}R \le r \le R. \end{cases} \tag{6.30}$$

For $0 \le r \le \frac{2}{3}R$ (regime A), Eq. (6.29) is valid. For $\frac{2}{3}R \le r \le R$ (regime C), substituting $-Y$ and 0 for M_r and M_θ, respectively, in Eq. (6.4) gives

$$(d^2/dr^2)(rY) = rp(r). \tag{6.31}$$

Since M_r is continuous and therefore vanishes at $r = \frac{2}{3}R$, and since the shear force is continuous and vanishes at $r = 0$ (no central concentrated force), integrating Eq. (6.31) gives

$$Y \equiv \sigma_0 \frac{H}{2} t = \begin{cases} \displaystyle\int_r^{2R/3} \frac{1}{\xi}\, d\xi \int_0^\xi \rho p(\rho) d\rho & \text{for } 0 \le r \le \frac{2}{3}R, \\[2ex] \displaystyle\frac{1}{r}\int_{2R/3}^r d\xi \int_0^\xi \rho p(\rho) d\rho & \text{for } \frac{2}{3}R \le r \le R. \end{cases} \tag{6.32}$$

For both types of support the conditions of equilibrium, yield, and optimality are satisfied. The functions $t(r)$ defined by Eqs. (6.29) and (6.32) thus furnish designs with absolute minimum volume of the face sheets for the considered loads. We now apply these results to two loading cases:

Case (a): *Circular loading.* $p(r) = p = \text{const}$ for $0 \le r \le a$, and $p(r) = 0$ for $a \le r \le R$.

For a *simple support* we obtain directly

$$\sigma_0 \frac{H}{2} t = \begin{cases} (P/4)(a^2 - r^2) + (pa^2/2)\ln(R/a) & \text{for } 0 \le r \le a, \\ (p/2)a^2 \ln(R/r) & \text{for } a \le r \le R. \end{cases} \tag{6.33}$$

The volume of the sheets is

$$V = \frac{1}{4}\left(2 - \frac{a^2}{R^2}\right)\frac{\pi p a^2 R^2}{\sigma_0 H}, \tag{6.34}$$

which we compare with the volume V_c of the sheets of the sandwich plate with constant sheet thickness, same radius and core, and same limit load. Because the limit load for this plate is $p = 6YR/a^2(3R - 2a)$ [see Save and Massonnet (1972)], we deduce that

$$V_c = \left(1 - \frac{2}{3}\frac{a}{R}\right)\frac{\pi p a^2 R^2}{\sigma_0 H}. \tag{6.35}$$

Saving in volume varies from 25% to 50% when a/R decreases from one to zero. With a concentrated central load P, one has

$$\sigma_0 \frac{Ht}{2} = \frac{P}{2\pi}\ln\frac{R}{r}, \tag{6.36}$$

and the volume of the sheets is

$$\sigma_0 \frac{Ht}{2} = \frac{P}{2\pi}\ln\frac{R}{r}, \tag{6.36}$$

For a *built-in plate* with $a < \frac{2}{3}R$, we obtain

$$\sigma_0 \frac{H}{2} t = \begin{cases} \dfrac{p}{4}(a^2 - r^2) + \dfrac{pa^2}{2}\ln\dfrac{2}{3}\dfrac{R}{a} & \text{for } 0 \le r \le a, \\[2ex] \dfrac{pa^2}{2}\ln\dfrac{2R}{3r} & \text{for } a \le r \le \frac{2}{3}R, \\[2ex] \dfrac{pa^2}{2}\dfrac{r - \frac{2}{3}R}{r} & \text{for } \frac{2}{3}R \le r \le R. \end{cases} \tag{6.38}$$

If $a > \frac{2}{3}R$,

$$\sigma_0 \frac{H}{2} t = \begin{cases} \dfrac{p}{4}\left(\dfrac{4}{9}R^2 - r^2\right) & \text{for } 0 \le r \le \frac{2}{3}R, \\ \dfrac{p}{6}\dfrac{r^3 - 8R^3/27}{r} & \text{for } \frac{2}{3}R \le r \le a, \\ \dfrac{p}{6}\dfrac{3a^2r - 8R^3/27 - 2a^3}{r} & \text{for } a \le r \le R. \end{cases} \tag{6.39}$$

With a concentrated load P at the center we have

$$\sigma_0 \frac{H}{2} t = \begin{cases} \dfrac{P}{2\pi} \ln \dfrac{2}{3}\dfrac{R}{r} & \text{for } 0 \le r \le \frac{2}{3}R, \\ \dfrac{P}{2\pi}\dfrac{r - \frac{2}{3}R}{r} & \text{for } \frac{2}{3}R \le r \le R, \end{cases} \tag{6.40}$$

and a sheet volume of

$$V = \frac{1}{3}\frac{PR^2}{\sigma_0 H}. \tag{6.41}$$

For a sandwich plate of uniform sheet thickness, the limit load is $P_l = 2\pi M_p$ (for both simple support and built-in edge), and the sheet volume is

$$V_c = \frac{PR^2}{\sigma_0 H}. \tag{6.42}$$

Hence, minimum-volume design saves 67% of sheet material. It must, however, be kept in mind that the design in Eq. (6.40) is purely theoretical because of its infinite thickness at the center.

The minimum-volume design for uniformly distributed load over the entire plate is obtained by substituting R for a in Eqs. (6.39). The volume of the sheet is

$$V = 0.117 \frac{\pi p R^4}{\sigma_0 H}, \tag{6.43}$$

whereas uniform thickness would have needed a volume

$$V_c = 0.178 \frac{\pi p R^4}{\sigma_0 H}. \tag{6.44}$$

The saving here is 34%.

Case (b): *Annular loading.* $p(r) = 0$ for $0 \le r \le a$, and $p(r) = p = \text{const}$ for $a \le r \le R$.

Application of the general formulas in Eqs. (6.29) and (6.32) gives $t(r)$ and shows that, for both the simply supported and the built-in plates, it is constant in the central region. In particular, $t = 0$ in the central region of the built-in plate when $a > \frac{2}{3}R$. Explicit relations for the simply supported plate are

$$Y = \frac{p}{2}\left[\frac{R^2 - a^2}{2} - a^2 \ln\left(\frac{R}{a}\right)\right], \qquad 0 \le r \le a,$$
$$Y = \frac{p}{2}\left[\frac{R^2 - r^2}{2} - a^2 \ln\left(\frac{R}{r}\right)\right], \qquad a \le r \le R, \tag{6.45}$$

and

$$V = \frac{Rp(R^2 - a^2)^2}{4\sigma_0 H}. \tag{6.46}$$

Equations (6.45) and (6.46), together with similar relations for line loads, are given in Hemp (1973, Chap. 3).

6.2.2.2. Because fabrication of a plate with continuously varying yield moment, having vanishing values at some radii, is most unlikely to be possible, our second step will be *to require the thickness of the face sheets to be piecewise constant in annular regions* defined by the radii $0 = r_0 < r_1 < \cdots < r_n = R$. Denote the thickness of the region $r_{i-1} < r < r_i$ by t_i and the area of its midsurface by A_i. The "cost" of the one ring is $A_i t_i = 2A_i Y_i/\sigma_0 H$, and the marginal cost is $2A_i/\sigma_0 H$ if we decide to take the yield moments Y_i of the rings as design variables.

In the absence of bounds on Y, the optimality condition in Eq. (6.9) will require that, in each ring, the dissipation per unit value of Y_i be equal to $2c^2 A_i/\sigma_0 H$. Because the radii r_i are given and c^2 is an arbitrary positive number, the optimality condition here is simply that the average dissipation per unit yield moment be the same in all rings. The design procedure is then as follows:

a. Make physically resonable assumptions on the stress regime (on the yield locus) for each ring and on the relative values of the yield moments.
b. Use the normality law to verify that the optimality condition can be satisfied; otherwise, modify the stress regimes.
c. Use the final stress regimes in the equilibrium equation, Eq. (6.4), which, upon integration, gives the stress field. The boundary conditions finally enable the design to be found explicitly.

Consider a simply supported plate that is uniformly loaded [Sheu and Prager (1969)]. When the thickness is constant, we know from limit analysis [Save and Massonnet (1972)] that collapse occurs in regime AB (Fig. 6.3). This regime will be used in each of the rings of the plate, assuming the presence of a hinge circle (point A in Fig. 6.3) at the boundary of the adjacent ring with larger t. Within regime AB, $\dot{\kappa}_r = 0$, $\dot{w}(r)$ is linear (conical mechanism), and

$$\dot{\kappa}_\theta = -\frac{\dot{w}_i - \dot{w}_{i-1}}{r(r_i - r_{i-1})}. \tag{6.47}$$

The resultant of the distributed dissipation for ring i is

$$Y_i k_i = \int_{r_{i-1}}^{r_i} Y_i \dot{\kappa}_\theta 2\pi r \, dr = Y_i 2\pi(\dot{w}_{i-1} - \dot{w}_i), \tag{6.48}$$

where $\dot{w}_i < \dot{w}_{i-1}$ in order to have positive values for k_i and $\dot{\kappa}_\theta$. The concentrated dissipation in a hinge circle of radius r is $2\pi r Y(r)|\dot{\varphi}|$, where $|\dot{\varphi}|$ is the absolute value of the relative rotation rate of the adjacent parts. For ring i the hinge circle occurs at $r = r_{i-1} + 0$, assuming $Y_i < Y_{i-1}$, which seems appropriate for the present problem. Hence, the dissipation in the hinge circle is

$$2\pi Y_i r_{i-1}(\dot{\varphi}_i - \dot{\varphi}_{i-1}), \tag{6.49}$$

where $\dot{\varphi}_i = (\dot{w}_{i-1} - \dot{w}_i)/(r_i - r_{i-1}) > 0$ is the absolute value of the slope of the conical mechanism of ring i (see Fig. 6.6).

We have assumed $0 \leq \dot{\varphi}_1 < \dot{\varphi}_2 < \cdots < \dot{\varphi}_n$ in accordance with $\dot{w}_i < \dot{w}_{i-1}$ and $0 = r_0 < r_1 < \cdots < r_n = R$. The optimality condition will thus state that

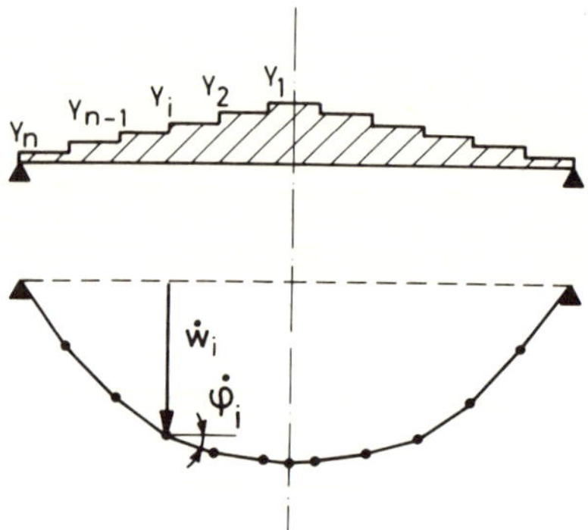

Figure 6.6. Circular supported plate with piecewise-constant yield moments and corresponding collapse mechanism.

the sum of the dissipations in Eqs. (6.48) and (6.49) is proportional to $Y_i\pi(r_i^2 - r_{i-1}^2)$. It gives

$$2(\dot{\varphi}_i r_i - \dot{\varphi}_{i-1} r_{i-1}) = \alpha(r_i^2 - r_{i-1}^2), \tag{6.50}$$

where α is the positive proportionality coefficient. The condition in Eq. (6.50) is satisfied by $\dot{\varphi}_i = \alpha r_i/2$ and furnishes the collapse mechanism $\dot{w}_{i-1} = \dot{w}_i + \alpha r_i(r_i - r_{i-1})/2$, $\dot{w}_n = 0$, compatible with the assumptions made on the variations of $\dot{w}_i$ and $\dot{\varphi}_i$.

Now, integrating Eq. (6.4), with p independent of r and $M_\theta = Y_i$, gives in ring i

$$M_{ri} = Y_i - pr^2/6 + C_i/r.$$

The constants C_i are determined from the boundary conditions $M_{ri} = Y_i$ at $r = r_{i-1}$, which yield $C_i = pr_{i-1}^3/6$. Hence,

$$M_{ri}(r) = Y_i - \frac{p(r^3 - r_{i-1}^3)}{6r}. \tag{6.51}$$

The design is obtained from the condition that

$$M_{ri}(r_i) = Y_{i+1} \quad \text{in Eq. (6.51)},$$

$$Y_{i+1} = Y_i - \frac{p(r_i^3 - r_{i-1}^3)}{6r_i}, \tag{6.52}$$

in agreement with our assumption $Y_i > Y_{i+1}$. The Y_i are found by using Eq. (6.52) iteratively from $i = n$ to $i = 0$. The starting value is $Y_{n+1} = 0$, since $M_r = 0$ at $r = R$; this gives

$$Y_n = \frac{p(R^3 - r_{n-1}^3)}{6R}. \tag{6.53}$$

When the number of rings tends to infinity, we may substitute dY for $Y_{i+1} - Y_i$, with $r_{i-1} = r$ and $r_i = r + dr$ in Eq. (6.52). Neglecting small quantities of higher order, we obtain

$$dY = -\frac{pr\,dr}{2} \quad \text{and} \quad Y = \frac{p(R^2 - r^2)}{4}$$

because $Y(R) = 0$. We see that we recover the design in Eq. (6.33) with $a = R$.

Table 6.1. Yield Moments Y_i—Simply Supported Circular Sandwich Plate Uniformly Loaded

Number of rings	Y_i/pR^2							
	$i = 1$	$i = 2$	$i = 3$	$i = 4$	$i = 5$	$i = 6$	$G = (\Gamma_c - \Gamma)/\Gamma_c$	G/G_{max}
1	0.1667						0.00	0.00
2	0.1875	0.1458					6.25	0.25
3	0.2006	0.1821	0.1173				11.10	0.44
4	0.2092	0.1988	0.1623	0.0960			14.10	0.56
5	0.2152	0.2086	0.1852	0.1430	0.0813		16.00	0.64
6	0.2197	0.2151	0.1988	0.1695	0.1627	0.0702	17.40	0.67
∞							25.00	1.00

Table 6.1 gives the yield moments Y_i up to 6 rings, together with the relative "savings" G, defined as

$$G = \frac{\Gamma_c - \Gamma}{\Gamma_c},$$

where $\Gamma_c = \pi R^4/6$ is the "cost" of the plate with constant Y and limit load p, and where, for minimum volume, we take $\gamma = Y$. We see that we obtain as much as 67% of the maximum possible savings already with $i = 6$.

The same design procedure can be used for the uniformly loaded *annular plate simply supported at the outer edge of radius R and free at the inner edge* of radius a with the same assumptions on the variations of Y_i and $\dot{\varphi}_i$ with i. The only particular point is that the innermost ring does not contain a hinge circle because it is not connected to a stronger ring. Hence, regime AB tends to A in all rings when the number n of rings increases, except for the innermost ring in which it tends to B; furthermore, the yield moment of that ring increases without bound. This "edge effect", first pointed out in a different way† by Megarefs (1966, 1967), is illustrated in Fig. 6.7, adapted from Sheu and Prager (1969).

The circular plate with a piecewise-varying thickness has been studied extensively by Lamblin (1975) in his thesis, some important results of which can be found in Lamblin and Guerlement (1976) and in Lamblin *et al.* (1985), where the problem at hand is treated with a supplementary line

† The stress variation method introduced by Megarefs (1966, 1967), Reiss and Megarefs (1971), and Ting and Reiss (1977) is a direct minimization method in which allowable stress variations are related to volume variations. It does not use an optimality criterion. For this reason it is not treated in this book.

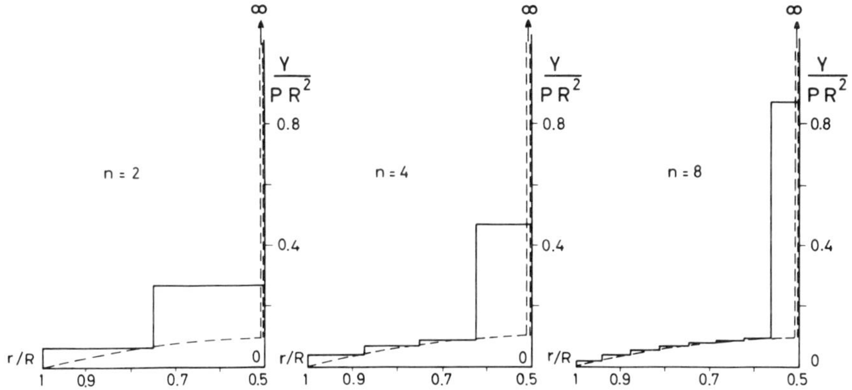

Figure 6.7. Optimal designs of annular plates.

load q per unit length along the inner edge. The optimal design is given by

$$Y_1 = \frac{Y_2}{1 - a/r_1} + \frac{p(r_1^2 + ar_1 - 2a^2)}{6} + qa \tag{6.54}$$

for the innermost ring ($i = 1$) and

$$Y_i = Y_{i+1} + p\left(1 - \frac{r_{i-1}}{r_i}\right)\frac{r_i^2 + r_i r_{i-1} + r_{i-1}^2 - 3a^2}{6} + qa\left(1 - \frac{r_{i-1}}{r_i}\right) \tag{6.55}$$

for $i = 2$ to n. The last equation, with $Y_{n+1} = 0$, gives

$$Y_n = p\left(1 - \frac{r_{n-1}}{R}\right)\frac{R^2 + r_{n-1}R + r_{n-1}^2 - 3a^2}{6} + qa\left(1 - \frac{r_{n-1}}{R}\right). \tag{6.56}$$

When n tends to infinity, Eq. (6.55) becomes

$$dY_i = -\frac{p(r^2 - a^2)dr}{2r} - qa\frac{dr}{r}.$$

Integration of this equation when $Y(R) = 0$ gives

$$Y(r) = \frac{p(R^2 - r^2)}{4} + \left(\frac{pa^2}{2} - qa\right)\ln\left(\frac{r}{R}\right). \tag{6.57}$$

The relation in Eq. (6.57) is valid in the range $[a + \varepsilon, R]$, where ε is an arbitrary, small, positive quantity. In the range $[a, a + \varepsilon]$, Eq. (6.54) must hold, with $Y_2 = Y(a + \varepsilon)$ given by Eq. (6.57) and $r_1 = a + \varepsilon$. We obtain

$$Y_1 = \frac{Y(a+\varepsilon)}{\varepsilon/(a+\varepsilon) + p(3a\varepsilon + \varepsilon^2)/6 + qa}, \tag{6.58}$$

showing that Y_1 tends to infinity when ε tends to zero, although the total volume of the sheets remains finite. This "edge effect" occurs for any value of a/R. On the contrary, if the outer edge were built-in and the inner edge were loaded by the line load q, the edge effect would appear only when $a/R < 0.5$.

Megarefs (1967) gives a detailed treatment of the various geometries and boundary conditions generating this phenomenon. To face that situation and arrive at a realistic design, it is necessary to set an upper bound on the yield moment and to possibly provide special stiffeners where the "edge effect" would occur. These cases will be treated at the end of this section.

We have so far considered the dividing radii r_i as given and have obtained the optimal values Y_i and the minimum cost Γ as functions of these radii. We can now regard them as parameters and minimize the cost with respect to them. This can be done either by solving the system of equations $\partial\Gamma/\partial r_i = 0$ or by a minimization procedure, such as the "grid" method used by Lamblin (1975). However, this problem of *optimization of the layout* can be solved in a much quicker way, at least for some geometries and loadings, by applying an *optimality criterion*, as we shall show by the example of a *uniformly loaded, simply supported plate.*

The criterion of optimal location of the steps in yield moment is

$$\gamma_i - \gamma_{i+1} = \dot{w}'_{i+1}(r_i)M'_{i+1}(r_i) - \dot{w}'_i(r_i)M'_i(r_i), \tag{6.59}$$

where the prime (′) denotes differentiation with respect to r. Derivation of the criterion in Eq. (6.59) is due to Cinquini *et al.* (1977). A similar derivation is given in Section 7.4.1 for beams. It is important to point out that the criterion requires the average dissipation per unit Y in a typical ring to be *equal* to, not simply proportional to, γ_Y in that ring.

For our example we remark that, having denoted by $\dot{\varphi}_i$ the absolute value of a negative slope, we have $\dot{w}'_i = -\dot{\varphi}_i = -\alpha r_i/2$. Also, from Eq. (6.51),

$$M'_i = -pr/3 - pr_{i-1}^3/6r^2. \tag{6.60}$$

Hence,

$$M'_i(r_i) = -pr_i/3 - pr_{i-1}^3/6r_i^2, \qquad M'_{i+1}(r_i) = -pr_i/2.$$

For minimum volume we take $\gamma = Y$. From Eq. (6.52) we have

$$Y_i - Y_{i+1} = \frac{p(r_i^3 - r_{i-1}^3)}{6r_i},$$

and, with $\alpha = 1$, the criterion, Eq. (6.59), becomes

$$\frac{r_i^3 - r_{i-1}^3}{6r_i} = \frac{r_i r_{i+1}}{4} - \frac{r_i^2}{6} - \frac{r_{i-1}^3}{12r_i}.$$

Dividing both sides by r_i, we find

$$r_i\left[\frac{1}{3} - \frac{(r_{i-1}/r_i)^3}{12}\right] = \frac{r_{i+1}}{4}, \tag{6.61}$$

to be used with $r_0 = 0$ and $r_n = R$. Using successively Eqs. (6.61), (6.52), and (6.53), we derive the expression for the cost:

$$\Gamma = \sum_{i=1}^{n} Y_i \pi (r_i^2 - r_{i-1}^2). \tag{6.62}$$

Evaluating the "savings" G as before, we obtain Table 6.2, in the last column of which we have given for comparison the savings obtained when the rings are of equal breadth (Table 6.1). Table 6.3 gives the corresponding values of the yield moments Y_i of the rings. Comparison with Table 6.1 shows that the four-ring plate with optimized radii is as economical as the plate with six equal rings, and that two optimized rings are enough to achieve more than 40% of the maximum possible saving.

Table 6.2. Optimal Division Radii and Corresponding Savings

Number of rings	r_i/R $i = 1$	$i = 2$	$i = 3$	$i = 4$	$i = 5$	$G - \Gamma_c - \Gamma/\Gamma_c$ %	G/G_{max}	G for equal rings %
2	0.750					10.6	0.42	6.25
3	0.629	0.838				14.7	0.59	11.10
4	0.552	0.737	0.878			17.0	0.68	14.10
5	0.499	0.665	0.793	0.903		18.5	0.74	16.00
6	0.460	0.612	0.730	0.831	0.920	19.5	0.78	17.40
∞						25.0	1.00	25.00

Table 6.3. Dimensionless Optimal Yield Moments

Number of rings	Y_i/pR^2					
	$i = 1$	$i = 2$	$i = 3$	$i = 4$	$i = 5$	$i = 6$
1	0.1667					
2	0.1901	0.0963				
3	0.2021	0.1362	0.0685			
4	0.2096	0.1586	0.1062	0.0532		
5	0.2148	0.1732	0.1304	0.0871	0.0436	
6	0.2187	0.1895	0.1473	0.1107	0.0739	0.037

When the uniformly loaded, circular, sandwich plate is *built in*, the stress profile for any ring is formed of regimes *AB* or *BC* or both (Fig. 6.3) on the hexagon for that ring. In Fig. 6.8 we show the yield hexagons of all rings superimposed on the same diagram. We assume that Y_i decreases with r in a central region $0 < i < k$ and then increases, $k < i < n$. In a typical ring of the central region the stress profile goes with increasing r from *A* to some point between *A* and *B*, possibly reaching *B*. In the outer region it goes from *B*, or some point between *B* and *C*, to point *C*. In the

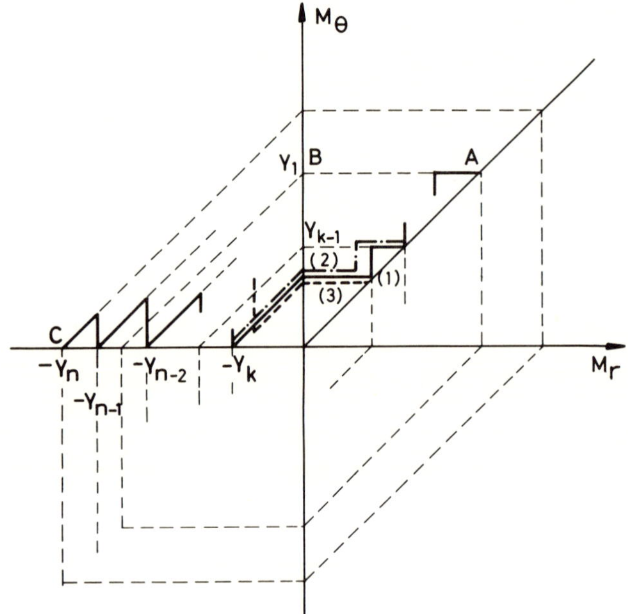

Figure 6.8. Stress profile for optimal built-in plates with piecewise-constant yield moment. (——) hinge circle at both ends; (—·—·—) hinge circle in r_k; (---) hinge circle in r_{k+1}.

intermediate ring k where the stress regime changes at radius r^* from AB to BC, three types of regimes can occur, as shown in Fig. 6.8. Either a positive hinge circle occurs at the inner radius of the ring and a negative hinge circle at the outer radius, or only one of these two hinge circles occurs. For each of these three cases an equation for r^* is obtained by combining equilibrium, continuity, and yield conditions. Since k is unknown, the solution is obtained by successive approximation by a numerical procedure implemented on an IBM 360 computer by Lamblin (1975).

For $n = 1$ (constant yield moment) we recover $Y = 0.089pR^2$ [Save and Massonnet (1972)] with $\Gamma_c = 0.089\pi pR^4$. The design in Eq. (6.39) with $a = R$ will be obtained by letting n tend to infinity, giving the absolute minimal cost $\Gamma_a = 0.058\pi pR^4$. Results for values of i from 1 to 6 are given in Table 6.4. Optimal values of the dividing radii have been obtained numerically by a grid method. We see that the saving increases very rapidly from $i = 2$ to $i = 5$ and then has a tendency to slow down strongly. Caution must be exercised in the numerical search for the optimal radii, because several local minima can exist [Lamblin and Guerlement (1976)].

Lamblin *et al.* (1985) give the solutions for various cases of loadings and boundary conditions for circular and annular plates. When an "edge

Table 6.4a. Optimal Division Radii

Number of rings	r_1/R	r_2/R	r_3/R	r_4/R	r_5/R	Gain G (%)
1						0
2	0.860					12.10
3	0.816	0.915				15.86
4	0.490	0.801	0.910			19.82
5	0.498	0.774	0.860	0.934		21.94
6	0.411	0.549	0.769	0.858	0.932	23.41
∞						34.00

Table 6.4b. Optimal Plastic Yield Moments

Number of rings	Y_1/pR^2	Y_2/pR^2	Y_3/pR^2	Y_4/pR^2	Y_5/pR^2	Y_6/pR^2
1	0.08900					
2	0.06565	0.11360				
3	0.05914	0.09150	0.12141			
4	0.08300	0.04300	0.07950	0.11190		
5	0.07945	0.03812	0.06627	0.09187	0.11587	
6	0.08971	0.06156	0.03240	0.06183	0.08769	0.11261

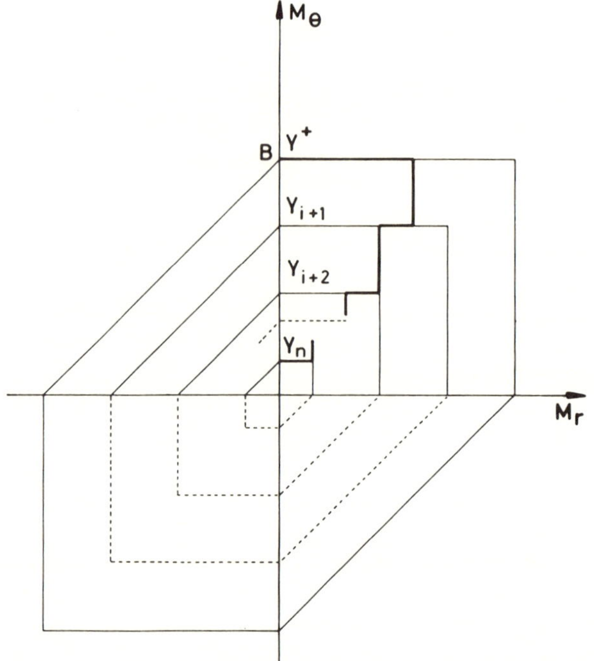

Figure 6.9. Stress profiles for optimal supported annular plate.

effect" is expected, *an upper bound Y^+ is set on the yield moment.* Obviously, Y^+ must be larger than the yield moment Y_c of the uniform-thickness plate, otherwise the plate could not support the load. On the other hand, for the condition $Y \leq Y^+$ to be relevant, the upper bound must be smaller than the maximum yield moment of the corresponding design without bound. Using the superscript * to denote, for comparison, elements pertaining to a design without upper bound, we must have

$$Y_c \leq Y^+ \leq Y^*_{\max}. \tag{6.63}$$

Consider, for example, an annular plate simply supported at the outer edge, free at the inner edge of radius a, and uniformly loaded. Assuming that Eq. (6.63) is satisfied and $M_{i+1} < M_i$, as in the absence of bounds, we choose the following stress profile (see Fig. 6.9):

a. in the central ring and possibly in some neighboring rings, $i = 1, \ldots, j$, where $Y = Y^+$, regime AB, from B to some intermediate point;
b. in the neighboring ring $j + 1$, regime AB between two intermediate points;
c. in all other rings, regime AB from A to an intermediate point as in the simply supported circular plate.

We then show that this stress profile is compatible with the optimality condition. The design is next obtained as usual from equilibrium, continuity, and yield conditions. Because the number j of rings where $Y = Y^+$ is unknown, the solution is obtained numerically by successive approximations starting from $j = 1$ and going up to the point where all conditions are satisfied. An example of results in the case of a plate made of four rings can be found in Fig. 6.10 and Table 6.5.

6.2.3. Assigned Limit Load—Sandwich von Mises Plate

Minimum-volume sandwich plates obeying the von Mises condition have been obtained by Freiberger and Tekinalp (1956), Eason (1960), and Reiss and Megarefs (1971) for circular and annular plates with various loadings and boundary conditions. For the simply supported plates, results coincide with those obtained with the yield condition of Tresca, Eqs. (6.29) and (6.33)–(6.37). Indeed, the plastic regime A used throughout the plate is common to both yield conditions, and the collapse mechanism, Eq. (6.28), that satisfies the optimality condition gives $\dot{\kappa}_r = \dot{\kappa}_\theta$, which is in agreement with the normality law applied to the von Mises criterion.

For the *built-in plate* subjected to circular loading, results differ only slightly from those obtained with the Tresca condition: $r_0 = 0.653R$ instead

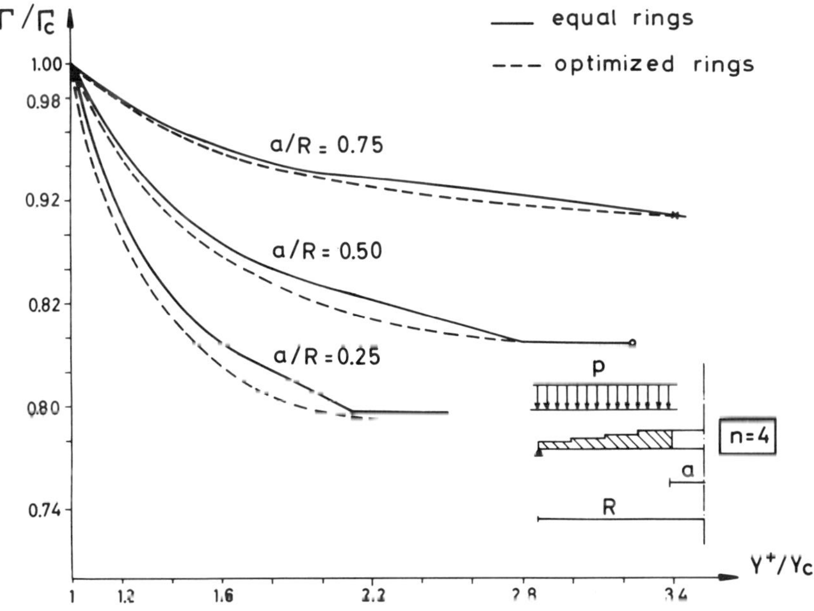

Figure 6.10. Variation of relative cost versus relative upper bound on Y.

Table 6.5. Four-Ring Plate: Optimal Radii and Yield Moments

	Y^+/Y_c						
$a/R = 0.25$	1	1.2	1.4	1.6	1.8	2	2.2
r_1/R		0.657	0.692	0.578	0.504	0.456	0.423
r_2/R		0.839	0.840	0.790	0.759	0.744	0.735
r_3/R		0.933	0.929	0.908	0.897	0.889	0.885
Y_1/pR^2	0.1875	0.2250	0.2625	0.3000	0.3375	0.3750	0.4124
Y_2/pR^2	0.1875	0.2250	0.1115	0.1368	0.1486	0.1545	0.1576
Y_3/pR^2	0.1875	0.0657	0.0653	0.0818	0.0913	0.0954	0.0974
Y_4/pR^2	0.1875	0.0292	0.0307	0.0390	0.0433	0.0462	0.0476
$G(\%)$	0	9.55	15.11	17.87	19.36	20.28	20.90
	Y^+/Y_c						
$a/R = 0.5$	1	1.2	1.6	1.8	2.2	2.4	2.8
r_1/R		0.906	0.773	0.730	0.673	0.654	0.626
r_2/R		0.947	0.876	0.863	0.843	0.836	0.827
r_3/R		0.976	0.946	0.940	0.932	0.929	0.925
Y_1/pR^2	0.1667	0.2000	0.2667	0.3000	0.3667	0.4000	0.4667
Y_2/pR^2	0.1667	0.0318	0.0657	0.0736	0.0819	0.0842	0.0870
Y_3/pR^2	0.1667	0.0188	0.0402	0.0437	0.0489	0.0506	0.0529
Y_4/pR^2	0.1667	0.0087	0.0189	0.0207	0.0233	0.0242	0.0254
$G(\%)$	0	5.46	11.3	12.8	14.8	15.6	16.551

of $2R/3$, and expressions for Y contain integrals that must be evaluated numerically. Under uniform load p the results of the computation are given in Table 6.6. Face-sheet volumes compare as follows: $0.111\pi pR^4/\sigma_0 H$ for von Mises and $0.117\pi pR^4/\sigma_0 H$ for Tresca. In all cases of loading and boundary conditions, the main features of the Tresca optimal plates are also present: edge effects and overall form of M_r and M_θ diagrams (except

Table 6.6. Built-in Uniformly Loaded Mises Plate: Distribution of Dimensionless Yield Moment

r/R	Y/pR^2	r/R	Y/pR^2
0	0.1067	0.653	0.00
0.1	0.1042	0.697	0.0135
0.2	0.0967	0.749	0.0290
0.3	0.0842	0.814	0.0482
0.4	0.0667	0.895	0.0740
0.5	0.0442	0.962	0.0970
0.6	0.0167	1.00	0.1112

for the discontinuity in M_θ). In view of the rather theoretical character of the designs considered, the supplementary mathematical difficulties connected with the condition of von Mises, and the small extra saving obtained, use of Tresca's condition gives sufficiently satisfactory results.

6.2.4. Assigned Limit Load—Sandwich Tresca Plate with Convex Cost Function

Let us assume that the specific cost is a piecewise-linear convex function of Y, and consider its simplest two-segment form:

$$\gamma = \alpha_1 Y, \qquad \alpha_1 > 0,\ 0 \le Y \le Y_0, \tag{6.64}$$

$$\gamma = \alpha_1 Y_0 + \alpha_2 (Y - Y_0), \qquad \alpha_2 > \alpha_1 > 0,\ Y_0 \le Y. \tag{6.65}$$

For a *uniformly loaded, simply supported plate*, according to what we know from Section 6.2.2, we may expect the following situations.

(a). If the load p is small, that is, if $M_r(0) = M_\theta(0) < Y_0$, the solution in Eq. (6.33) with $a = R$ will hold:

$$Y = \frac{p(R^2 - r^2)}{4}. \tag{6.66}$$

This is valid for

$$p < p^* = 4Y/R^2. \tag{6.67}$$

(b) For loads p larger than p^*, the optimality condition, Eq. (6.23), applied at regime A is, according to Eq. (6.14d) with positive curvatures,

$$\dot\kappa_r + \dot\kappa_\theta = \alpha_2, \qquad 0 \le r \le r_0, \tag{6.68}$$

$$\dot\kappa_r + \dot\kappa_\theta = \alpha_1, \qquad r_0 \le r \le R. \tag{6.69}$$

Here r_0 is the unknown radius separating the central region, where $Y > Y_0$, from the external region, where $Y < Y_0$.

Because the curvature rates must be equal at the center by isotropy, we have, as in Section 6.2.2, $\dot\kappa_r = \dot\kappa_\theta = \alpha_2/2$ throughout the central region. At the boundary $r = r_0$ we must have continuity of $\dot\kappa_\theta$ in order to avoid concentrated dissipation arising from slope discontinuity in the collapse mechanism. Hence, at $r = r_0$ in the external region, we have $\dot\kappa_\theta = \alpha_2/2$ and $\dot\kappa_r = \alpha_1 - \alpha_2/2$ in order to satisfy the optimality condition, Eq. (6.69). Positiveness of $\dot\kappa_r$ will occur only if $\alpha_2 < 2\alpha_1$, a condition that we temporarily assume to be satisfied. With the initial conditions just found at $r = r_0$, the condition in Eq. (6.69), and the relations in Eq. (6.3), the collapse mechanism can be obtained in the external region, in which it should be noted that both $\dot{\kappa}_r$ and $\dot{\kappa}_\theta$ vary with r. In any case regime A prevails in the whole plate and the design in Eq. (6.66) remains valid as long as $\alpha_2 < 2\alpha_1$. Using $Y = Y_0$ at $r = r_0$ in this relation, we obtain $r_0/R = [1 - 4Y/pR^2]^{1/2}$.

(c) When $\alpha_2 > 2\alpha_1$, the two regions of the mechanism considered in (b) must be connected by an intermediate region $r_0 \le r \le r_1 < R$, in which $\dot{\kappa}_\theta$ varies from $\alpha_2/2$ to $\alpha_1 < \alpha_2/2$, and $\dot{\kappa}_r$ vanishes. This conical part of the mechanism is obtained with $Y = Y_0$ throughout the region; γ_Y varies continuously from α_2 to α_1, enabling regime AB to be used. With these regimes we integrate the equation of equilibrium in the three regions and use the boundary conditions and the continuity of M_r at r_0 and r_1 to obtain

$$\begin{aligned} &Y = M_r = M_\theta = Y_0 + \frac{p(r_0^2 - r^2)}{4}, && 0 \le r \le r_0, \\ &Y = M_\theta = Y_0, \qquad M_r = Y_0 - \frac{p[r^2 - (r_0^3)/r)]}{6}, && r_0 \le r \le r_1, \\ &Y = M_r = M_\theta = Y_0 + p[r_1^2 + 2r_0^3/r_1 - 3r^2], && r_1 \le r \le R. \end{aligned} \tag{6.70}$$

In the central conical mechanism we have $\dot{\kappa}_\theta = C/r$, where C is a positive constant, $\dot{\kappa}_\theta(r_0) = \alpha_2/2$, and $\dot{\kappa}_\theta(r_1) = \alpha_1$. Hence, $r_0/r_1 = 2\alpha_1/\alpha_2$ and, with $M_r(R) = 0$, we obtain

$$\frac{r_1}{R} = \left\{\frac{3[1 - p^*/p]}{1 + 2(2\alpha_1/\alpha_2)^3}\right\}^{1/2}, \tag{6.71}$$

where $p^* = 4Y/R^2$. Obviously, we must have $p > p^*$ and $r_1 < R$, which requires

$$1 < \frac{p}{p^*} < \frac{1.5}{1 - (2\alpha_1/\alpha_2)^3}. \tag{6.72}$$

When this condition is satisfied, the design is given by Eq. (6.70) with r_1 evaluated from Eq. (6.71) and $r_0 = 2\alpha_1 r_1/\alpha_2$.

For greater values of p/p^*, only the two first lines of Eq. (6.70) are valid; when $M_r(R) = 0$ in the second line, r_0 is determined to be

$$r_0/R = [1 - 3p^*/2p]^{1/3}. \tag{6.73}$$

Figures 6.11 and 6.12 give the radii r_0 and r_1 and the nondimensional cost

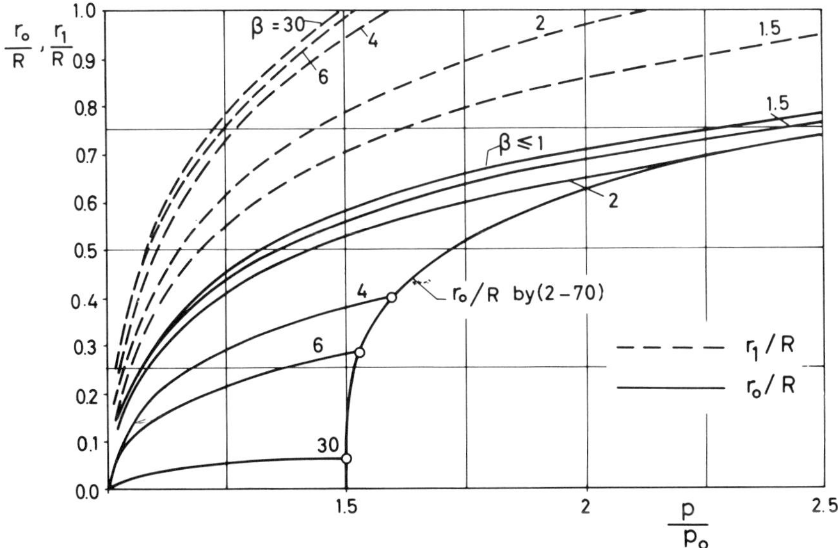

Figure 6.11. Supported circular plate with convex cost function: dimensionless division radii versus dimensionless uniform load.

$\Gamma/\pi R^2\alpha_1 Y_0$, respectively, as functions of p/p^* with curves labeled in values of $\beta \equiv (\alpha_2 - \alpha_1)/\alpha_1$. The preceding results were obtained by Marçal and Prager (1964) in a somewhat less direct manner.

If we let α_1 tend to zero, the specific cost vanishes for all values of Y not larger than Y_0; therefore, these yield moments can all be increased to

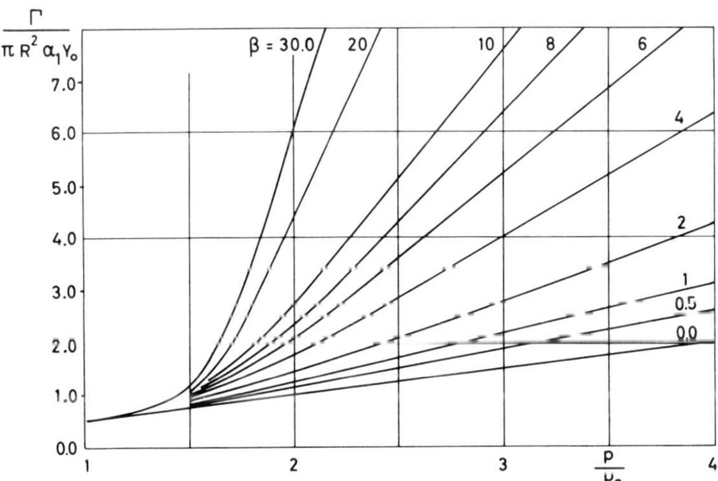

Figure 6.12. Supported circular plate with convex cost function: dimensionless cost versus dimensionless uniform load.

Y_0 without additional cost. We are in the same situation as for a linear cost with a lower bound $Y \equiv Y_0$ on Y. The load must be larger than the collapse load $p_c = 6Y_0/R^2$ of the plate with constant Y equal to Y_0; otherwise, this plate is optimal. Then, since $p_c = 1.5p^*$ and $\alpha_1 = 0$, Eq. (6.72) will never be satisfied. The design will thus be given by the first two lines of Eq. (6.70), $r_1 = R$, and r_0 given by Eq. (6.73).

The optimality condition, Eqs. (6.68) and (6.69), has been used by Marçal (1967) for the optimal design of the *uniformly loaded built-in plate.* The solution can be found in a manner very similar to that used above, with the supplementary difficulty that more regimes must be considered: A and C, with Y possibly smaller or larger than Y_0 and AB and BC associated with $Y = Y_0$. Numerical solution of the corresponding equations turns out to be necessary.

6.2.5. Assigned Limit Load—Minimum-Volume Solid Tresca Plate

It is convenient to first consider a plate with piecewise-constant yield moment. Because $Y_i = \sigma_0 t_i^2/4$, for a solid ring i of thickness t_i, minimum volume will be obtained by minimizing $\Sigma_{i=1}^n A_i(Y_i)^{1/2}$. Hence, $\gamma = Y^{1/2}$ and $\gamma_Y = 0.5Y^{-1/2}$. Note that the specific cost is a *concave* function of the design variable Y. Consequently, the optimality criterion is only a *necessary* condition for a *local* optimum.

Let us first study a *simply supported* circular or annular plate transversely loaded in one direction only. The procedure of Section 6.2.2.2 can be repeated completely, substituting only $0.5Y_i^{-1/2}$ for the constant α. The optimality condition is

$$\dot{\varphi}_i = 0.25r_iY_i^{-1/2}, \tag{6.74}$$

from which we obtain

$$\dot{\varphi}_{i+1} = \frac{\dot{\varphi}_i(r_{i+1}Y_i^{1/2})}{r_iY_{i+1}^{1/2}}. \tag{6.75}$$

The relation in Eq. (6.75) can be satisfied within the framework of the assumptions $r_{i+1} > r_i$, $\dot{\varphi}_{i+1} > \dot{\varphi}_i$, $Y_{i+1} < Y_i$. We conclude that the optimal designs Y_i ($i = 1, \ldots, n$) for these solid plates are identical to the minimum-volume designs of the corresponding sandwich plates, *provided the division radii are given.* Obviously, identical designs Y_i give different volumes for the solid and sandwich plates and, hence, different sets of optimal division radii. For example, in the case of the uniformly loaded, circular plate made of two equal rings, we have $Y_1 = 0.1875pR^2$ and $Y_2 = 0.1458pR^2$. But if we optimize the division radius r_1, we find $r_1 = 0.75R$, yielding $Y_1 = 0.1901pR^2$ and $Y_2 = 0.0963pR^2$ for the sandwich plate, whereas minimization of the volume of a solid plate with respect to r_1 gives $r_1 = 0.806R$ and corresponding yield moments $Y_1 = 0.1877pR^2$ and $Y_2 = 0.0794pR^2$ [Hopkins and Prager (1955)].

If the number of rings increases without limit, we recover the corresponding functions $Y(r)$ obtained for sandwich plates, but the saving with respect to the constant-thickness plate is different: for example, it is 18% for the uniformly loaded solid plate, as compared to 25% for the corresponding sandwich plate. Examples of circular plates with 2 to 6 rings, subjected to circular or annular loadings, can be found in Lamblin *et al.* (1985).

As we emphasized at the beginning of this section, the designs considered here furnish relative optima. Hence, it may be expected that other designs exist that exhibit smaller costs. For example, a two-ring plate uniformly loaded and simply supported can be made of a very narrow, strong external ring in which the internal ring is "built in." As shown by Kozlowski and Mroz (1969), if one is prepared to accept $r_1 > 0.97R$ and $Y_2 > 39.5\,Y_1$ (!), a volume smaller than that of the classical design of Hopkins and Prager (1955) can be achieved. If one recalls that this latter design saves only 6.87% of the volume with respect to the constant thickness plate, whereas transforming the simple support of this plate into a built-in edge multiplies its limit load by 1.88 [see Save and Massonnet (1972)], the result of Kozlowski and Mroz was to be expected, even with some cost assigned to the fixed support so obtained.

Another way of decreasing the volume is to use stiffeners, as shown by these authors in the same paper. The volume can even be made to vanish theoretically with an infinite number of infinitely thin and infinitely high stiffeners, as first pointed out by Brotchie (1967). In order to take advantage of this possibility, we must, on the one hand, admit the use of functions with an unlimited number of discontinuities and, on the other hand, set an upper bound on the height of the stiffeners. Such a procedure is discussed by Rozvany *et al.* (1982) and applied to the simply supported, uniformly loaded, circular Tresca plate. For example, setting the upper bound $Y^+ = 3\,Y_c$, where Y_c is the yield moment of the solid plate with constant thickness and the same limit load, one finds that the design with the smallest volume consists of a solid central region ($0 < r < 0.5014R$) with a variable thickness, such that $Y(r) < Y^+$, surrounded by an infinity of radial beams (stiffeners without plate) of maximum height $t = (4Y^+/\sigma_0)^{1/2}$ and variable breadth, such that $Y(r) = M_r(r)$. The saving obtained is 35%, compared to 18% given by the design with a smooth function $Y(r)$. However, this saving of 35% is likely to be reduced by the need to approximate the optimal design by a finite number of radial beams and, for application of the load, a nonvanishing plate, even in the outer region.

Now consider the uniformly loaded *built-in* plate. Plastic regimes A and C (Fig. 6.3) will be used, as for the corresponding sandwich plate, but the radius r_0 separating the regions of application of these two regimes is different because it now depends on the type of loading. Indeed, it is determined from kinematic continuity conditions applied to a collapse mechanism satisfying the optimality criterion. But the latter contains $Y(r)^{-1/2}$, which in turn depends on the type of loading. Hence, the optimal designs for corresponding sandwich and solid plates are not identical any more. A successive approximation procedure similar to that used for the sandwich plate could be applied. In order to avoid these supplementary lengthy and tedious calculations, we shall satisfy ourselves with the approximations given by the optimal designs Y_i $(i = 1, \ldots, n)$ of the sandwich plates. They are able to support the load, but, as a rule, do not satisfy the optimality criterion of the solid plate and, hence, have a volume in excess of the (relative) minimum.

The goodness of these approximations to the optimal designs will be evaluated by referring to the exact, minimum-volume design obtained for continuous, smooth variation of Y with r by Onat *et al.* (1957). For circular loading p over a central circle of radius a, they have found that (a) when $a < r_0$,

$$
\begin{aligned}
Y(r) &= \frac{p(a^2 - r^2)}{4} + pa^2 \frac{\ln(r_0/a)}{2}, && 0 \le r \le a, \\
&= \frac{pa^2 \ln(r_0/r)}{2}, && a \le r \le r_0, \\
&= \frac{pa^2(r - r_0)}{2r}, && r_0 \le r \le R;
\end{aligned}
\tag{6.76}
$$

(b) when $a > r_0$,

$$
\begin{aligned}
Y(r) &= \frac{p(r_0^2 - r^2)}{4}, && 0 \le r \le r_0, \\
&= \frac{p(r^3 - r_0^3)}{6r}, && r_0 \le r \le a, \\
&= \frac{p(3a^2 r - r_0^3 - 2a^3)}{6r}, && a \le r \le R.
\end{aligned}
\tag{6.77}
$$

The radius r_0 is given in Fig. 6.13 as a function of a/R.

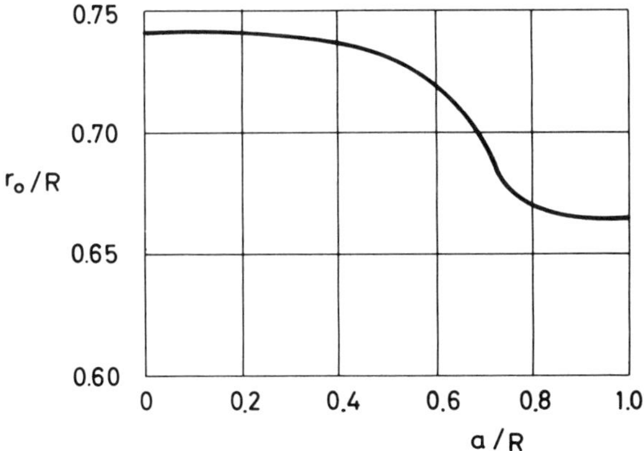

Figure 6.13. Built-in solid circular plate: dimensionless division radius for stress profile versus dimensionless radius of loaded area.

Uniform loading corresponds to $a = R$, Eq. (6.77), with $r_0 = 0.664R$. The volume is evaluated by integration, with $t = 2(Y/\sigma_0)^{1/2}$, and the saving is found to be 23%. Tables 6.7 and 6.8 show that the design of approximate optimality (i.e., yield moments of the sandwich plate) very rapidly gives a large amount of the maximum saving, either with the optimal division radii of the sandwich plate or upon separate optimization of these radii by the grid method for the solid plate, a computation that thus proves not useful for practical application.

6.2.6. Assigned Elastic Compliance—Sandwich Plate

Consider a *simply supported circular sandwich plate* that is to be designed for minimum volume of the face sheets with an upper bound $\mathscr{C}$ set on the

Table 6.7. Built-In Solid Plate, Uniformly Loaded. $r_1, r_2, \ldots, r_{n-1}$ are Optimal Division Radii for Sandwich Plate

	Number of rings					
	1	2	3	4	5	∞
r_1/R		0.86	0.816	0.490	0.498	
r_2/R			0.915	0.801	0.774	
r_3/R				0.910	0.860	
r_4/R					0.934	
r_5/R						
$G(\%)$	0	6.95	9.26	11.9	13.1	23

Table 6.8. Built-In Solid Plate, Uniformly Loaded. Optimal Division Radii Obtained by the Grid Method

		Number of rings					
		1	2	3	4	5	∞
	x_1		0.831	0.507	0.517	0.448	
	x_2			0.826	0.763	0.573	
$x_i = r_i/R$	x_3				0.886	0.763	
	x_4					0.885	
	x_5						
$G(\%)$		0	7.1	9.33	12.15	13.19	23

elastic compliance. We take the bending stiffness $B = EH^2t/4$ both as the design variable and as specific cost, since it is porportional to the thickness t. We know from Section 3.3 that, in the absence of bounds on B, the optimal design exhibits a deflection field with constant specific strain energy per unit stiffness.

As was pointed out in Section 6.2.3, the design with Eqs. (6.29) and (6.33)-(6.37) is also valid for the von Mises yield condition. Moreover, the state of stress is homogeneous because regime A only has been used throughout the plate: $M_r = M_\theta = Y = \sigma_0 Ht/4$. Hence, this design is, except for a scalar positive coefficient, the optimal design for assigned elastic compliance *for any value of Poisson's ratio* ν.

Consider first a *uniform load* p. We have

$$B = fp(R^2 - r^2) \tag{6.78}$$

and

$$w = \frac{w_0(R^2 - r^2)}{R^2}, \tag{6.79}$$

where the positive constant factor f and the deflection w_0 at the center must be related to the assigned compliance $\mathscr{E}$ and the load p. Indeed, we must have $\int pw\, dA = \mathscr{E}$, giving for constant p, from Eq. (6.79),

$$w_0 = 2\mathscr{E}/p\pi R^2. \tag{6.80}$$

But we also have $2\int \phi\, dA = \mathscr{E}$. With ϕ given by Eq. (6.9), in which we let $\kappa_r = \kappa_\theta = 2w_0/R^2$ from Eq. (6.79), we obtain

$$8w_0^2 \int B\, dA = \mathscr{E}R^4(1 - \nu). \tag{6.81}$$

From Eqs. (6.81) and (6.80) we deduce the cost $\Gamma = \int B\,dA$ (or the volume $V = 4\Gamma/EH^2$):

$$\Gamma = \frac{p^2\pi^2 R^8(1-\nu)}{32\mathscr{E}}. \tag{6.82}$$

Evaluating $\Gamma = \int B\,dA$, using Eq. (6.78) for B, and comparing it with Eq. (6.82) gives

$$f = \frac{\pi p R^4(1-\nu)}{16\mathscr{E}};$$

hence,

$$B = \frac{\pi p^2 R^4(1-\nu)(R^2 - r^2)}{16\mathscr{E}}. \tag{6.83}$$

For this design to make sense it must remain elastic under the applied load. We leave it to the reader to show that this condition is satisfied when $\mathscr{E}/p < \pi R^4(1-\nu)\sigma_0/2HE$.

Other downward axisymmetric loadings can be treated in the same manner. For example, in the case of *annular loading* over $a \leq r \leq R$ we find

$$w_0 = \frac{\mathscr{E}}{\pi p a^2(1 - a^2/2R^2)}, \qquad \Gamma = \frac{\pi^2 p^2 a^4(2R^2 - a^2)^2(1-\nu)}{32\mathscr{E}},$$

and

$$f = \frac{\pi p(2R^2 - a^2)^2(1-\nu)a^2}{16\mathscr{E}(3a^2 - 2R^2)},$$

if B is to be given by the right-hand sides of Eq. (6.33) multiplied by $4f$.

For the *built-in plate* the state of stress is no longer homogeneous. Hence, we are only allowed to take advantage of the designs obtained with the von Mises yield condition and use them for an elastically incompressible material ($\nu = 1/2$). Under uniform load such a design is given in Table 6.6. We assume that the assigned compliance $\mathscr{E}$ is small enough to ascertain elastic behavior under the applied load p. Because the moment field $M_r = \sigma_r Ht/2$, $M_\theta = \sigma_0 Ht/2$ is the field of optimal plastic design, and thus well defined, the thickness t must be amplified by a factor $f > 1$ in order that $\sigma_r^2 + \sigma_\theta^2 - \sigma_r\sigma_\theta < \sigma_0^2$. Furthermore, this must be done in order to give a constant value C^2 to the specific complementary energy $(\sigma_r^2 + \sigma_\theta^2 -$

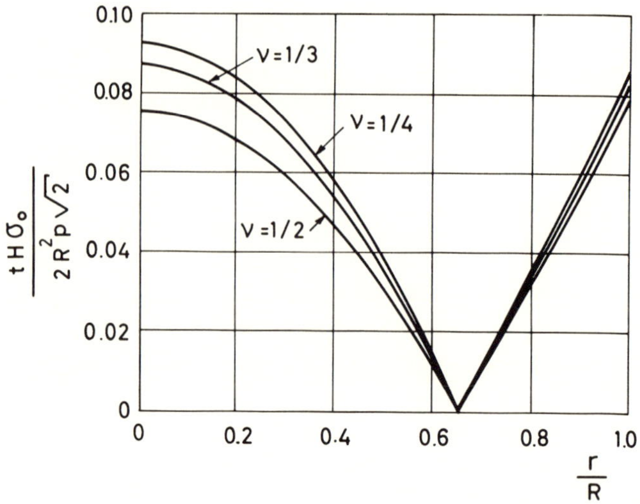

Figure 6.14. Optimal designs of built-in circular plates for assigned elastic compliance.

$2\nu\sigma_r\sigma_\theta)/2E$ corresponding to the given compliance. This is achieved with $f = \sigma_0/C(2E)^{1/2}$, since $\nu = 0.5$, and C is related to $\mathscr{E}$ by $2\int \phi\, dA = \mathscr{E}$. With $\phi/t = C^2$, we obtain $2C^2V = \mathscr{E}$. But V is the product of f and the volume $0.111p\pi R^4/\sigma_0 H$ of the plastic optimal design, as given in Section 6.2.3. Hence, $C = H\mathscr{E}(2E)^{1/2}/0.222\pi pR^4$ and

$$f = \frac{0.111\pi pR^4\sigma_0}{EH\mathscr{E}}. \tag{6.84}$$

When the material is elastically compressible ($\nu \neq 1/2$), solutions by Taylor (1969) can be used if one is prepared to limit the compliance by assigning an upper bound on the specific strain energy per unit of thickness. The thickness distributions for various values of Poisson's ratio are given in Fig. 6.14. They give the minimum volume of the face sheets of uniformly loaded, built-in, circular plates subjected to an upper bound $\sigma_0^2/2E$ on the "stress intensity" defined by the magnitude of the elastic strain energy per unit volume of face sheet.

According to the discussion in the last three paragraphs of Section 6.1, these designs exhibit minimum volume for an assigned compliance $\mathscr{E}$ related to $\sigma_0^2/2E$ by $\mathscr{E} = V\sigma_0^2/E$, where V is the (minimum) volume obtained.

6.2.7. Other Problems for Circular Elastic Plates

Consider first a simply supported *solid* circular plate of radius R, under a uniform load p, to be designed for minimum volume and prescribed upper bound kR on the central deflection. The optimality criterion arising from stationariness of the cost is that the specific mutual strain energy per unit bending stiffness be proportional to the derivative of the specific cost with respect to the bending stiffness as shown by Eq. (3.142) for beams in bending. In the present case the thickness $t(r)$ is the unknown, the bending stiffness is $B \equiv t^3E/12$, and the specific cost is $\gamma \equiv t$. The specific mutual complementary energy may be substituted for the corresponding strain energy because we deal with linear elasticity. It is given by Eq. (1.10), provided we use r and θ for both j and k, and substitute for Q_j the bending moments M_r, M_θ under the given load, and for Q_k the bending moments $\bar{M}_r$, $\bar{M}_\theta$ caused by a unit central concentrated load. From comparison with Eq. (6.10) we have $\tilde{e}_{11} = \tilde{e}_{22} = 1$ and $\tilde{e}_{12} = \tilde{e}_{21} = -\nu$. The specific complementary mutual strain energy $\bar{\psi}$ is thus given by

$$\bar{\psi} = \{M_r\bar{M}_r + M_\theta\bar{M}_\theta - \nu(M_r\bar{M}_\theta + M_\theta\bar{M}_r)\}/2B.$$

Hence, the optimality condition is

$$(1/t^4)[M_r\bar{M}_r + M_\theta\bar{M}_\theta - \nu(M_r\bar{M}_\theta + M_\theta\bar{M}_r)] = \alpha. \tag{6.85}$$

In Eq. (6.85) the positive constant α is related to the assigned central deflection by applying the theorem of Betti (see Section 1.2.5) to the given load and the central unit load:

$$kR = \int (\bar{\psi}/t)t\,dA.$$

Substituting the above expression for $\bar{\psi}$ and using Eq. (6.85) in the last equation result in

$$kR = 12\alpha V/E. \tag{6.86}$$

The other governing equations are the equation of equilibrium, Eq. (6.4), and the equation of compatibility in the linear elastic range, relating the bending moments by using Hooke's law, Eq. (6.5), in Eqs. (6.22).

These last two equations hold for the functions $\bar{M}_r$ and $\bar{M}_\theta$ as well as for M_r and M_θ. With Eq. (6.85) we thus have a system of five equations for the five unknown functions M_r, M_θ, $\bar{M}_r$, $\bar{M}_\theta$, and t, with boundary conditions $M_r(R) = \bar{M}_r(R) = 0$, $M_r(0)$ finite, and $M_r(0)$ finite. The constant α is given by Eq. (6.86). This set of equations has been solved by Erbatur and Mengi (1977a), who used a numerical procedure based on iteration and the method of finite differences. They extended their solution (1977b) to take into account the weight of the plate. Figures 6.15 and 6.16 summarize the results for $\nu = 0.2$ using loading parameters $\beta_1 = F/12p$, $\beta_2 = R\delta/p$,

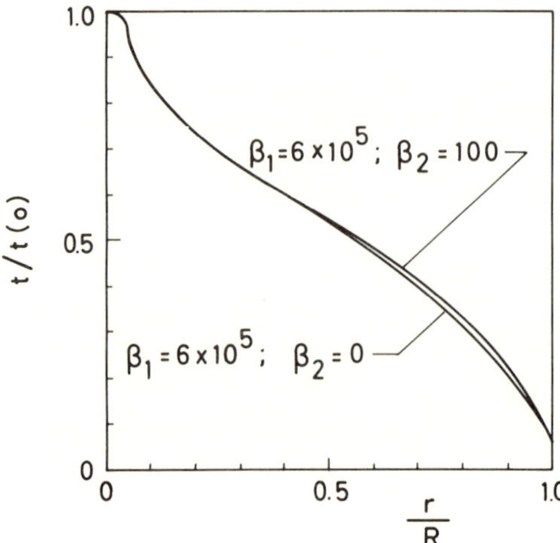

Figure 6.15. Optimal dimensionless profiles for solid supported plates for assigned central elastic deflection.

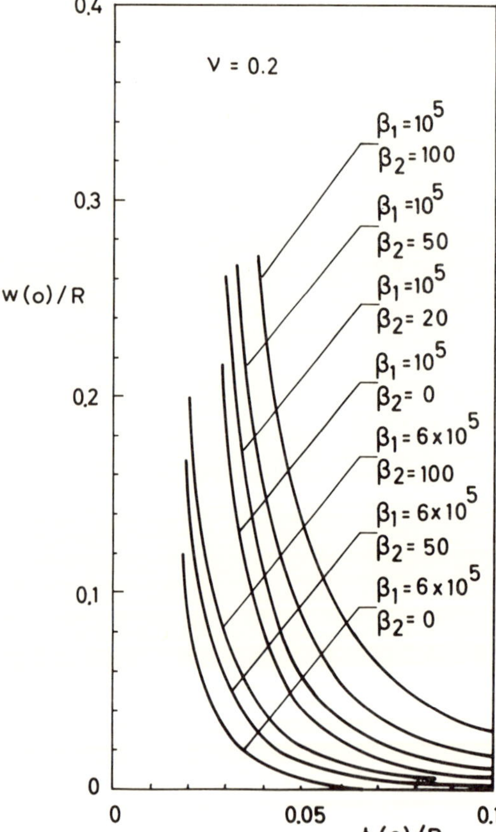

Figure 6.16. Dimensionless central deflection versus dimensionless central thickness.

where δ is the weight per unit volume. The optimum dimensionless profiles with and without the plate's weight show very little difference.

Similar problems for circular and annular plates with various support conditions and upper bounds on deflections or rotations have been treated by Lamblin *et al.* (1980), where the function $t(r)$ is assumed to be piecewise constant. Since the division radii are regarded as given, each annular region of constant thickness is used as a "natural" finite element. The relevant equations, including the optimality criterion, are then formulated in matrix form and solved by an iterative procedure.

The circular solid plate of minimum volume for assigned elastic compliance has been considered by Reiss (1976), using the "statical" approach. It is not possible to give here a detailed treatment of this problem or of other optimal design problems for circular plates. We mention the works of Olhoff (1970), for an assigned lower bound on the first natural frequency, and Niordson (1981), on a combination of upper bound on compliance and a constraint on the derivative of the thickness function. In all of these problems the mathematical difficulties, as a rule, necessitate lengthy numerical solutions. A general discussion of such problems is given in Chapter 7.

6.3. Other Plate Problems

The mathematical difficulty of most optimal design problems for noncircular plates precludes, as a rule, the hope of achieving a closed-form solution, no matter what the behavioral constraints, cost function, and design variable may be. A noteworthy exception, however, is the minimum-volume elliptic sandwich plate under assigned uniform pressure at plastic collapse, which was solved by Prager (1955) for simple support and by Shield (1960) for built-in boundary. In the first case, if the plate boundary has the equation $x^2/a^2 + y^2/b^2 = 1$, the optimal distribution of the yield moment Y is simply

$$Y = \frac{pa^2b^2(1 - x^2/a^2 - y^2/b^2)}{2(a^2 + b^2)}. \tag{6.87}$$

For the built-in plate, numerical solutions of the relevant equations are necessary.

Square and rectangular solid elastic plates with constraints on natural frequency, stress intensity, deflection, and bounds on thickness are treated (by numerical procedures) in the book of Haug and Arora (1979) for both static and time-varying loads. It should be emphasized that these solutions are *local* optima, as are the solutions for vibrating rectangular plates by

Olhoff (1974) and for limited static deflection by Armand and Lodier (1978), as these authors point out [see Olhoff (1975)]. Minimum volume of solid plates of general shape for assigned load factor at plastic collapse has been considered by Cinquini (1981), who treated an example of a simply supported, square plate subjected to a centrally concentrated force.

6.4. Cylindrical Shell under Axisymmetric Loads

6.4.1. Basic Relations

We consider a circular cylindrical thin shell for which the Kirchhoff–Love assumptions of shell theory apply. The system of cylindrical coordinates is shown in Fig. 6.17a, where positive directions of displacement u, v, w of the midsurface are indicated. Positive internal forces (per unit of length) and applied forces (per unit area) are shown in Fig. 6.17b in the particular case of axial symmetry and zero longitudinal force.

6.4.1.1. Kinematics. Under general loading the components of the strain tensor and of the tensor of the curvature changes of the midsurface are [see, for example, Timoshenko and Woinowsky-Krieger (1959, p. 512)]

$$\varepsilon_x = \frac{\partial u}{\partial x}, \qquad \varepsilon_\theta = \frac{1}{R}\frac{\partial v}{\partial \theta} + \frac{w}{R}, \qquad \gamma_{x\theta} = \frac{1}{R}\frac{\partial u}{\partial \theta} + \frac{\partial v}{\partial x}, \tag{6.88a}$$

$$\kappa_x = +\frac{\partial^2 w}{\partial x^2}, \qquad \kappa_\theta = \frac{1}{R^2}\left(\frac{\partial v}{\partial \theta} + \frac{\partial^2 w}{\partial \theta^2}\right), \qquad \kappa_{x\theta} = \frac{1}{R}\left(\frac{\partial v}{\partial x} + \frac{\partial^2 w}{\partial x\,\partial \theta}\right). \tag{6.88b}$$

If the loading is axisymmetric ($p_\theta = 0$ and p_x, p_r independent of θ), all displacements v and all derivatives with respect to the coordinate θ vanish, and Eqs. (6.88) reduce to

$$\varepsilon_x = \partial u/\partial x, \qquad \varepsilon_\theta = +w/R, \qquad \gamma_{x\theta} = 0, \tag{6.89a}$$

$$\kappa_x = +\partial^2 w/\partial x^2, \qquad \kappa_\theta = 0, \qquad \kappa_{x\theta} = 0. \tag{6.89b}$$

Equations (6.89) can be also regarded as relating velocities and generalized strain rates.

6.4.1.2. Statics. For general loading, the static variables (generalized stresses) are the axial forces N_x, N_θ, the in-plane shear $N_{x\theta}$, the bending moments M_x, M_θ, and the twisting moment $M_{x\theta}$. *If the loading is axisymmetric*, longitudinal and circumferential directions are principal axes by symmetry, and hence $N_{x\theta} = M_{x\theta} = 0$. Moreover, M_θ is a reaction because

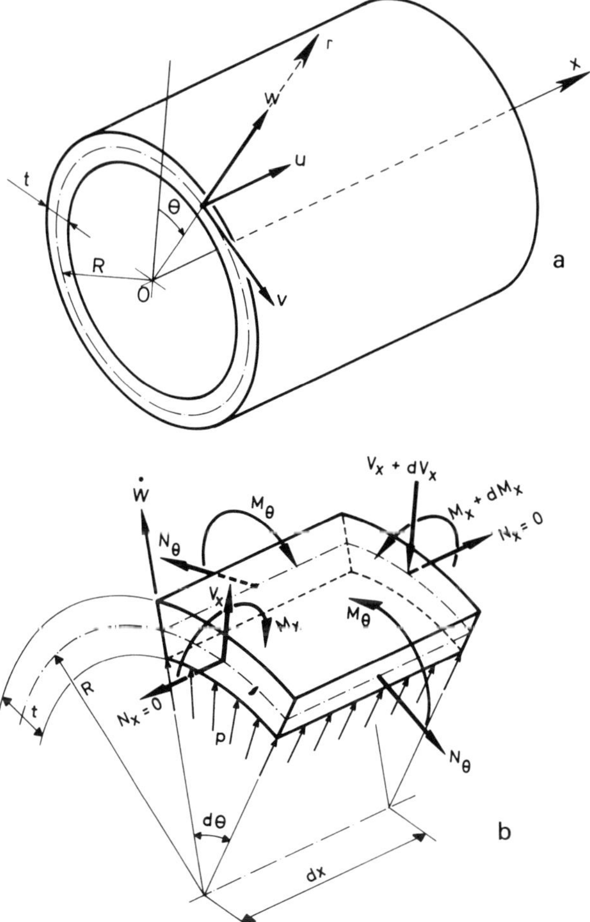

Figure 6.17. (a) Coordinates and displacements in cylindrical shell. (b) Cylindrical shell element.

$\kappa_\theta = 0$. Indeed, because $v = 0$ and w is independent of θ, M_θ does zero work in the deformation of a shell element. The generalized stresses reduce to N_x, N_θ, and M_x (the transverse shear V_x being also a reaction, as in the case of plates). In the absence of longitudinal forces ($p_x = 0$, $N_x = 0$), the equation of equilibrium is

$$\frac{d^2 M_x}{dx^2} + \frac{N_\theta}{R} + p_r = 0. \tag{6.90}$$

6.4.1.3. Linear Elasticity. In the axisymmetric case Eqs. (6.6) hold, with the expressions of Eq. (6.89b) for the curvatures. In addition, the

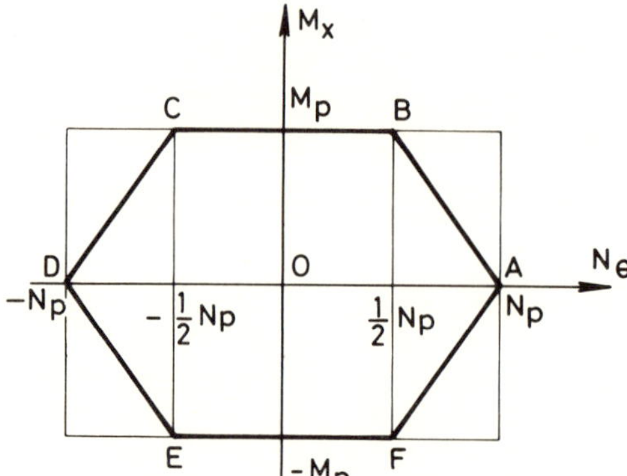

Figure 6.18. Hodge's yield hexagon for cylindrical shell.

elastic relations for the membrane forces are, after substituting x for r,

$$N_x = \frac{Et}{1-\nu^2}(\varepsilon_x + \nu\varepsilon_\theta), \qquad N_\theta = \frac{Et}{1-\nu^2}(\varepsilon_\theta + \nu\varepsilon_x), \tag{6.91}$$

where t is the thickness of the solid shell or the sum of the thicknesses of the face sheets of the sandwich shell.

6.4.1.4. Perfect Plasticity. Because the generalized stresses have different dimensions (some are moments per unit arc length of the middle surface, and others are forces per unit length), the yield strength of a shell element must be described by a locus depending on the values of the fully plastic moment M_p and the fully plastic force N_p. The value of M_p is given by the right-hand side of Eqs. (6.11) or (6.12), whereas

$$N_p = t\sigma_0. \tag{6.92}$$

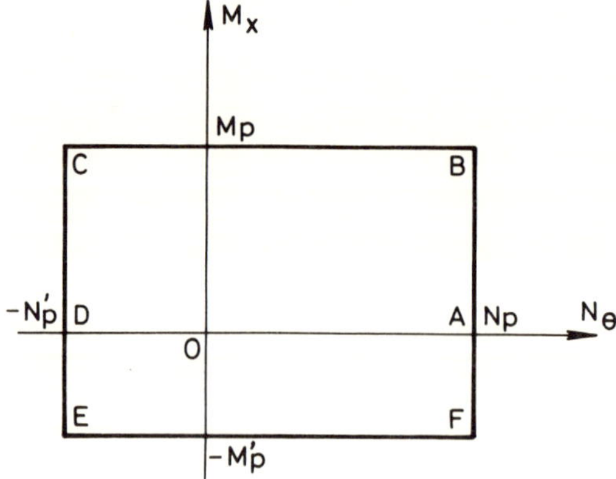

Figure 6.19. Yield condition for reinforced concrete cylindrical shell.

The examples to be treated will use either the yield locus of a Tresca sandwich shell without axial force, called Hodge's hexagon (Fig. 6.18), or the analogous yield locus for reinforced concrete (Fig. 6.19). The reader interested in the derivation of these two curves and in other yield conditions for shells is referred to Save and Massonnet (1972).

6.4.2. Assigned Limit Load—Sandwich Tresca Cylinder

6.4.2.1. As we did in Section 6.2.2.1 for plates, we first want to design a sandwich Tresca cylinder for minimum volume of the face sheets with no bound on the thickness $t(x)$ [see Onat and Prager (1956), Freiberger (1956), and Shield (1960c)]. The thickness t, to which both M_p and N_p are proportional, will be our design variable. With specific cost $\gamma = t$, the optimality condition is $D/t = c^2$ or

$$\frac{M_x\dot{\kappa}_x + N_\theta\dot{\varepsilon}_\theta}{t} = c^2. \tag{6.93}$$

It is easily shown that only corner regimes may be used with Eq. (6.93). Indeed, regimes BC (or EF) imply $\dot{\varepsilon}\theta = 0$ and, hence, by Eq. (6.89), $\dot{w} = 0$.

Regime AB (and similar ones) implies $\dot{\kappa}_x/\dot{\varepsilon}_\theta = 1/H$ by the normality law. The equation of AB is

$$N_\theta/N_p + M_x/2M_p = 1 \qquad \text{or} \qquad N_\theta + M_x/H = \sigma_0 t.$$

Hence, Eq. (6.93) can be written

$$(1/t)(M\dot{\varepsilon}_\theta/H + N_\theta\dot{\varepsilon}_\theta) = c^2 \qquad \text{or} \qquad (1/t)\dot{\varepsilon}_\theta(M_x/H + N_\theta) = c^2$$

and, finally, as $\dot{\varepsilon}_\theta = c^2/\sigma_0$. This last condition implies that $\dot{w} = Rc^2/\sigma_0$ and, hence, $\dot{\kappa}_x = 0$, which contradicts $\dot{\kappa}_x = \dot{\varepsilon}_\theta/H$.

Let L be the length and R the radius of a *simply supported* cylinder subjected to a general, outward radial load $p(x)$ (Fig. 6.20). For relatively *short shells*, bending will be important and, hence, the plastic regime of point F on the yield locus of Fig. 6.18 will be relevant, with the correct signs for M_x, N_θ, $\dot{\kappa}_x$, and $\dot{\varepsilon}_\theta$. The coordinates of point F are

$$M_x = -\sigma_0 Ht/2, \qquad N_\theta = \sigma_0 t/2. \tag{6.94}$$

Substituting Eq. (6.94) in the optimality condition, Eq. (6.93), gives

$$-d^2w/dx^2 + w/RH = 2c^2/\sigma_0 H. \tag{6.95}$$

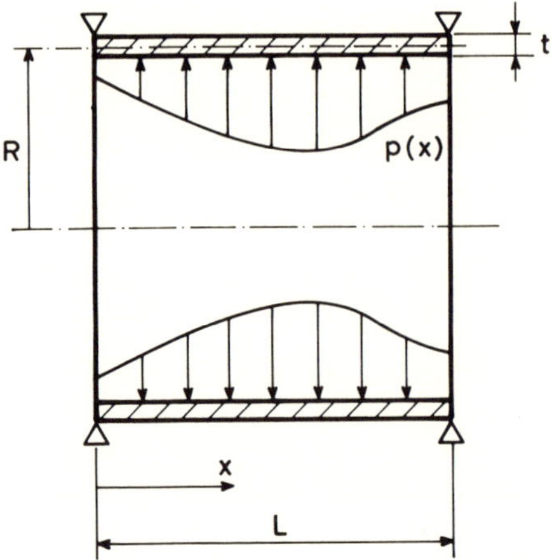

Figure 6.20. Simply supported cylindrical shell and loading.

Integrating Eq. (6.95), with boundary conditions $\dot{w}(0) = 0$ and $\dot{w}(L) = 0$, we obtain the collapse mechanism

$$\dot{w} = \frac{2c^2R}{\sigma_0}\left\{1 - \frac{\cosh[(L-2x)/2\sqrt{RH}]}{\cosh(L/2\sqrt{RH})}\right\}. \tag{6.96}$$

This mechanism is valid as long as the strain rate vector with components $\dot{\kappa}_x$, $\dot{\varepsilon}_\theta$ lies in the angle made by the outward normals to sides AF and FE at point F (Fig. 6.18). This condition is satisfied when $0 \le \dot{w}/R \le -Hd^2\dot{w}/dx^2$ or, from Eq. (6.96) when

$$L \le 2.634(RH)^{1/2}. \tag{6.97}$$

The inequality of Eq. (6.97) defines the set of short shells.

For *long shells* ($L > 2.634(RH)^{1/2}$) we may expect a central part to be in a pure membrane state with the stress regime of point A (Fig. 6.18), whereas the shell regions near the supports will remain in regime F. In the central part where $M_x = 0$ and $N_\theta = \sigma_0 t$, it follows that $\dot{w} = c^2R/\sigma_0$ because of Eqs. (6.93) and (6.89). Continuity of $\dot{w}$ and of $d\dot{w}/dx$ (to avoid concentrated dissipation) at the junction of the regions in regimes A and F along with the boundary conditions $\dot{w}(0) = \dot{w}(L) = 0$ give the collapse mechanism [see Shield (1960c)]. The mechanism is given by the $\dot{w}$ of Eq. (6.96) for $0 \le x \le 1.317(RH)^{1/2}$, by $\dot{w} = c^2R/\sigma_0$ for $1.317(RH)^{1/2} \le x \le L/2$, and by $\dot{w}(x)$ symmetric with respect to the central cross section $x = L/2$. It is worth pointing out that, for a shell of given parameter $\omega = L/(RH)^{1/2}$, the same collapse mechanism applies regardless of the distribution of outward radial load $p(x)$. Hence, by the superposition method of Save and Shield (1966), already quoted in Section 3.8, if $t_i(x)$ is the optimal design for $p_i(x)$, the optimal design for $\Sigma\, a_i p_i$, $a_i \geq 0$, is simply $\Sigma\, a_i t_i$.

Consider now the particular case of a *uniform load p*. Because of the symmetry with respect to the central cross section, we need only consider $0 \leq x \leq L/2$. Substituting the expressions in Eq. (6.94) for M_x and N_θ in the equation of equilibrium, Eq. (6.90), and integrating, with boundary conditions $M_x(0) = M_x(L) = 0$ (and hence $t(0) = t(L) = 0$), we obtain for a *short shell* that

$$t(x) = \frac{2pR}{\sigma_0}\left\{\frac{(\cosh \omega L - 1)\sinh \omega x}{\sinh \omega L} - \cosh \omega x + 1\right\}, \tag{6.98}$$

valid for $0 < \omega L \equiv L/(RH)^{1/2} \leq 1.317$. The volume of the sheets is

$$\Gamma = \frac{8\pi R^2 p}{\sigma_0}\left(\frac{1 - \cosh \omega L}{\omega \sinh \omega L} + \frac{L}{2}\right). \tag{6.99}$$

For *long shells* ($\omega L \geq 2.634$) the optimal design is

$$t(x) = \frac{2pR}{\sigma_0}\left\{\frac{(\cosh \omega L^* - 1)\sin \omega x}{\sin \omega L^*} - \cosh \omega x + 1\right\},$$

$$0 \leq x \leq L^* = 1.317(RH)^{1/2}, \tag{6.100}$$

simply obtained by substituting L^* for L in Eq. (6.98), and

$$t(x) = pR/\sigma_0, \qquad L^* \leq x \leq L/2. \tag{6.101}$$

At $x = L^*$, continuity of M_x is satisfied (with $M_x = 0$), but the shear force is discontinuous across $x = L^*$, falling from

$$\begin{aligned} V_x(L^*) &= \left(\frac{dM_x}{dx}\right)_{x=L^*} \\ &= -p\left[\frac{(\cosh \omega L^* - 1)\cosh \omega L^*}{\sinh \omega L^*} \quad \sinh \omega L^*\right], \end{aligned} \tag{6.102}$$

obtained from Eq. (6.100) with $M_x = \sigma_0 t H/2$, down to zero. This discontinuity requires at $x = L^*$ a reinforcing ring with mean radius R and area

$$A - V_x(L^*)/R\sigma_0 \tag{6.103}$$

that is subjected to the same circumferential strain rate $\dot{\varepsilon}_\theta$ as the central part in a pure membrane state. A typical solution for a long shell is shown

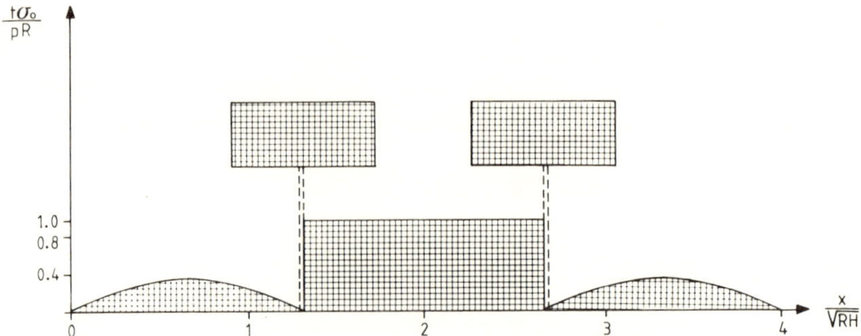

Figure 6.21. Solution for a long shell.

in Fig. 6.21. The reinforcing rings are supposed to be connected to the shell by vanishingly thin webs and to be placed symmetrically with respect to the mid-surface. In the limiting case where $L = 2L^*$, both designs, with or without a reinforcing ring, are equally acceptable and give the same volume, which is to be expected.

Figure 6.22 shows the variation of the ratio Γ/Γ_c of the minimum-volume (cost) design to the design with constant t and same limit load and boundary conditions, given by Guerlement (1975):

$$t_c = 2pR\frac{1 - \cos(\omega L/2)}{2 - \cos(\omega L/2)} \qquad \text{if } \omega L \leq \pi$$

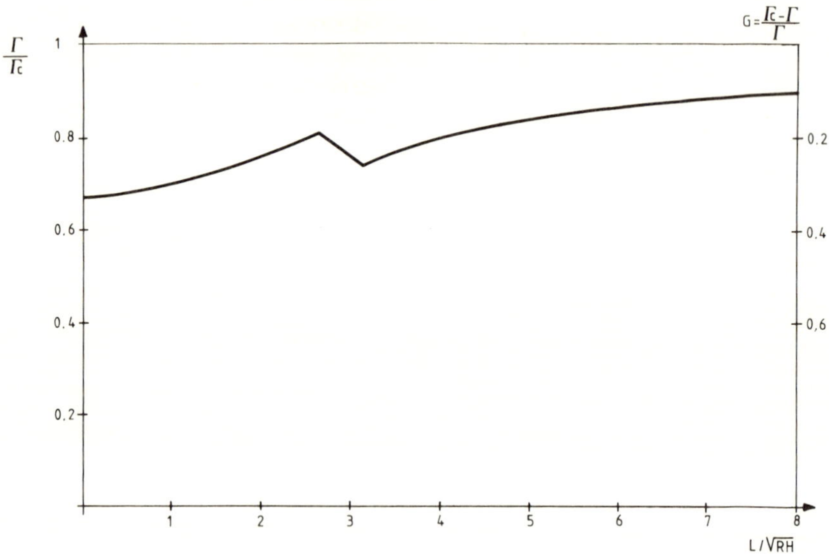

Figure 6.22. Variation of the ratio of the minimum volume design to the design with constant t.

and

$$t_c = pR/\sigma_0 \qquad \text{if } \omega L > \pi.$$

Considering now a *hydrostatic pressure*

$$p(x) = p(0)(1 - x/L), \tag{6.104}$$

where $p(0)$ is given, we proceed in a similar manner [see Igic (1980)] and obtain the following:

(a) for a short shell ($\omega L \leq 2.634$),

$$t(x) = \frac{2p(0)R}{\sigma_0}\left[\coth(\omega L) \quad \sinh(\omega x) - \cosh(\omega x) - \frac{x}{L} + 1\right]; \tag{6.105}$$

(b) for a long shell ($\omega L \geq 2.634$),

$$t(x) = \frac{2p(0)R}{\sigma_0}\left[\coth(\omega L^*)\sinh(\omega x) - \cosh(\omega x) - \frac{x}{L} + 1 - \frac{(\sinh \omega x)[1 - L^*/L]}{\sinh wL^*}\right] \tag{6.106}$$

for $0 \leq x \leq L^* = 1.317(RH)^{1/2}$,

$$t(x) = \frac{p(0)R(1 - x/L)}{\sigma_0} \tag{6.107}$$

for $L^* \leq x \leq L - L^*$, and

$$t(z) = \frac{2p(0)R}{\sigma_0}\frac{z}{L} - \frac{L \sinh \omega z}{L \sinh \omega L^*} \tag{6.108}$$

for $0 \leq z \leq L^*$, with $z = L - x$.

Reinforcing rings with cross-sectional areas A_1 and A_2 must be placed at the abscissas L^* and $L - L^*$, respectively. We find

$$A_1 = \frac{p(0)R}{\omega\sigma_0}\left[\frac{1}{\sinh \omega L^*} - \frac{1}{\omega L} - (\coth \omega L^*)\left(1 - \frac{L^*}{L}\right)\right] \tag{6.109}$$

and

$$A_2 = \frac{p(0)R}{\omega\sigma_0}\left(\frac{1}{\omega L} - \frac{2L^* \cosh \omega L^*}{L \sinh \omega L^*}\right). \tag{6.110}$$

6.4.2.2. A shell *built in at both ends* can be treated in a similar manner. The collapse mechanism is again symmetric with respect to the central cross section, and we need only consider $0 \le x \le L/2$. *Short shells* will have no pure membrane zone: They will exhibit positive rate of curvature $\dot{\kappa}_x$ near the supports (regime B) for $0 < x < x_1$ and negative curvature (regime F) in the central region. Integrating the optimality criterion with the boundary conditions that $\dot{w}$ and $d\dot{w}/dx$ are continuous at $x = x_1$, $\dot{w}$ and $d\dot{w}/dx$ vanish at the built-in ends, and $d\dot{w}/dx$ vanishes for $x = L/2$ gives the four constants of integration and the value of x_1. We obtain

$$\dot{w} = 2c^2R(1 - \cos \omega x)/\sigma_0, \qquad 0 \le x \le x_1, \tag{6.111}$$

$$\dot{w} = \frac{2c^2R}{\sigma_0}\left[1 - \frac{\cos \omega x_1 \cosh \omega(L/2 - x)}{\cosh \omega(L/2 - x_1)}\right], \qquad x_1 \le x \le L/2, \tag{6.112}$$

$$\tanh \omega x_1 = \tanh[\omega(L/2 - x_1)]. \tag{6.113}$$

For the strain rate vectors $(\dot{\kappa}_x, \dot{\varepsilon}_\theta)$ at F to be acceptable according to the normality law, we must have

$$\cosh \omega(L/2 - x_1) \le 2 \cosh \omega x_1. \tag{6.114}$$

The limiting case is obtained by setting $\omega x_1 = l_1$, $\omega L/2 = l_2$, l_1 and l_2 being dimensionless abscissas, in both Eqs. (6.113) and (6.114), the latter relation being taken with the equality sign. The resulting system of two equations in L_1 and L_2 gives $l_1 = 0.65907$ and $l_2 = 1.69078$. Hence, the longest short shell will correspond to $\omega L = 3.38156$. For larger values of ωL we deal with *long shells* that will exhibit, with increasing x,

(1) a zone with length l_1/ω in regime B,
(2) a zone with length $(l_2 - l_1)/\omega$ in regime F,
(3) a zone with length $L/2 - l_2/\omega$ in regime A (pure membrane).

The boundary conditions are $\dot{w} = d\dot{w}/dx = 0$ for $x = 0$, $\dot{w}$ and $d\dot{w}/dx$ continuous for $x = L_1/\omega$ and $x = L_2/\omega$. When applied to the integrated optimality condition, they give the following collapse mechanisms:

$$\dot{w} = \frac{2c^2R(1 - \cos \omega x)}{\sigma_0}, \qquad 0 \le x \le \frac{l_1}{\omega}, \tag{6.115}$$

$$\dot{w} = \frac{2c^2R}{\sigma_0}\left[1 - \frac{\cosh(l_2 - \omega x)}{2}\right], \qquad \frac{l_1}{\omega} \le x \le \frac{l_2}{\omega}, \tag{6.116}$$

and

$$\dot{w} = c^2 R/\sigma_0, \qquad l_2/\omega \le x \le L/2. \tag{6.117}$$

These mechanisms were first given by Shield (1960c). We apply them to the case of a hydrostatic pressure

$$p(x) = p(L)(x/L) \tag{6.118}$$

where $p(L)$ is given. Despite the symmetry of the collapse mechanism, the design $t(x)$ will not be symmetric. For a *short shell*, integration of the equation of equilibrium, Eq. (6.90), with the use of regimes B and F give, respectively,

$$t(x) = C_1 \sin \omega x + C_2 \cos \omega x + p(x)R/\sigma_0, \qquad 0 \le x \le x_1,$$

$$t(x) = C_3 \sin \omega x + C_4 \cos \omega x + p(x)R/\sigma_0, \qquad L - x_1 \le x \le L,$$

$$t(x) = D_1 \sinh \omega x + D_2 \cosh \omega x + p(x)R/\sigma_0, \qquad x_1 \le x \le L - x_1.$$

The static boundary conditions are $M_x = 0$ at $x = 0$ and $x = L$, and at $x = x_1$ and $x = L - x_1$, because M_x must change continuously from positive to negative values at these latter positions. Also the shear force V_x is continuous at $x = x_1$ and $x = L - x_1$. Since M_x (and V_x) is related to t by the plastic regimes, the boundary conditions above yield C_1, C_2, C_3, C_4, D_1, and D_2, and the corresponding design is

$$t = \frac{2p(L)R}{\sigma_0 L}\left\{\frac{x + x_1 \sinh \omega(x - L - x_1) - (L - x_1)\sinh \omega(x - x_1)}{\sinh(L - 2x_1)}\right\},$$

$$x_1 \le x \le L - x_1, \tag{6.119}$$

$$t = \frac{2p(L)R}{\sigma_0 L}\left\{x - x_1 \cos \omega(x - x_1)\right.$$

$$\left. + \left[\frac{x_1 + x_1 \cosh \omega(L - 2x_1) - L}{\sinh \omega(L - 2x_1)} + 2\right] \sin \omega(x_1 - x)\right\},$$

$$0 \le x \le x_1, \tag{6.120}$$

and

$$t = \frac{2p(L)R}{\sigma_0 L}\left\{x - (L - x_1)\cos\omega(x - L + x_1)\right.$$

$$\left. - \left[\frac{x_1 - (L - x_1)\cosh\omega(L - 2x_1)}{\sinh(L - 2x_1)} + 2\right]\sin\omega(x - L + x_1)\right\},$$

$$L - x_1 \le x \le L. \tag{6.121}$$

The volume of the face sheets is obtained by integration to be

$$\Gamma = \frac{4\pi R^2 p(L)}{\omega\sigma_0}\left\{\frac{\omega L}{2} - \sinh\omega x_1 - \frac{[\cosh\omega(L - 2x_1) - 1]\cos\omega x_1}{\sinh\omega(L - 2x)}\right\}. \tag{6.122}$$

We recall that x_1 is given by Eq. (6.113).

An analogous solution can be obtained for a *long shell*, the difference being that there exists a central region in pure membrane state (regime A), at the boundaries of which reinforcing rings must be placed to apply the necessary shear forces to the neighboring regions subjected to bending. We refer the reader to Igic (1980), for details.

6.4.3. Assigned Limit Load—Reinforced Concrete Cylinder

6.4.3.1. Optimality Criterion. The thickness of the shell is considered to be determined by a limit on the elastic deflections or by cracking conditions. *We want to minimize the total volume of the reinforcing steel.* The generalized variables of importance are M_x, N_θ, $\dot{\kappa}_\theta$, and $\dot{\varepsilon}_\theta$. Reinforcement obviously should be placed longitudinally only at the top or bottom faces, depending on the sign of M_x. Let A_x be the cross-sectional area per unit length of this reinforcement. Circumferential reinforcement of area A_θ per unit length should be placed at the midsurface wherever N_θ is positive. The regions in pure circumferential compression are costless and need not be considered, provided the strength of the concrete is checked to be sufficiently large. The yield force and moments are $N_{p\theta} = A_\theta\sigma_0$ and $M_{px} = A_x\sigma_0 H$, where the moment arm H of the internal forces is assumed to be a known constant fraction of the total thickness of the shell. The shell can then be regarded as a *grid* of longitudinal sandwich beams of unit width in bending and circumferential rings of unit width in pure tension. In complete analogy with Sections 4.1 and 5.1, the optimality criterion for the absolute minimum of

$$\Gamma = \int_A (A_x + A_\theta)\,dA \tag{6.123}$$

is found to be

$$|\dot{\kappa}_x| \le c^2/H \quad \text{and} \quad \dot{\varepsilon}_\theta \le c^2 \qquad \text{when } |\dot{\kappa}_x| \ne 0,\ \dot{\varepsilon}_\theta > 0, \tag{6.124}$$

$$|\dot{\kappa}_x| \le c^2/H \qquad \text{when } \dot{\kappa}_x \ne 0,\ \dot{\varepsilon}_\theta < 0, \tag{6.125}$$

$$\dot{\varepsilon}_\theta \le c^2 \qquad \text{when } \dot{\kappa}_x = 0,\ \dot{\varepsilon}_\theta > 0, \tag{6.126}$$

where the equality sign must be used when the lower bounds A_x^- and A_θ^- set on the reinforcement areas are *not* reached, and inequality signs must be used when they are reached. Note that in the absence of bounds, since A_x and A_θ must physically be nonnegative, the optimality criterion indicates that A_x vanishes when $|\dot{\kappa}_x| < c^2/H$, and A_θ vanishes when $|\dot{\varepsilon}_\theta| < c^2$.

6.4.3.2. Consider a cylindrical reinforced concrete shell that is *simply supported* at both ends and subjected to internal radial pressure $p(x)$. No bound is set on the reinforcement. For sufficiently *short shells* we may expect that bending will be prominent, and we may use a collapse mechanism analogous to that in Section 6.4.2.1, with $\dot{\kappa}_x < 0$, $\dot{\varepsilon}_\theta > 0$, and $H|\dot{\kappa}_x| \ge \dot{\varepsilon}_\theta$. Hence, the optimality condition, Eq. (6.124), applies and gives

$$-Hd^2\dot{w}/dx^2 = c^2. \tag{6.127}$$

From this, with boundary conditions $w(0) = w(L) = 0$, we obtain the collapse mechanism

$$\dot{w} = (c^2/2H)(Lx - x^2), \tag{6.128}$$

symmetric with respect to the central cross section. The validity of Eq. (6.128) is restricted by the fact that $\dot{\varepsilon}_\theta$ must not become larger than c^2 for $0 < x < L/2$. The corresponding limit for the parameter $\omega L \equiv L/(RH)^{1/2}$ is 2.8184, which defines the range of short shells.

For *long shells,* with $L/(RH)^{1/2} > 2.8184$, we take the two regions of length $L^* = 1.4142(RH)^{1/2}$ adjacent to the supports, with $\dot{\kappa}_x < 0$, and a central region in pure circumferential extension (membrane state).

The boundary conditions for the regions in bending are $w(0) = 0$ and $(d\dot{w}/dx)_{x=L^*} = 0$ to avoid concentrated specific dissipation. Equation (6.127) then gives

$$\dot{w} = (c^2/H)[L^*x - x^2/2], \qquad 0 \le x \le L^*, \tag{6.129}$$

from which we immediately obtain, from $\dot{\varepsilon}_\theta = c^2$,

$$\dot{w} = c^2 R, \qquad L^* \le x \le L/2. \tag{6.130}$$

Note that the value of L^* is such that continuity of $\dot{w}$ is satisfied at $x = L^*$.

We now apply the results obtained above to a *short shell* subjected to a *hydrostatic pressure*

$$p(x) = p(0)(1 - x/L). \tag{6.131}$$

The equation of equilibrium is simply $d^2M_x/dx^2 = p(x)$. We substitute $M_x = -A_x\sigma_0 H$ and obtain, with boundary conditions $M_x(0) = M_x(L) = 0$,

$$A_x = \left[\frac{p(0)x}{\sigma_0 H}\right](x/2 - x^2/3L - xL/6). \tag{6.132}$$

We must recall that, because M_x is everywhere negative, the reinforcement is placed everywhere at the external face of the shell.

If we increase L to reach the range of *long shells* ($L > 2.8184(RH)^{1/2}$), $\dot{\kappa}_x < 0$ holding for $0 \le x \le L^*$ will give, with $M_x(0) = M_x(L^*) = 0$,

$$A_x(x) = \left[\frac{p(0)x}{2\sigma_0 H}\right]\left(x - L^* - \frac{x^2 - L^{*2}}{3L}\right), \qquad 0 \le x \le L^*. \tag{6.133}$$

In the central region $L^* \le x \le L - L^*$ in pure circumferential extension, $N_\theta = p(x)R$ and, with $N_\theta = A_\theta\sigma_0$, we obtain

$$A_\theta(x) = \left[\frac{p(0)R}{\sigma_0}\right](1 - x/l), \qquad L^* \le x \le L. \tag{6.134}$$

Finally, the end region $L - L^* \le x \le L$ is in bending; equilibrium with end conditions $M_x(L - L^*) = M_x(L) = 0$ leads to

$$A_x(x) = \frac{p(0)}{2\sigma_0 H}\left\{\left(1 - \frac{x}{3L}\right)x^2 + \frac{(L - L^*)^2(2L + L^*)(x - L)}{3LL^*} - \frac{2L^2(x - L + L^*)}{3L}\right\},$$

$$L - L^* \le x \le L. \tag{6.135}$$

At the abscissas $x = L^*$ and $x = L - L^*$, discontinuities in shear force are

$$V_1 = V(L^*) = p(0)L^*(1 - 2L^*/3L)/2$$

and

$$V_2 = V(L - L^*) = p(0)\{[(L - L^*)^3 - L]/3LL^* + L - L^*\},$$

respectively, and necessitate concentrated circumferential reinforcement with areas $A_{\theta 1} = V_1 R/\sigma_0$ and $A_{\theta 2} = V_2 R/\sigma_0$.

Exercise 6.1. Find the design with minimum volume of the face sheets of a sandwich cylinder of length L and radius R simply supported at the two ends and subjected to a load $p(x) = a + b(1 - x/L)$, where a and b are positive constants.

Exercise 6.2. For the shell of Exercise 6.1, with loading $p(x) = a$ $(b = 0)$ and $\omega L = 2.634$, verify that the volumes of the optimal designs with and without reinforcing ring are equal.

Exercise 6.3. Find the design with minimum volume of the face sheets of a sandwich cylinder of length L and radius R built in at both ends and subjected to a uniformly distributed internal pressure p.

Exercise 6.4. Replace one built-in end of the shell of Exercise 6.3 by a free end and determine the various collapse mechanisms, together with the corresponding ranges of the parameter ωL able to satisfy the optimality criterion.

Exercise 6.5. Consider a *long* cylinder, as defined in the solution of Exercise 6.4, and find its minimum-volume design under uniformly distributed load p.

Exercise 6.6. Show that consideration of the normality law for regime B in Section 6.4.2.2 does not add a relevant constraint on l_1.

Exercise 6.7. Find the distribution of the minimum-volume reinforcement of a *long* cylinder in reinforced concrete simply supported at both ends and subjected to a hydrostatic pressure $p(x) = p(0)(1 - x/L)$.

Exercise 6.8. Find the minimum-volume distribution of reinforcement of a *short* cylinder in reinforced concrete subjected to a uniformly distributed, internal, radial pressure p. Verify that it can be obtained by applying the superposition method of Save and Shield to the design of Eq. (6.132). Apply the same method to obtain the optimal design for *long* shells under uniform load.

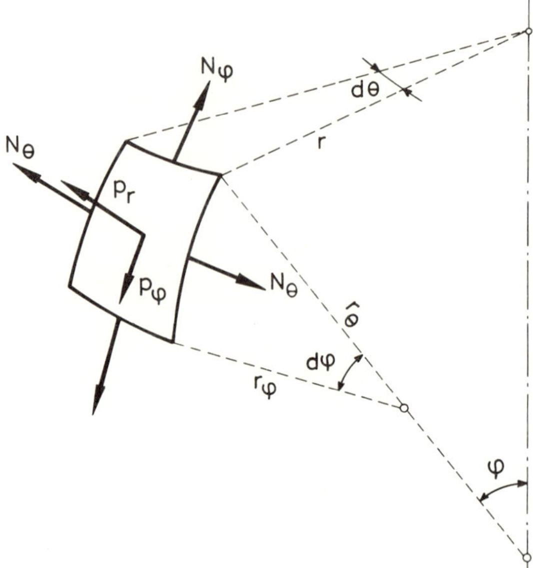

Figure 6.23. Element of membrane of revolution.

6.5. Other Shell Problems

Optimal design solutions for shells other than cylindrical are very rare† (even by purely numerical procedures), except for shells of revolution under axially symmetric loading, which are assumed to be in a pure membrane state. Because such shells are statically determinate, an obvious necessary condition for minimum volume of material is that they be "fully stressed." When the mid-surface, support conditions compatible with the membrane state and the loads are given (ignoring the shell's own weight and, more generally, all body forces), the thickness at every point is determined straightforwardly by the condition of limited stress intensity in equality form. If $t(\varphi)$ denotes the unknown thickness function, the well-known equations of equilibrium are (see Fig. 6.23 for notations)

$$d(rN_\varphi)/d\varphi - r_\varphi \cdot \cos\varphi \cdot N_\theta = -p_\varphi r \cdot r_\varphi, \tag{6.136}$$

$$N_\varphi/r_\varphi + N_\theta/r_\theta = p_r, \tag{6.137}$$

where $N_\varphi = \sigma_\varphi t$, $N_\theta = \sigma_\theta t$, and r_φ and r_θ are the principal radii of curvature of the mid-surface.

† Conical sandwich shells of minimum volume for assigned limit loads have been treated by Reiss (1974) by the statical method.

Using Eqs. (6.136) and (6.137) with either the Tresca yield condition,

$$\max\{|\sigma_\varphi|, |\sigma_\theta|, |\sigma_\varphi - \sigma_\theta|\} = \sigma_0, \tag{6.138}$$

or the von Mises condition,

$$\sigma_\varphi^2 + \sigma_\theta^2 - \sigma_\varphi\sigma_\theta = \sigma_0, \tag{6.139}$$

in plane stress, Issler (1964) has obtained the following thickness functions:

(a) uniform surface loading p ($p_r = -p \cos \varphi$, $p_\varphi = p \sin \varphi$), von Mises spherical shell with radius a:

$$t(\varphi) = 2t_0(1 + \cos \varphi)^{-1} \times (\cos^4 \varphi + 2 \cos^3 \varphi - 2 \cos^2 \varphi - 3 \cos \varphi + 3)^{1/2}, \tag{6.140}$$

with

$$t_0 \equiv t(0) = ap/2\sigma_0; \tag{6.141}$$

(b) uniform surface loading p, Tresca spherical shell:

$$t(\varphi) = 2t_0(1 + \cos \varphi)^{-1}, \qquad 0 \leq \varphi \leq \varphi_1, \tag{6.142}$$

$$t(\varphi) = 2t_0(1 - \cos \varphi)(1 + \cos \varphi)^{-1}(2 + \cos \varphi), \qquad \varphi_1 \leq \varphi \leq \varphi^*, \tag{6.143}$$

where

$$\varphi_1 = \operatorname{arc} \cos\{(\sqrt{5} - 1)/2\} \tag{6.144}$$

and t_0 is given by Eq. (6.141);

(c) snow load q ($p_r = -q \cos^2 \varphi$, $p_\varphi = q \sin \varphi \cdot \cos \varphi$), von Mises spherical shell:

$$t(\varphi) = t_0(3 - 6 \cos^2 \varphi + 4 \cos^4 \varphi)^{1/2}, \tag{6.145}$$

with t_0 again given by Eq. (6.141) with p replaced by q.

(d) snow load, Tresca spherical shell:

$$t(\varphi) = qa/2\sigma_0 \equiv t_0, \qquad 0 \leq \varphi \leq \pi/4, \tag{6.146}$$

$$t(\varphi) = 2t_0 \sin^2 \varphi, \qquad \pi/4 \leq \varphi \leq \pi/2. \tag{6.147}$$

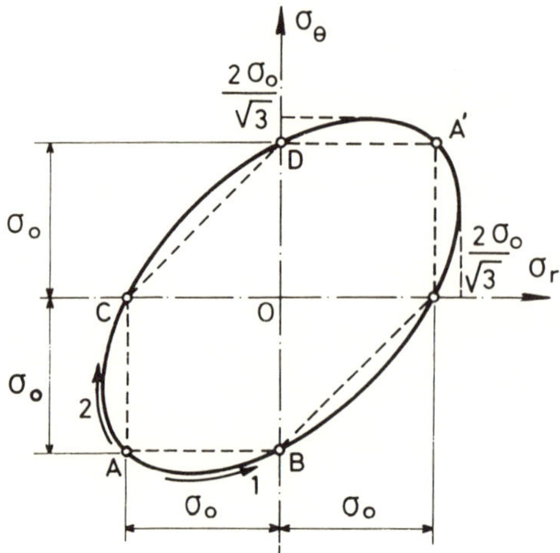

Figure 6.24. von Mises's yield condition in plane stress.

In the case of nonspherical shells there exist, in general, three intervals for φ, corresponding to the three ways of satisfying Eq. (6.138) and the three corresponding expressions for $t(\varphi)$. When the loading is the *weight of the shell* itself, the problem is slightly more complicated. Because of the symmetry of revolution, the state of stress at the vertex of the shell is represented by point A, Fig. 6.24 (or A' if the shell is suspended instead of being supported). For a given shape of the mid-surface, two uniform strength designs exist, corresponding to the two different stress profiles AB or AC lying on both sides of the yield curve with respect to point A, Fig. 6.24. If δ is the weight per unit volume of the shell material, the two equilibrium equations for a *spherical shell* of radius a are

$$(d/d\varphi)(t\sigma_\varphi \sin\varphi) - t\sigma_\theta \cos\varphi + a\delta t \sin^2\varphi = 0, \tag{6.148}$$

$$\sigma_\varphi + \sigma_\theta + a\delta \cos\varphi = 0. \tag{6.149}$$

At the vertex, $\sigma_\theta = \sigma_\varphi = -\sigma_0$ (point A, Fig. 6.22) and $\varphi = 0$.

We obtain $a = 2\sigma_0/\delta$ from Eq. (6.149), whereas Eq. (6.148) is identically satisfied. Hence, the thickness t_0 at the vertex remains arbitrary, and the weight of the shell can be made arbitrarily small. The function $t(\varphi)/t_0$ is obtained by integrating Eq. (6.148), using Eq. (6.149) and the appropriate stress profile 1 or 2 of Fig. 6.2 on the Tresca hexagon or on the von Mises ellipse. The four solutions are given in Fig. 6.25. We see that profile 2 gives lower weights: The corresponding collapse mechanism will exhibit a constant difference between the dissipation per unit volume and the power of the body force per unit volume, in analogy with the criterion in Eq. (3.122),

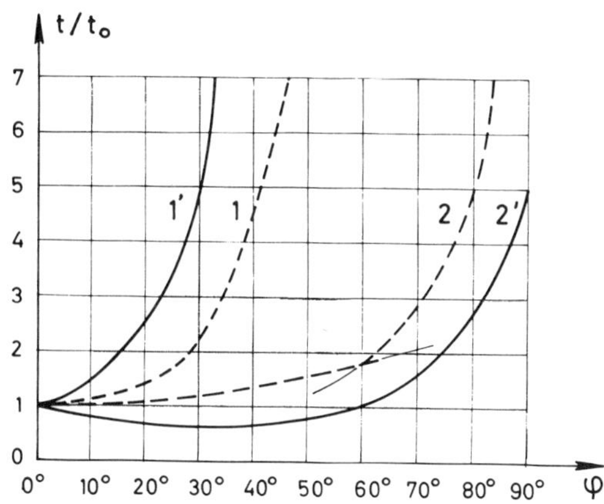

Figure 6.25. Optimal dimensionless thickness for spherical membranes.

whereas the other will not. Curves 1 and 2, corresponding to the Tresca condition, were first obtained by Ziegler (1958). That t_0 and the volume remain arbitrary is due to the fact that the only load acting is the weight of the shell itself. This unlikely and paradoxical situation is avoided by considering the combination of the structural weight of the shell and the weight p per unit area of a cover sheet of given uniform thickness. This was done by Prager and Rozvany (1980) for spherical cupolas made of materials without tensile strength (plain concrete, for example). The limiting stress condition is then part BAC of the Tresca hexagon in Fig. 6.24:

$$\max(|\sigma_\varphi|, |\sigma_\theta|) \le \sigma_0, \qquad \sigma_\varphi \le 0, \sigma_\theta \le 0. \tag{6.150}$$

Using regime AC, we obtain the uniform strength designs of Fig. 6.26, where the following dimensionless variables are used:

$$\bar{t} = t(\varphi)/t_0, \qquad \bar{q} = p/\delta t_0, \qquad \bar{\alpha} = \sigma_\theta/\sigma_\varphi, \tag{6.151}$$

with $t_0 = p/q\delta$

The subscript 1 corresponds to the base of the cupola; the subscript 0 to the apex. The first three or four digits of $t(\varphi)$ are given by the approximate formula

$$\bar{t}(\varphi) \cong (\cos(\varphi/2))^{-(4+3.8\bar{q})/(1+1.9\bar{q})}. \tag{6.152}$$

We can minimize either the total weight W (total load carried by the supports) or only the structural weight W^* (excluding the weight of the

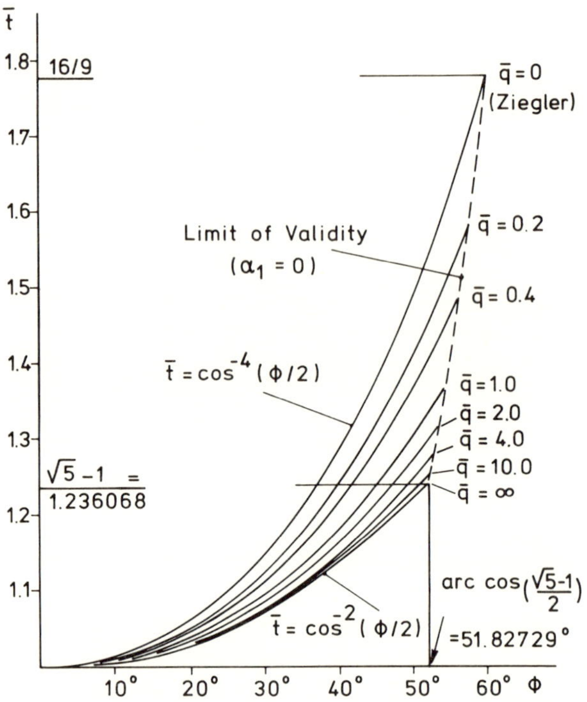

Figure 6.26. Spherical cupolas with no tensile strength: thickness variation of uniform-strength designs.

cover sheet) by proper choice of the radius ρ of the sphere for given base radius r_0. If we let

$$\bar{\rho} = \frac{\rho\gamma}{\sigma_0}, \qquad \bar{r} = \frac{r_0\gamma}{\sigma_0}, \qquad \bar{W} = \frac{W}{\pi p r_0^2}, \qquad \bar{W}^* = \frac{W^*}{\pi p r_0^2}, \tag{6.153}$$

we can summarize the results in Figs. 6.27 and 6.28 for

$$\bar{\rho} = \frac{2}{1 + \bar{q}}. \tag{6.154}$$

The same kind of optimization has been achieved by Nakamura *et al.* (1981) for spherical cupolas obeying the Tresca yield condition (hence, removing the limitation of vanishing tensile strength) and by Dow *et al.* (1982) for von Mises spherical cupolas and Tresca shells with meridional section defined by a quadratic function, both for a variety of applied loads, all including the weight of the shell.

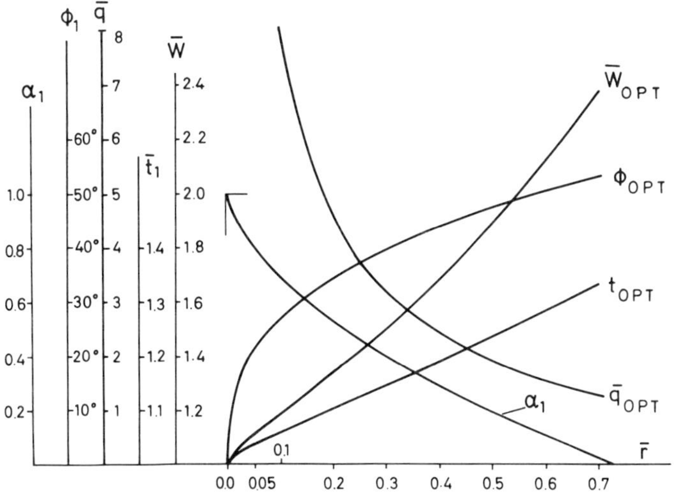

Figure 6.27. Designs of spherical cupolas of uniform strength for minimum total weight.

The uniform strength condition is sometimes formulated [see Flugge (1960), Milankovic (1908), Dokmeci (1966)] as

$$\sigma_\varphi = \sigma_\theta = \pm\sigma_0. \tag{6.155}$$

Both principal stresses must be of the same sign and equal to the limit stress

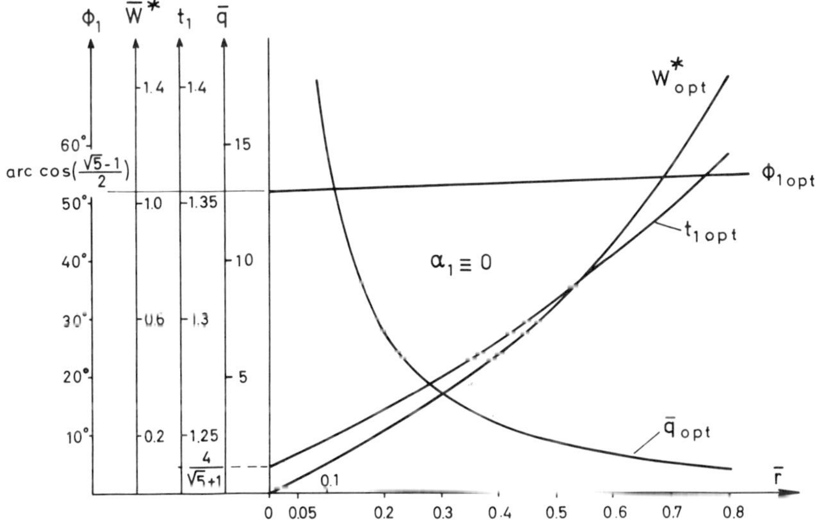

Figure 6.28. Designs of spherical cupolas of uniform strength for minimum structural weight.

at every point of the shell. Because the condition of Eq. (6.155) completely defines the stress tensor when the principal directions are known, the two equilibrium equations will give both the meridian curve and the thickness distribution.

It is worth recalling that all designs based on membrane forces exclusively imply that the supports must permit the corresponding membrane displacements and furnish only reactions of the membrane type.

6.6. Circular Disks in Plane Stress

6.6.1. Basic Relations

For the sake of simplicity we restrict the explicit treatment of disk problems to the *circular or annular disk with axially symmetric loading.* The radial and circumferential directions are then those of principal stresses, strains, and strain rates. We assume that the thickness $t(r)$ will remain small enough for plane stress conditions to be used. We shall nevertheless regard the disk as a three-dimensional body and take σ_r, σ_θ, ε_r, and ε_θ as our generalized variables. With the notations of Fig. 6.29 we have the following basic relations:

6.6.1.1. Kinematics. The strains are given by

$$\varepsilon_r = du/dr, \qquad \varepsilon_\theta = u/r, \tag{6.156}$$

where $u(r)$ is the displacement field. Obviously, similar relations hold for strain rates $\dot{\varepsilon}_r$, $\dot{\varepsilon}_\theta$, and velocity $\dot{u}(r)$.

6.6.1.2. Statics. The equation of radial equilibrium is (Fig. 6.29)

$$\frac{d(\sigma_r t)}{dr} + \frac{(\sigma_r - \sigma_\theta)t}{r} + ft = 0, \tag{6.157}$$

where $t(r)$ is the (unknown) thickness of the disk and f is the radial body force per unit volume.

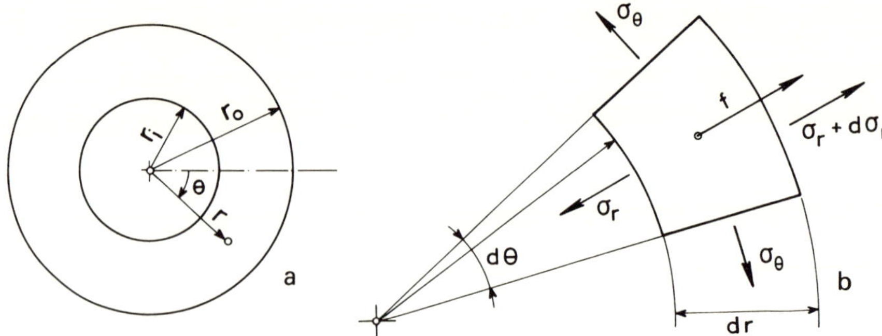

Figure 6.29. (a) Annular disk. (b) Element of annular disk.

6.6.1.3. Linear Elasticity. For plane stress, Hooke's law is given by the relations

$$\varepsilon_r = \frac{\sigma_r - \nu\sigma_\theta}{E}, \qquad \varepsilon_\theta = \frac{\sigma_\theta - \nu\sigma_r}{E}, \tag{6.158}$$

or

$$\sigma_r = \frac{E(\varepsilon_r + \nu\varepsilon_\theta)}{1 - \nu^2}, \qquad \sigma_\theta = \frac{E(\varepsilon_\theta + \nu\varepsilon_r)}{1 - \nu^2}, \tag{6.159}$$

where ν is Poisson's ratio. The strain energy per unit volume is

$$\phi = \frac{E(\varepsilon_r^2 + \varepsilon_\theta^2 + 2\nu\varepsilon_r\varepsilon_\theta)}{2(1 - \nu^2)}, \tag{6.160}$$

and the complementary energy per unit volume is

$$\psi = (\sigma_r^2 + \sigma_\theta^2 - 2\nu\sigma_r\sigma_\theta)/2E. \tag{6.161}$$

As we pointed out in Section 1.2.1, $\phi = \psi$ for solution states. Note that Eqs. (6.158)–(6.161) can be obtained from Eqs. (6.5)–(6.10) by mere substitution of σ_r, σ_θ, ε_r, ε_θ, and E for M_r, M_θ, κ_r, κ_θ, and B, respectively.

6.6.1.4. Perfect Plasticity. In a state of plane stress ($\sigma_x = 0$) with in-plane principal stresses σ_r and σ_θ, the yield conditions of Tresca and von Mises for a volume element of material are known to be as follows† [see Save and Massonnet (1972)]. The yield condition of Tresca is

$$\max\{|\sigma_r|, |\sigma_\theta|, |\sigma_r - \sigma_\theta|\} = \sigma_0. \tag{6.162}$$

The corresponding yield locus is shown in Fig. 6.30. The specific dissipation $D = \sigma_r\dot{\varepsilon}_r + \sigma_\theta\dot{\varepsilon}_\theta$ is given by the following:

$$\begin{aligned}
&\text{For sides } AB \text{ or } DE: && D = \sigma_0|\dot{\varepsilon}_\theta|, && \text{(a)} \\
&\text{For sides } BC \text{ or } EF: && D = \sigma_0|\dot{\varepsilon}_\theta| = \sigma_0|\dot{\varepsilon}_r| \text{ because } \dot{\varepsilon}_r = -\dot{\varepsilon}_\theta && \text{(b)} \\
& && \quad \text{and } |\sigma_r - \sigma_\theta| = \sigma_0, && \\
&\text{For sides } AF \text{ and } DC: && D = \sigma_0|\dot{\varepsilon}_r|, && \text{(c)} \quad (6.163) \\
&\text{For corners } A \text{ or } D: && D = \sigma_0|\dot{\varepsilon}_r + \dot{\varepsilon}_\theta|, && \text{(d)} \\
&\text{For corners } B \text{ or } E: && D = \sigma_0|\dot{\varepsilon}_\theta|, && \text{(e)} \\
&\text{For corners } C \text{ or } F: && D = \sigma_0|\dot{\varepsilon}_r|. && \text{(f)}
\end{aligned}$$

† Equations (6.162)–(6.169) can also be obtained from Eqs. (6.13)–(6.20) by the same substitution as noted in Section 6.6.1.3.

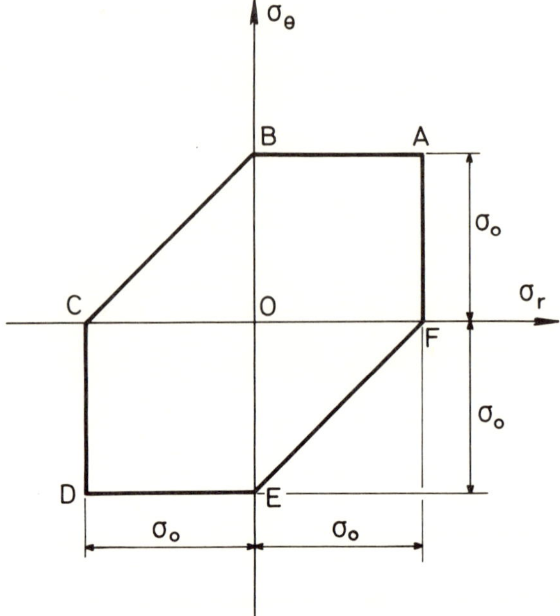

Figure 6.30. Yield condition of Tresca in plane stress.

The yield condition of von Mises is (see Fig. 6.31)

$$\sigma_r^2 + \sigma_\theta^2 - \sigma_r\sigma_\theta = \sigma_0^2. \qquad (6.164)$$

The strain rates are

$$\dot{\varepsilon}_r = \alpha(2\sigma_r - \sigma_\theta), \qquad \dot{\varepsilon}_\theta = \alpha(2\sigma_\theta - \sigma_r), \qquad (6.165)$$

from which we obtain

$$\sigma_r = \frac{2\dot{\varepsilon}_r + \dot{\varepsilon}_\theta}{3\alpha}, \qquad \sigma_\theta = (2\dot{\varepsilon}_\theta + \dot{\varepsilon}_r)/3\alpha, \qquad (6.166)$$

where α is a nonnegative scalar function of r. From Eqs. (6.164) and (6.166) we obtain

$$\alpha = \frac{\sigma_0}{3(\dot{\varepsilon}_r^2 + \dot{\varepsilon}_\theta^2 + \dot{\varepsilon}_r\dot{\varepsilon}_\theta)}. \qquad (6.167)$$

The dissipation per unit volume $D = \sigma_r \dot{\varepsilon}_r + \sigma_\theta \dot{\varepsilon}_\theta$ is given by

$$D = (2/\sigma_0)[3(\dot{\varepsilon}_r^2 + \dot{\varepsilon}_\theta^2 + \dot{\varepsilon}_r \dot{\varepsilon}_\theta)^{1/2}]. \tag{6.168}$$

A parametric form of Eq. (6.164) can be given as follows:

$$\sigma_r = \frac{2\sigma_0 \sin(\beta + \pi/6)}{\sqrt{3}}, \qquad \sigma_\theta = \frac{2\sigma_0 \sin(\beta - \pi/6)}{\sqrt{3}}, \tag{6.169}$$

with $\tan \beta = \sqrt{3} \cdot \sin \varphi$, φ being the parameter $(0 \leq \varphi \leq \pi)$.

6.6.2. Assigned Limit Load for a Tresca Disk

6.6.2.1. We want to find the thickness distribution $t(r)$ of an annular disk made of a Tresca material in order to achieve *minimum volume* of the disk. In the *absence both of bounds on t and of body forces f*, the optimality criterion is, by analogy with Eq. (3.9), that the specific† dissipation Dt per unit strength $t\sigma_0$ be a positive constant throughout the disks, as first given by Drucker and Shield (1956). Because the constant is arbitrary, it can include σ_0, and the optimality condition may be written as

$$D = \sigma_r \dot{\varepsilon}_r + \sigma_\theta \dot{\varepsilon}_\theta = \alpha > 0. \tag{6.170}$$

We leave it to the reader (Exercise 6.1) to prove that, in view of Eq. (6.156) applied to $\dot{\varepsilon}_r$, $\dot{\varepsilon}_\theta$, and the velocity $\dot{u}(v)$, the condition in Eq. (6.170) can be satisfied only with corner regimes C (or F) or D (or A) (Fig. 6.30).

With regime C (or F), $\sigma_\theta = 0$ and $|\sigma_r| = \sigma_0$, so that, with $f = 0$, the equilibrium equation, Eq. (6.157), gives

$$rt = \text{const.} \tag{6.171}$$

With regime D (or A) we have $\sigma_r = \sigma_\theta = \pm\sigma_0$, and the same equation yields

$$t = \text{const} \tag{6.172}$$

Using Eq. (6.156) in Eq. (6.170), we successively obtain the following:

(a) For regime C, $-\sigma_0 d\dot{u}/dr = \alpha$. Hence $\dot{u}$ is a linear function of r that can be written as

$$\dot{u} = a(b - r). \tag{6.173}$$

† That is, per unit midsurface area of the disk.

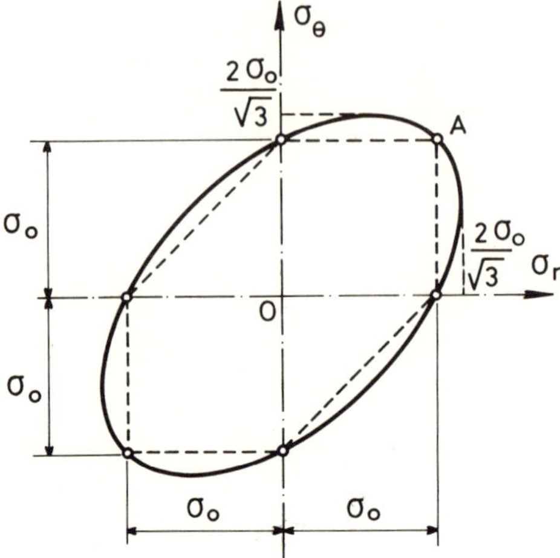

Figure 6.31. Yield condition of von Mises in plane stress.

The normality law at C requires that $\dot{\varepsilon}_\theta = \dot{u}/r \geq 0$, $\dot{\varepsilon}_r = d\dot{u}/dr < 0$, and $|\dot{\varepsilon}_r| \geq \dot{\varepsilon}_\theta$. With nonvanishing $\dot{u}$ and $r \geq 0$, these conditions result in $a > 0$, $b > 0$, and

$$b/2 \leq r \leq b. \tag{6.174}$$

In regime F the relations in Eqs. (6.173) and (6.174) still hold, but with $a < 0$.

(b) For regime D, the optimality condition is

$$-\sigma_0(d\dot{u}/dr + \dot{u}/r) = \alpha.$$

Hence

$$\dot{u} = a(r \pm c^2/r)/2. \tag{6.175}$$

The normality law at D implies that both $\dot{\varepsilon}_r$ and $\dot{\varepsilon}_\theta$ are nonpositive, resulting in $a < 0$ and

$$c^2 \leq r^2. \tag{6.176}$$

In regime A the relations in Eqs. (6.175) and (6.176) continue to hold, provided a is positive.

The regimes C and D, or F and A, can be matched together in only one way: The partial design, Eq. (6.172), with constant thickness must be external to the partial design with Eq. (6.171). The reader will easily show that this is a consequence of Eqs. (6.174) and (6.176) in conjunction with the continuity of $\dot{u}$.

6.6.2.2. We are now able to construct the various minimum-volume designs corresponding to *radial line loads of magnitudes P and Q per unit of length applied to the inner radius r_i and the outer radius r_0 of the annular disk, respectively.* We first remark that, from equilibrium, we must have

$$Pr_i = Qr_0 \tag{6.177}$$

for a design of the type of Eq. (6.171), where $\sigma_\theta = 0$, and

$$P = Q \tag{6.178}$$

for a design of the type of Eq. (6.172). In Eqs. (6.177) and (6.178) we obviously assume P and Q the same sign.

If we start with $|P| > |Q|$, the design in Eq. (6.171) will hold alone only if $r_0/r_i = P/Q$. Assuming $Pr_i < Qr_0$, we must end the design in Eq. (6.171) at an intermediate radius $r^* = r_i(P/Q)$, where a resulting load per unit of length Q will be applied by an external design of the type in Eq. (6.172) in order to satisfy $Pr_i = Qr^*$, according to Eq. (6.177). Because of Eqs. (6.174) and (6.171), we must have $r^* \leq 2r_i$—that is, $P < 2Q$. Nevertheless, if the radii r_i and r_0 are modified so that $Pr_i > Qr_0$, the external ring with constant thickness is not necessary anymore, but the radial equilibrium equation, Eq. (6.177), is not satisfied. Hence, a flange with vanishing thickness and infinite height, but finite cross-sectional area, is needed at $r = r_i$ to resist the unbalanced radial distributed forces $(Pr_i - Qr_0)/r_i$. This flange collapses in pure circumferential extension $\dot{\varepsilon}_\theta = \dot{u}(r_i)/r_i$, with $\sigma_\theta = \sigma_0$. Its cross-sectional area is given by $A_f = (Pr_i - Qr_0)/\sigma_0$, and its specific dissipation is $D_f = \sigma_0\dot{\varepsilon}_\theta = \sigma_0 a(b - r_i)/r_i$, equal to the constant specific dissipation in the disk $\sigma_0|\dot{\varepsilon}_r| = a\sigma_0$ with $b = 2r_i$, a condition in agreement with Eq. (6.174).

The other cases of relative values and signs of P and Q can be discussed in a similar manner. The results are summarized in Fig. 6.32, whereas the saving achieved is given in Fig. 6.33.

6.6.2.3. Now consider a *complete disk* ($r_i = 0$) *in rotation subjected to a uniform outward radial force T per unit of length at its boundary* $r = r_0$. We want to design it for minimum volume when a lower bound ω capable of producing plastic collapse is set on its angular velocity. If m is the mass per unit volume of the material, the corresponding specific body force is $f = m\omega^2 rt$. In complete analogy with Eq. (3.122) the optimality condition is that the difference between the dissipation per unit volume and the power $f\dot{u}/t$ of the body force per unit thickness be a constant throughout the disk [Drucker and Shield (1957)]. Hence, we must have

$$\sigma\dot{\varepsilon}_r + \sigma_\theta\dot{\varepsilon}_\theta - m\omega^2 r\dot{u} = \alpha > 0. \tag{6.179}$$

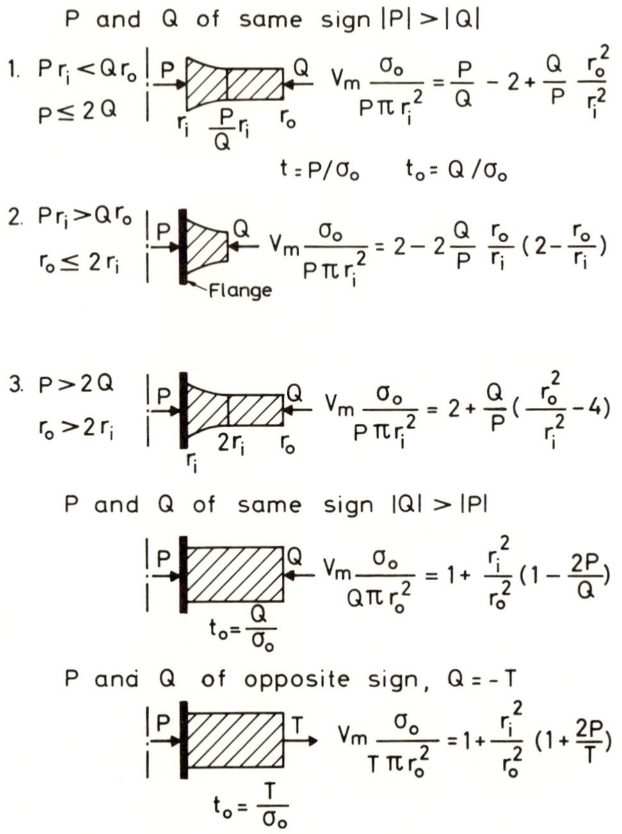

P and Q of same sign $|P| > |Q|$

1. $P r_i < Q r_o$, $P \leq 2Q$

r_i $\quad \frac{P}{Q} r_i$ $\quad r_o$

$$V_m \frac{\sigma_o}{P \pi r_i^2} = \frac{P}{Q} - 2 + \frac{Q}{P} \frac{r_o^2}{r_i^2}$$

$t = P/\sigma_o \qquad t_o = Q/\sigma_o$

2. $P r_i > Q r_o$, $r_o \leq 2 r_i$

Flange

$$V_m \frac{\sigma_o}{P \pi r_i^2} = 2 - 2 \frac{Q}{P} \frac{r_o}{r_i} \left(2 - \frac{r_o}{r_i}\right)$$

3. $P > 2Q$, $r_o > 2 r_i$

r_i $\quad 2r_i$ $\quad r_o$

$$V_m \frac{\sigma_o}{P \pi r_i^2} = 2 + \frac{Q}{P} \left(\frac{r_o^2}{r_i^2} - 4\right)$$

P and Q of same sign $|Q| > |P|$

$$V_m \frac{\sigma_o}{Q \pi r_o^2} = 1 + \frac{r_i^2}{r_o^2} \left(1 - \frac{2P}{Q}\right)$$

$t_o = \frac{Q}{\sigma_o}$

P and Q of opposite sign, $Q = -T$

$$V_m \frac{\sigma_o}{T \pi r_o^2} = 1 + \frac{r_i^2}{r_o^2} \left(1 + \frac{2P}{T}\right)$$

$t_o = \frac{T}{\sigma_o}$

Figure 6.32. Optimal designs of annular disks.

In view of Eq. (6.156) (used for strain rates and velocity) and the fact that both principal stresses are tensile, regime A (Fig. 6.30) must be used to satisfy Eq. (6.179). With $\sigma_r = \sigma_\theta = \sigma_0$ and $f = m\omega^2 r$, integrating the equilibrium equation gives, with boundary condition $\sigma_r(r_0) = \sigma_0$,

$$t = t_0 \exp\left[\frac{K(1 - \rho^2)}{2}\right], \tag{6.180}$$

where

$$\rho = r/r_0, \qquad K = m\omega^2 r_0^2/\sigma_0, \tag{6.181}$$

and the thickness t_0 at the outer radius is

$$t_0 = T/\sigma_0.$$

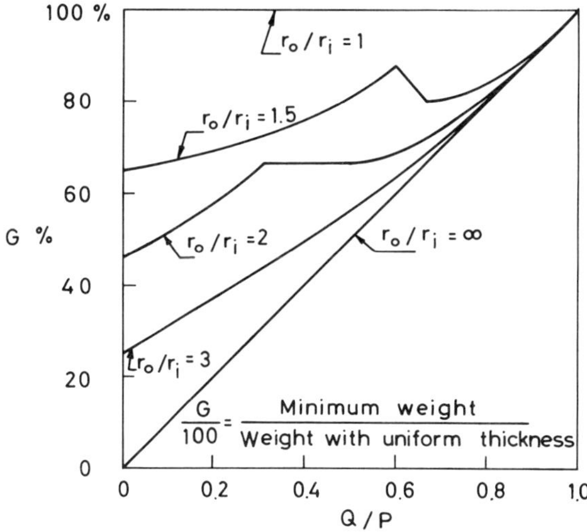

Figure 6.33. Weight savings in optimal annular disks.

In the case of an *annular disk* ($r_i > 0$) the boundary condition $\sigma_r(r_i) = 0$ cannot be satisfied by using the plastic regime A. In order to avoid placing a somewhat unrealistic line flange at the inner radius, an upper bound t^+ can be placed on the thickness t. As shown by Drucker and Shield (1957), the upper bound t^+ is reached for $r_i \leq r < r^*$, and the exponential profile, Eq. (6.180), applies for $r^* \leq r \leq r_0$. The internal ring with constant thickness is in regime AB (Fig. 6.28), and its dissipation per unit volume is larger than that of the external ring, in accordance with the optimality condition in the presence of an upper bound on the design variable. The division radius r^* is given by the equation

$$t^+\left(1 - K\rho^{*2}/3 - \frac{3\rho_i - K\rho_i^3}{3\rho^*}\right) = t_0 \exp\left[\frac{K(1-\rho^{*2})}{2}\right], \qquad (6.182)$$

expressing the continuity of $\sigma_r t$ at r^*.

It is finally worth noting that very little economy in volume is achieved by the exponential profile over a tapered profile, as can be seen from numerical examples given by Heyman (1958).

6.6.3. Assigned Radial Elastic Displacement of the Outer Edge of a Rotating Disk

We again consider the rotating circular disk of Section 6.6.2.3, and we set an upper bound u^* on the radial elastic displacement of its edge $r = r_0$, loaded with the uniformly distributed traction T. We know from Section 3.5.6 that we must place outward radial forces $\bar{T}$ with unit intensity at the edge and consider the mutual elastic strain energy per unit volume $\varepsilon_\theta\bar{\varepsilon}_\theta + \varepsilon_r\bar{\varepsilon}_r$ of the elastic strain fields ε_θ, ε_r and $\bar{\varepsilon}_\theta$, $\bar{\varepsilon}_r$ due to T and $\bar{T}$, respectively. The optimality condition is that the difference between this mutual energy and the work per unit thickness $f\bar{u}$ of the body force ft in the displacement $\bar{u}$ be a positive constant [Chern and Prager (1970)]. From Eqs. (6.156) and (6.158), we have $\bar{u} = r(\bar{\sigma}_\theta - \nu\bar{\sigma}_r)/E$, and the optimality condition is written, in analogy with Eq. (6.85), as

$$\sigma_r\bar{\sigma}_r + \sigma_\theta\bar{\sigma}_\theta - \nu(\sigma_r\bar{\sigma}_\theta + \bar{\sigma}_r\sigma_\theta) - m\omega^2r^2(\bar{\sigma}_\theta - \nu\bar{\sigma}_r) = E\alpha > 0. \tag{6.183}$$

Because $\sigma_r = \sigma_\theta$ and $\bar{\sigma}_r = \bar{\sigma}_\theta$ at $r = 0$, Eq. (6.183) gives

$$E\alpha = 2(1 - \nu)\sigma_r(0)\bar{\sigma}_r(0). \tag{6.184}$$

We eliminate u from Eq. (6.156) to obtain the compatibility condition

$$d(r\varepsilon_\theta)/dr = \varepsilon_r. \tag{6.185}$$

Using Hooke's law, Eq. (6.158), the relations in Eqs. (6.157) and (6.185) are transformed into one differential equation on σ_r only and one differential equation relating σ_θ to σ_r. The boundary conditions are that $\sigma_r(r_0) = T/\sigma_0$ and $\sigma_r(0)$ and $\sigma_\theta(0)$ are equal and finite. The same equations and boundary conditions hold for $\bar{\sigma}_r$ and $\bar{\sigma}_\theta$, with $m\omega^2r = 0$ and $\bar{T}$ substituted for T. Because $t(r)$ remains to be found, the internal forces $S_\theta = \sigma_\theta t$ and $S_r = \sigma_r t$, rather than the stresses, are taken as unknowns. When the force fields S_r, S_θ and $\bar{S}_r$, $\bar{S}_\theta$ are obtained, the optimality condition, Eq. (6.183), used with Eq. (6.184), gives $t(r)$ except for the positive factor $t(0)$. It remains to evaluate the assigned radial displacement u^* at $r = r_0$ by the theorem of virtual work applied to $\bar{T}$, $\bar{S}_\theta$, $\bar{S}_r$ and to actual elastic deformations of the rotating disk to determine $t(0)$. In the static case ($\omega = 0$), it is readily verified that the solution has constant thickness $h(r) = h(0)$, and that $\sigma_r = \sigma_\theta = \sigma_0$ for all r. That the optimal design be fully stressed agrees with the discussion of Section 6.1, because assigned radial displacement coincides with assigned compliance in the absence of body forces.

For comparison we now want to derive the elastic uniform strength design of the rotating disk. We may immediately use the solution, Eq. (6.180), of the plastic, minimum-volume, design problem. Indeed, the statically admissible stress field $\sigma_\theta = \sigma_r = \sigma_0$, being homogeneous, is an elastic stress field because the compatibility equation is satisfied (for any value of Poisson's ratio). Because the strains are constant, $u(r_0) = \varepsilon_r r_0$. From the

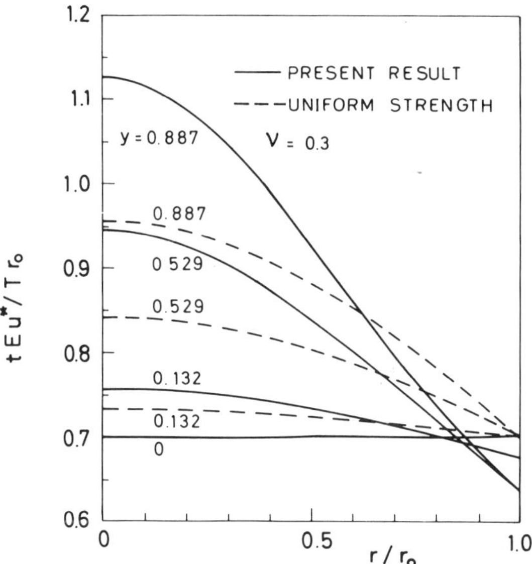

Figure 6.34. Optimal designs of rotating circular disk for assigned elastic displacement of its outer boundary.

assigned value u^* of $u(r_0)$ and $\sigma_r(r_0) = \sigma_0 = T/t(r_0)$, we have $t_0 \equiv t(r_0) = T(1-\nu)r_0/Eu^*$. The comparison of the profiles obtained is given in Fig. 6.34 for $\nu = 0.3$ from Chern and Prager (1970), who found that, despite the strong difference between the functions $t(r)$, the resulting volumes of the uniform-strength designs are less than 1% larger than the minimum-volume designs.

6.7. Other Optimal Disk Problems

We consider hereafter the minimum-volume design of a disk of both *unknown shape and thickness distribution*, capable of supporting at the verge of plastic collapse, a set of loads in equilibrium or of transmitting given loads to given rigid foundation arcs. Assuming that we deal with an isotropic Tresca material, we adopt the coinciding trajectories of principal stresses (S_1 and S_2) and strain rates ($\dot{\varepsilon}_1$ and $\dot{\varepsilon}_2$) for the coordinate system. If s_1 and s_2 are the coordinates of a generic point where the lines of principal stresses have the radii of curvature ρ_1 and ρ_2, we may use the yield condition of Fig. 6.30, Eq. (6.162), and the optimality criterion, Eq. (6.170), by mere substitution of subscripts 1 and 2 for r and θ.

The equilibrium equations of a disk element are known to be

$$\frac{\partial S_1}{\partial s_1} + \frac{S_1 - S_2}{\rho_2} = 0, \qquad \text{(a)}$$
$$\frac{\partial S_2}{\partial s_2} + \frac{S_2 - S_1}{\rho_1} = 0, \qquad \text{(b)} \tag{6.186}$$

where

$$S_1 = \sigma_1 t, \ S_2 = \sigma_2 t. \tag{6.187}$$

Inspection of Fig. 6.30 and Eq. (6.163) (with 1 and 2 substituted for r and θ, respectively) shows that the optimality criterion will be satisfied by the following types of strain rate fields:

(a) Regime AB:

$$\sigma_2 = \sigma_0, \quad \sigma_0 > \sigma_1 > 0, \quad \dot{\varepsilon}_2 = \alpha/\sigma_0, \quad \dot{\varepsilon}_1 = 0; \tag{6.188}$$

(b) Regime BC:

$$\sigma_2 - \sigma_1 = \sigma_0, \qquad -\dot{\varepsilon}_1 = \dot{\varepsilon}_2 = \alpha/\sigma_0; \tag{6.189}$$

(c) Corner B:

$$\sigma_2 = \sigma_0, \quad \sigma_1 = 0, \quad \dot{\varepsilon}_2 = \alpha/\sigma_0, \quad -\dot{\varepsilon}_2 \leq \dot{\varepsilon}_1 \leq 0; \tag{6.190}$$

(d) Corner C:

$$\sigma_1 = -\sigma_0, \quad \sigma_2 = 0, \quad -\dot{\varepsilon}_1 = \alpha/\sigma_0, \quad 0 \leq \dot{\varepsilon}_2 \leq -\dot{\varepsilon}_1. \tag{6.191}$$

Referring to Sections 4.2–4.4, we recognize that the strain rates in (a) and (c) form a Michell field of type R^+; those in (b), of type T; those in (d), of type R^-. Similar conclusions would be obtained with regimes DE, EF, E, and F. For corner regimes A and D, we obtain, with $\sigma_1 = \sigma_2 = \pm\sigma_0$, $\dot{\varepsilon}_1 + \dot{\varepsilon}_2 = \pm\alpha/\sigma_0$. Because a Michell field of type S, if matched with fields of other types, would have $\dot{\varepsilon}_1 + \dot{\varepsilon}_2 = 2\alpha/\sigma_0$ (as $|\dot{\varepsilon}_1| = \alpha/\sigma_0$, $|\dot{\varepsilon}_2| = \alpha/\sigma_0$), we can draw the following conclusion: *All Michell strain (rate) fields constructed for truss-like continua satisfy the optimality criterion, Eq.* (6.170) *for a disk made of a Tresca material, provided it does not contain fields of type S or is formed exclusively of one such field.*

Once we know the strain rate field for optimality, integration of the equilibrium equations, Eqs. (6.186), will give the field of forces S_1 and S_2, and the thickness $t(s_1, s_2)$ will be obtained from Eq. (6.187), taking into account the proper stress regime, Eqs. (6.188)–(6.191). The resulting volume of material will be identical to that of the corresponding truss-like continuum, because it is equal in both cases to the quotient by α of the power of the loads in the common collapse mechanism. The designs of Figs. 4.3, 4.13, and 4.14 are examples that can be regarded as minimum-volume disks.

A graphical procedure for determining the thickness distribution has been given by Prager (1958), which consists of mapping the physical plane of the disk on the force plane with coordinates S_1, S_2. For the problem of Fig. 4.3, Hu and Shield (1961) have given an analytical solution for $t(s_1, s_2)$. They have also pointed out the example of Fig. 6.35, in which the solution for the disk differs from that of the truss-like continuum, which here is any frame that consists entirely of bars in compression [Michell (1904)]; for example, bars along the sides of the triangle *MNG*. The corresponding minimum volume is $6.93Ta/\sigma_0$ (see Fig. 6.35). The minimum-volume disk, given by Cox (1958), is formed of circular arcs in compression with cross section $A = T/\sigma_0$, and the region between the arcs is of constant thickness $T/2\sigma_0 a$ in plane hydrostatic compression $-\sigma_0$. The total volume of this design is $6.60Ta/\sigma_0$.

Kozlowski and Mroz (1970) have derived the minimum-volume criterion for a Tresca disk with upper and lower bounds on its thickness and limits on its free boundaries. It is found, which can be expected from the various criteria already derived in the present book, that the dissipation

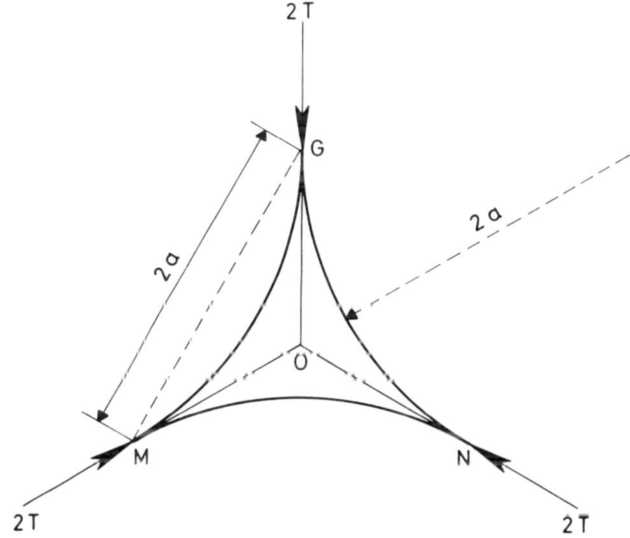

Figure 6.35. Optimal disk for three equal loads with 120° angles.

per unit volume should be a constant where the bounds are not reached, not larger than this constant where the lower bound of the thickness or the internal limit of the boundaries are reached, and not smaller than this constant where the upper bound and the external limit are attained. Because bounds on the thickness render a complete solution very difficult, the authors present what they call near-optimum solutions, obtained by statically admissible modifications of stress fields that do not satisfy the bounding constraints. Examples of minimum-volume disks can also be found in the book by Hemp (1973).

When the material of the disk has vanishing yield stress in tension, which can be approximately taken for concrete, one may think of reinforcing it with ductile cords in the directions of tensile stresses and design it for minimum total volume (or cost) of disk and reinforcing cords. This problem has been treated by Prager (1958).

Exercise 6.9. Prove that elementary volumes with bounding arcs ds_1 and ds_2 of a Michell truss-like continuum and of a minimum-volume Tresca disk have the same value except for fields of type S and corner regimes A or D, respectively.

Exercise 6.10. Explain by physical arguments why the strain rate field of the design of Fig. 6.35 is not adequate for a truss-like continuum, even for producing a good upper bound to the minimum volume.

7

Lagrange Multiplier Methods for Optimization with Constraints

7.1. Introduction

From the preceding chapters it is clear that structural optimization essentially consists of finding a function $t(x)$ that minimizes an integral of the type $\int_v \gamma(t)\,dx$ under a set of equality and inequality constraints. For example, if we wish to minimize the volume of a structure subject to assigned loads at plastic collapse, conditions of equilibrium are equality constraints, whereas the yield condition is expressed by weak inequality constraints.

The general mathematical formulation is as follows. Find

$$\inf_t \Gamma = \inf_t \int_v \gamma(t)\,dx \tag{7.1}$$

under state constraints (e.g., equilibrium equations)

$$G(Q, q, t, p) = 0 \tag{7.2}$$

and behavioral constraints (e.g., material behavior relations)

$$F(Q, q, t) \leq 0, \tag{7.3}$$

where γ is the specific cost function, Q stands for the generalized stresses, q are generalized strains or strain rates, t is the design variable, and p stands for the applied loads. Note that both Eqs. (7.2) and (7.3) are, as a rule, a *set* of relations.

In this general form the problem is quite complex. It is difficult to decide whether or not the variable sets defined by Eqs. (7.2) and (7.3) are convex, which is required for many formal mathematical proofs of existence of optima. The constraints usually involve variables that do not appear in the cost functional. Boundary conditions on the mechanical state functions are not prescribed everywhere on the boundary. Despite these basic difficulties, this problem has been successfully formulated and solved by proper discretization into a finite number of variables and use of numerical procedures, as described in Volume 2 of this treatise. On the other hand, it is also useful to study some aspects of this problem as continuous optimization: though mathematical proofs of existence and uniqueness seem out of reach, one can at least achieve the formulation of necessary conditions of optimality by a simple and systematic procedure.

The form of the variational formulation depends upon the corresponding physical problem. Two classical forms in structural optimization are the statical and the kinematical formulations. In classical solid mechanics, variational principles such as minimum complementary energy and the lower-bound theorem of limit analysis in which stresses subject to equilibrium are the primary variable functions can be said to be of *statical* form. Those principles such as minimum potential energy and the upper-bound theorem of limit analysis in which strains subject to compatibility constraints are the primary variables have *kinematical* form. In structural optimization these same statical and kinematical forms appear with additional design variables occurring.

In the statical formulation the unknowns are the design function $t(x)$ and the generalized stresses, which are subjected to equilibrium and behavioral constraints. The kinematical formulation considers $t(x)$ and the generalized strains (or strain rates, or possibily deflections or velocities) as the unknowns, subjected to kinematic compatibility conditions. In the first case, equilibrated variations are given to the generalized stresses, whereas compatible strain variations are used in the second approach.

As later sections will show, though some formulations may be basically statical or kinematical, the necessary use of the constitutive equations most often couples the static and kinematic variables. This is the general rule in elastic optimality problems. On the contrary, when the behavioral constraint is a lower bound on the load factor at plastic collapse, perfect plasticity enables us to deal with completely uncoupled static and kinematic formulations that are mathematically dual problems. That is, they involve variables that show a reciprocal or dual nature—dual kinematic for primal static, and conversely.

The *Lagrange multiplier technique* has proved especially fruitful, as we shall see in specific examples later on. It furnishes a reasonably unified frame for optimal design of structures subject to a variety of physical constraints [see, for example, Haug and Cea (1981)]. Also, from the engineer's point of view, the fact that the optimality conditions so obtained are most often only necessary is not to be regarded as too much of a handicap.

The reader interested in existence and uniqueness conditions for the solution of an optimal design problem regarded as the minimization of a functional is referred to Hestenes (1966). In the case of the minimization of a function of several variables subject to constraints, it can be shown easily how the Lagrange multiplier technique transforms a problem of constrained minimization into an unconstrained minimization in a larger space.

Indeed, let us decide to find

$$\inf f(x), \qquad x \in R^1, \tag{7.4}$$

with

$$g(x) \leq b, \tag{7.5}$$

where f and g are functions, R^1 is the set of real numbers, and b is a given real number. This problem is equivalent to the following one. Find

$$\inf_x \sup_\lambda \mathscr{L}, \qquad \mathscr{L} = f(x) + \lambda\{g(x) - b\}, \tag{7.6}$$

with

$$\phi \equiv \sup_\lambda\{\lambda[g(x) - b]\} = \begin{cases} 0 & \text{if } g(x) \leq b, \\ \infty & \text{if } g(x) > b. \end{cases} \tag{7.7}$$

The real number λ is a Lagrange multiplier.

Clearly, for Eq. (7.7) to hold, λ must necessarily be nonnegative for all x. When the constraint of Eq. (7.5) is an equality, λ is no longer sign-restricted and is nonvanishing only when the equality is satisfied. The original minimum problem is thus transformed into a search for a minimax with no constraint except the nonnegativity of the Lagrange multiplier λ.

The extension of the design space is clearly shown in Fig. 7.1, where the nonanalytical minimum of the linear function $y = 1 - x$ for $0 \leq x \leq 0.8$ occurs at point A' with the same abscissa, $x = 0.8$, as the analytical saddle point A of the nonlinear function $\mathscr{L}(x, \lambda)$. The coordinates of A satisfy $\partial\mathscr{L}/\partial x = \partial\mathscr{L}/\partial\lambda = 0$.

As a further example, suppose we wish to minimize

$$y = 1 + (x^4 - x),$$

subject in turn to each of the following constraints (see Fig. 7.2):

$0.36 \leq x \leq 0.86,$ (a) (convex)

$0 \leq x \leq 0.36$ and $0.86 \leq x \leq 1,$ (b) (nonconvex)

$0 \leq x \leq 0.36.$ (c) (convex)

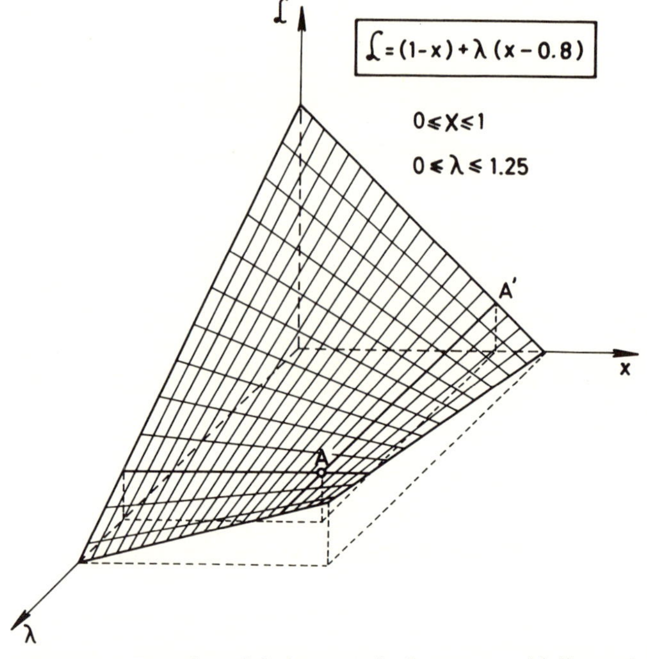

Figure 7.1. Extension of design space by Lagrange multiplier method.

Constraint (a) defines a convex set in x. We let

$$\mathscr{L} = 1 + (x^4 - x) + \lambda(0.36 - x) + \mu(x - 0.86) \tag{7.8}$$

and seek (x, λ, μ) to solve the minimax problem

$$\inf_{x} \sup_{\substack{\lambda \geq 0 \\ \mu \geq 0}} \mathscr{L}.$$

Stationariness of $\mathscr{L}$ implies $\partial\mathscr{L}/\partial x = \partial\mathscr{L}/\partial\lambda = \partial\mathscr{L}/\partial\mu = 0$, giving

$$4x^3 - 1 - \lambda + \mu = 0, \tag{7.9a}$$

$$0.36 - x = 0, \tag{7.9b}$$

$$-0.86 + x = 0. \tag{7.9c}$$

These three equations clearly have no common root, so the minimum of $\mathscr{L}$ cannot occur at a stationary point, but it must occur when at least one of the variables is on the boundary of its admissible set. We first show that x

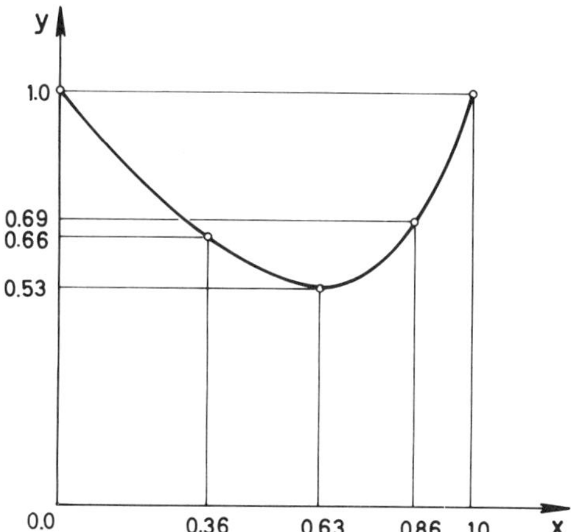

Figure 7.2. Function y to minimize with constraints on variable x.

cannot be at either of its boundary values 0.36 or 0.86. Suppose $x = 0.36$. Then $\mathscr{L}$ has the value $1 + (0.36^4 - 0.36) - 0.5\mu$ independent of the value of λ. The supremum $\mathscr{L}^* \equiv 0.66$ of this for $\mu \geq 0$ occurs at $\mu = 0$. Similarly, for $x = 0.86$, the value of $\mathscr{L}$ is independent of μ, and the supremum is $\mathscr{L}^{**} = 0.69$, occurring at $\lambda = 0$.

Both $\mathscr{L}^*$ and $\mathscr{L}^{**}$ are greater than the value of $\mathscr{L}$ occurring when $\lambda = \mu = 0$, and a stationary value is taken with respect to x alone; this occurs when $x = \sqrt[3]{1/4} \simeq 0.63$, with $\mathscr{L}_{\min} = 0.53$. Indeed, setting $\lambda = 0$ or $\mu = 0$ first and then seeking a minimum by stationary conditions with respect to the other two variables leads to the same difficulty. The other Lagrange multiplier must be negative at the stationary point and thus lies outside the admissible set. For $x = 0.36$, $\mu = 0$, Eq. (7.9) gives $\lambda = -0.81$; for $x = 0.86$, $\lambda = 0$, Eq. (7.9) gives $\mu = -1.54$ for stationariness.

Constraints (b) define a nonconvex set. With

$$\mathscr{L} = 1 + (x^4 - x) + \lambda(x - 0.36) + \mu(-x + 0.86) + \eta(0 - x) + \nu(x - 1),$$

we obtain, in the same manner as with constraints (a), two candidate solutions:

$$x = 0.36, \quad \mu = \eta = \nu = 0, \quad \lambda = 0.81 > 0, \quad y = 0.66,$$

and

$$x = 0.86, \quad \lambda = \eta = \nu = 0, \quad \mu = 1.54 > 0, \quad y = 0.69,$$

of which the first gives the solution to $\inf_x \sup_{\lambda,\mu,\nu,\eta \geq 0} \mathscr{L}$. We remark that the smallest value of y is *not* obtained at the abscissa $x = 0.63$ of the unconstrained minimum, contrary to an assumption commonly made.

Constraints (c) define a convex set. With

$$\mathscr{L} = 1 + (x^4 - x) + \lambda(x - 0.36) + \mu(0 - x),$$

the solution to $\inf_x \sup_{\lambda,\mu\geq 0} \mathscr{L}$ is

$$x = 0.36, \quad \mu = 0, \quad \lambda = 0.81 > 0, \quad \textit{and} \quad y = 0.66.$$

A further special case may occur. Consider the constraints $0 \leq x \leq 0.63$. It is easily seen that the solution $x = 0.63$, $\lambda = \mu = 0$, $y = 0.53$ is an analytical minimum occurring at the boundary of the admissible domain.

7.2. Beams in Bending

7.2.1. Assigned Load Factor at Plastic Collapse

We begin with the *statical formulation* based on the corresponding theorem of limit analysis. Using the notations of Chapter 3 and a superscript $'$ to denote derivatives with respect to x, we must have

$$p + M'' = 0 \qquad \text{(equilibrium)}, \tag{7.10}$$

and

$$-Y^- \leq M \leq Y^+ \qquad \text{(yield condition)}. \tag{7.11}$$

We construct the functional

$$\mathscr{L} = \int_0^l \gamma(t)\,dx + \int_0^l \lambda(p + M'')\,dx + \int_0^l \mu_1(M - Y^+)\,dx + \int_0^l \mu_2(-M - Y^-)\,dx, \tag{7.12}$$

where t is the cross-sectional design variable, and state the minimax problem

$$\inf_{\substack{t\geq 0\\ M}} \sup_{\substack{\lambda\\ \mu_1,\mu_2\geq 0}} \mathscr{L}. \tag{7.13}$$

The stationary conditions for $\mathscr{L}$ furnish the relations in Eqs. (7.10) together with the supplementary relations

$$\mu_1(M - Y^+) = 0, \tag{7.14}$$

$$\mu_2(-M - Y^-) = 0, \tag{7.15}$$

$$\gamma_t - \mu_1 Y_t^+ - \mu_2 Y_t^- \geq 0, \tag{7.16}$$

$$\lambda'' + \mu_1 - \mu_2 = 0, \tag{7.17}$$

where the subscript t indicates partial derivation with respect to t. The inequality sign in Eq. (7.16) must be introduced because t is physically sign-constrained. The reader may verify that adding the explicit constraint $t \geq 0$ requires adding the term $\int_0^l \nu(x)t\,dx$ to $\mathscr{L}$ with nonpositive Lagrange multiplier ν. This leads to Eq. (7.16) as the stationary condition, in which the inequality sign holds only when $t = 0$.

Note that Eq. (7.17) is obtained by integration by parts, using the boundary conditions

$$(M\lambda')_{\substack{x=0\\x=l}} = 0, \qquad (M'\lambda)_{\substack{x=0\\x=l}} = 0. \tag{7.18}$$

Indeed, the stationary condition with respect to M is

$$\delta\mathscr{L}_M \equiv \int_0^l [\lambda(\delta M)'' + \mu_1\delta M - \mu_2\delta M]\,dx = 0.$$

The first term of the integrand can be written as $[\lambda(\delta M)']' - (\lambda'\delta M)' + \lambda''\delta M$. With the boundary conditions in Eq. 7.18), valid for λ and δM, we obtain

$$\delta\mathscr{L}_M \equiv \int_0^l \delta M(\lambda'' + \mu_1 - \mu_2)\,dx = 0.$$

The mechanical interpretation of Eqs. (7.14)-(7.18) follows. The multipliers $-\mu_1$ and μ_2 are rates of positive and negative plastic curvature, respectively, and Eqs. (7.14) and (7.15) show that only when one yield condition is satisfied does the corresponding rate of curvature have a nonvanishing value; the other is zero. The relation in Eq. (7.16) is the optimality criterion of Eq. (3.9) when no bound is set on the design variable t except the physical bound of nonnegativeness. The relation in Eq. (7.17) expresses the rate of curvature as a function of the velocity λ of the collapse mechanism.

If we adopt the *kinematical formulation*, we consider a velocity field $\dot{v}(x)$, the corresponding rates of curvature $-\dot{v}''$, and define the specific dissipation as

$$D = \begin{cases} -\dot{v}''Y^+ & \text{if } \dot{v}'' < 0, \\ \dot{v}''Y^- & \text{if } \dot{v}'' > 0, \\ 0 & \text{if } \dot{v}'' = 0. \end{cases} \tag{7.19}$$

The unknown functions $\dot{v}(x)$ and $t(x)$ (or $Y(x)$) are subject to the constraint

$$\int_0^l (p\dot{v} - D)\, dx \le 0, \tag{7.20}$$

derived from the kinematic theorem of limit analysis, when we require that the design $t(x)$ must be able to support the load, and $\dot{v}$ is kinematically admissible. Hence, the relevant functional is

$$\mathcal{L} = \int_0^l \gamma(t)\, dx + \alpha \int_0^l (p\dot{v} - D)\, dx,$$

and the optimization problem is expressed as

$$\inf_{\substack{t\ge 0\\ \dot{v}}} \sup_{\alpha \ge 0} \mathcal{L}. \tag{7.21}$$

The stationary condition for $\mathcal{L}$ furnishes Eq. (7.20) and

$$\gamma_t - \alpha D_t \ge 0, \tag{7.22}$$

$$p + M'' = 0. \tag{7.23}$$

Recalling that a collapse mechanism is defined except for a positive constant, we see that Eq. (7.22) is identical to Eq. (7.16). The equilibrium equation, Eq. (7.23), is obtained from the stationariness of $\mathcal{L}$ with respect to $\dot{v}$ (or $\dot{v}''$) by applying integration by parts with natural boundary conditions of the type in Eq. (7.18) and by taking into account the fact that the structure is completely plastic, $D = -\dot{v}''M$. We see that we recover the results of Section 3.1: namely, the optimality criterion and, by using the Lagrange multipliers, the coupling of the static and kinematic aspects of the solution.

7.2.2. Assigned Bounds on Elastic Deflections

Let us assign upper and lower bounds on the elastic deflection $v(x)$ of a beam with stiffness $B[t(x)]$:

$$-v^- \le v \le +v^+, \tag{7.24}$$

where v^- and v^+ are given nonnegative numbers.

The only other constraint is the equation of elastic equilibrium

$$p - (Bv'')'' = 0. \tag{7.25}$$

The functional $\mathscr{L}$ is readily written as

$$\mathscr{L} = \int_0^l \gamma(t)\,dx + \int_0^l \eta\{p - (Bv'')''\}\,dx + \int_0^l \mu_1(v - v^+)\,dx + \int_0^l \mu_2(-v - v^-)\,dx. \tag{7.26}$$

The optimization problem is then

$$\inf_{\substack{t\geq 0\\ v}} \sup_{\substack{\eta\\ \mu_1,\mu_2\geq 0}} \mathscr{L}. \tag{7.27}$$

Besides Eq. (7.25) the stationary conditions for $\mathscr{L}$ are

$$\mu_1(v - v^+) = 0, \tag{7.28}$$

$$\mu_2(-v - v^-) = 0, \tag{7.29}$$

$$\gamma_t - B_t \cdot v''\eta'' \geq 0, \tag{7.30}$$

$$-(B\eta'')'' + \mu_1 - \mu_2 = 0. \tag{7.31}$$

They have the following mechanical significance. The multipliers μ_1 and μ_2 are distributions of concentrated forces at the points where v reaches the bounds. They cannot both be nonvanishing at the same abscissa because of Eqs. (7.28) and (7.29), a physically obvious fact since v cannot simultaneously attain v^+ and $-v^-$. The concentrated forces μ_1 and μ_2 produce elastic deflections η and curvatures η'' governed by the equation of elastic equilibrium, Eq. (7.31). Boundary conditions for v and η are identical. The necessary optimality condition in Eq. (7.30) shows that the mutual specific strain energy per unit design variable t is to be proportional to the marginal specific cost.† This is a generalization of the condition of Eq. (3.142), which is easily recovered by simply substituting $v(x_0) \leq v^+$ for Eq. (7.24) [see Cinquini and Sacchi (1980)].

Equations (7.28)-(7.31) can be given a discretized finite difference form and be solved numerically by successive approximations [Cantu and Cinquini (1979)]. In the procedure, the optimality criterion in Eq. (7.30) serves to define the successive improvements in the design $t(x)$.

7.2.3. Multipurpose Optimal Design

In multipurpose optimal design the variational approach directly furnishes the relevant optimality criterion together with the governing field

† Where $t = 0$, the inequality sign may hold.

equations. Let us consider the minimum-cost design of a beam that must exhibit bounded deflections under the service load $p(x)$ and an assigned minimum value sp of its plastic collapse load. Under load p, elastic behavior is assumed† and the constraints in Eqs. (7.24) and (7.25) hold. On the other hand, the constraints in Eqs. (7.10) and (7.11) are valid under the load sp. Hence,

$$\begin{aligned}\mathscr{L} = &\int_0^l \gamma(t)\,dx + \int_0^l \eta\{p - (Bv'')''\}\,dx + \int_0^l \mu_1(v - v^+)\,dx \\ &+ \int_0^l \mu_2(-v - v^-)\,dx + \int_0^l \lambda(sp + M'')\,dx + \int_0^l \nu_1(M - Y^+)\,dx \\ &+ \int_0^l \nu_2(-M - Y^-)\,dx, \end{aligned} \tag{7.32}$$

with μ_1, μ_2, ν_1, and ν_2 nonnegative.

The stationary conditions for $\mathscr{L}$ are written as usual; in particular, the condition with respect to t gives the optimality criterion

$$\gamma_t - B_t v''\eta'' - \nu_1 Y^+ - \nu_2 Y^- \geq 0. \tag{7.33}$$

We recall that the inequality sign may hold only where $t = 0$. With the physical meanings given to v, η, μ_1, and μ_2 in Section 7.2.2 and to λ, μ_1, and μ_2 in Section 7.2.1, the optimality criterion in Eq. (3.219) is recovered when Eq. (7.33) is particularized to the case with a deflection limitation at a given point and cost identified with volume.

7.3. Plates in Bending

7.3.1. Elastic Sandwich Plates with Limited Stress Intensity

In order to achieve generality with respect to possible plate geometries, we hereafter use tensorial notation. The generalized stresses are the moments M_{ik} $(i, k = 1, 2)$, bending moments corresponding to $i = k$, and twisting moments to $i \neq k$ (with $M_{ik} = M_{ki}$). If w is the deflection field, the generalized strains are the curvatures $w_{,ik}$, where the comma subscript indicates partial derivation with respect to the following subscripts. Bending curvatures are obtained with $i = k$; twisting curvatures, with $i \neq k$. The summa-

† This assumption is usually verified a posteriori. Introducing the supplementary constraint of elastic behavior (limited effective stress) from the very beginning would greatly complicate the problem.

tion convention on repeated subscripts is used [see Synge and Schild (1949)]: for example,

$$a_{ik}b_{ik} = a_{11}b_{11} + a_{12}b_{12} + a_{21}b_{21} + a_{22}b_{22}.$$

The general relation between moments and curvatures for sandwich plates is then

$$M_{ik} = -tC_{ikrs} \cdot w,_{rs}, \tag{7.34}$$

where the fourth-order tensor C_{ikrs} depends only on the elastic constants of the material.

The stress intensity constraint is

$$f(M_{ik}, t) \le 0, \tag{7.35}$$

and the equation of equilibrium reads

$$(tC_{ikrs}w,_{rs}),_{ik} + p = 0. \tag{7.36}$$

Hence, the relevant functional is

$$\mathscr{L} = \int_A \gamma(t)\, dx + \int_A \mu\{(tC_{ikrs}w,_{rs}),_{ik} + p\}\, dA$$
$$+ \int_A \beta_{ik}(M_{ik} + tC_{ikrs}w,_{rs})\, dA - \int_A \lambda f(M_{ik}, t)\, dA, \tag{7.37}$$

where A stands for the mid-surface of the plate, dA for an element of area of this, and μ, β_{ik}, and λ are Lagrange multipliers. The stationary conditions for $\mathscr{L}$ are

$$\partial\gamma/\partial t + \mu,_{ik}C_{ikrs}w,_{rs} + \beta_{ik}C_{ikrs}w,_{rs} - \lambda\, \partial f/\partial t \ge 0, \tag{7.38}$$

$$\beta_{ik} - \lambda\, \partial f/\partial M_{ik} = 0, \tag{7.39}$$

$$(\mu,_{ik} + \beta_{ik})t = 0. \tag{7.40}$$

Note that the second term or the left-hand side of Eq. (7.38) is obtained, as usual, by applying the Green–Ostrogradsky formula.

Substitution of Eq. (7.40) into Eq. (7.38) results in

$$\lambda\, \partial f/\partial t \le \partial\gamma/\partial t. \tag{7.41}$$

Assume now that the optimal design for the single loading p is "fully stressed"—that is,

$$f(M_{ik}, t) = 0 \tag{7.42}$$

all over the plate—and that f is homogeneous in M_{ik} and t. Then from Euler's theorem on homogeneous functions we may write

$$(\partial f/\partial M_{ik})M_{ik} + (\partial f/\partial t)t = 0. \tag{7.43}$$

Multiplying both sides of eq. (7.43) by γ and using eqs. (7.39)-(7.41), we obtain the necessary optimality condition

$$\frac{\mu_{,ik}M_{ik}}{t} \leq \frac{\partial \gamma}{\partial t}. \tag{7.44}$$

We conclude that, if there is an optimal design, there must exist a fictitious deflection field μ related to a fully stressed, elastic moment field M_{ik} by the normality law (as shown by Eqs. (7.39) and (7.40)) applied to the limiting homogeneous function $f(M_{ik}, t)$. Moreover, the fictitious specific energy per unit thickness of the two fields $\mu_{,ik}$ and M_{ik} is equal to the derivative to the specific cost with respect to the thickness where $t > 0$, and smaller than that derivative where $t = 0$.

It can be verified that μ can indeed be regarded as a kinematically admissible deflection field by applying the Green–Ostrogradsky formula to the second integral of Eq. (7.37) and regarding the resulting expression as though it were derived from the theorem of virtual work for the statically admissible (elastic) moment field.

Note that, when $f(M_{ik}, t)$ is the specific elastic strain energy of the plate element, $\mu_{,ik}$ coincides with $w_{,ik}$.

The results of the present section have been presented by Shield (1960b) and Save (1968) as sufficient conditions for the absolute minimum volume of sandwich structures (see Section 6.1).

7.3.2. Reinforced Concrete Plates with Assigned Limit Load.

For the sake of simplicity we restrict ourselves to considering reinforcement in two orthogonal directions x and y at the lower face of the plate (positive bending only). We aim at minimizing the total volume of reinforcement

$$V = \int_A (t_x + t_y)\, dA, \tag{7.45}$$

where t_x, t_y are the cross-sectional areas of the reinforcement per unit length in the x- and y-directions, respectively. Assuming a known constant value H for the moment arm of the internal forces in the collapse state, the yield moments are

$$M_{px} = t_x\sigma_0 H, \qquad M_{py} = t_y\sigma_0 H. \tag{7.46}$$

Hence, we may use the cost function

$$\Gamma = \int_A (M_{px} + M_{py})\, dA. \tag{7.47}$$

The generalized stresses are the bending moments M_x and M_y and the twisting moment M_{xy}. The constraint of plastic admissibility is known to be [see Save and Massonnet (1972)]:

$$f = (M_{px} - M_x)(M_{py} - M_y) - M_{xy}^2 \geq 0. \tag{7.48}$$

With the well-known equilibrium equation

$$\frac{\partial^2 M_x}{\partial x^2} + \frac{\partial^2 M_y}{\partial y^2} + 2\frac{\partial^2 M_{xy}}{\partial x\, \partial y} + p = 0, \tag{7.49}$$

the relevant functional is

$$\mathscr{L} = \int_A (M_{px} + M_{py})\, dA + \int_A \mu\left(\frac{\partial^2 M_x}{\partial x^2} + \frac{\partial^2 M_y}{\partial y^2} + 2\frac{\partial^2 M_{xy}}{\partial x\, \partial y} + p\right) dA$$
$$+ \int_A \lambda f(M_x, M_y, M_{xy}, M_{px}, M_{py})\, dA.$$

The minimax problem is then

$$\inf_{\substack{M_x, M_y, M_{xy} \\ M_{px}, M_{py} \geq 0}} \ \sup_{\substack{\mu \\ \lambda \geq 0}} \mathscr{L}.$$

Let M_1 and M_2 be the principal bending moments, and let α be the angle of the first principal direction with the x-axis. Then

$$\begin{aligned} M_x &= M_1 \cos^2 \alpha + M_2 \sin^2 \alpha, \\ M_y &= M_1 \sin^2 \alpha + M_2 \cos^2 \alpha, \\ M_{xy} &= (M_2 - M_1) \sin \alpha \cdot \cos \alpha. \end{aligned} \tag{7.50}$$

We now use the Green-Ostrogradsky formula to transform the second integral of $\mathscr{L}$ in such a way as to express equilibrium by a virtual work equation, substitute Eqs. (7.50) for M_x, M_y, and M_{xy} in the resulting expression for $\mathscr{L}$, and express the stationary conditions for the functional so obtained (with variables M_x, M_y, and M_{xy} replaced by M_1, M_2, α). The resulting necessary conditions of optimality are satisfied if

$$\alpha = 0, \qquad (\partial^2\mu/\partial x\, \partial y)(M_2 - M_1) = 0, \tag{7.51}$$

$$\lambda(M_{px} - M_x) \geq -1, \qquad \lambda(M_{py} - M_y) \geq -1, \tag{7.52}$$

$$\partial^2\mu/\partial x^2 - \lambda(M_{py} - M_y) = 0, \qquad \partial^2\mu/\partial y^2 - \lambda(M_{px} - M_x) = 0. \tag{7.53}$$

Equations (7.53) show that μ can be regarded as a plastic field of deflection rates, because its curvatures ($\kappa_x = -\partial^2\mu/\partial x^2$, $\kappa_y = -\partial^2\mu/\partial y^2$) are obtained by applying the normality law to the yield condition of Eq. (7.48).

Equations (7.51) indicate that in the general case, when $M_2 \neq M_1$, the twisting curvature vanishes and the reinforcing bars must be placed along the coincident principal directions of bending moments and curvatures. Finally, Eqs. (7.53) and (7.52) give

$$-\partial^2\mu/\partial x^2 \leq 1, \qquad -\partial^2\mu/\partial y^2 \leq 1. \tag{7.54}$$

Inequality signs hold only where $t_x = 0$ or $t_y = 0$, respectively. The conditions in Eqs. (7.54) were first obtained by Morley (1966) and later discussed by Sacchi and Save (1969).

7.4. Optimal Segmentation and Optimal Location of Supports

The variational formulation is especially convenient for obtaining optimality criteria for segmentation in the case of a piecewise-constant design function and for support locations. For the sake of simplicity we consider beams in bending with at most one discontinuity in the design variable and one unknown support location. The results are valid for several discontinuities and support locations and can be extended to plates and shells [see Masur (1975a), Dems and Mroz (1980), Cinquini *et al.* (1977)]. Plastic and elastic behaviors are treated successively.

7.4.1. Assigned Limit Load—Optimal Segmentation

Consider again the problem treated in Section 7.2.1, and let a be the unknown abscissa of a discontinuity in the design variable t. Let us use subscripts (1) and (2) to indicate that functions are taken in the intervals $0 \le x \le a$ and $a \le x \le l$, respectively. In the functional $\mathcal{L}$, dependence on a is made explicit by splitting into two the integral over the whole interval $(0, l)$: over $(0, a)$ and (a, l).

The stationary conditions for $\mathcal{L}$ are Eqs. (7.14)-(7.17), to which we must now add $\partial\mathcal{L}/\partial a = 0$. We recall that the partial derivative of an integral with respect to a limit of integration is nothing but the value of the integrand at that limit, and we take Eqs. (7.14) and (7.15) into account in order to use the simplified functional

$$\mathcal{L}^* = \int_0^a \gamma_{(1)}\,dx + \int_a^l \gamma_{(2)}\,dx + \int_0^a \lambda_{(1)}(p_{(1)} + M''_{(1)})\,dx + \int_a^l \lambda_{(2)}(p_{(2)} + M''_{(2)})\,dx. \tag{7.55}$$

Using integration by parts, we can rewrite Eq. (7.55) as

$$\mathcal{L}^* = \int_0^a \gamma_{(1)}\,dx + \int_a^l \gamma_{(2)}\,dx + \int_0^a \lambda_{(1)}p_{(1)}\,dx + \int_0^a (\lambda_{(1)}M'_{(1)})'\,dx - \int_0^a \lambda'_{(1)}M'_{(1)}\,dx + \int_a^l \lambda_{(2)}p_{(2)}\,dx + \int_a^l (\lambda_{(2)}M'_{(2)})'\,dx - \int_a^l \lambda'_{(2)}M'_{(2)}\,dx.$$

Because the loading p, the velocity λ of the collapse mechanism, and the shear force M' are continuous at $x = a$, the stationary condition $\partial\mathcal{L}^*/\partial a = 0$ furnishes

$$\gamma_{(1)}(a) - \gamma_{(2)}(a) = [\lambda'_{(1)}(a) - \lambda'_{(2)}(a)]M'(a). \tag{7.56}$$

We thus recover Eq. (3.229).

Equation (7.56) is a so-called transversality condition for $\mathcal{L}$ [see Korn and Korn (1961, Section II-5-4b)].

7.4.2. Assigned Limit Load—Optimal Support Location

The presence of a support at $x = a$ renders $M'(x)$ discontinuous at that section. Hence, in the absence of discontinuity of t at the same abscissa, Eq. (7.56) becomes

$$\lambda'_{(2)}(a)M'_{(2)}(a) - \lambda'_{(1)}(a)M'_{(1)}(a) = 0. \tag{7.57}$$

If λ' is continuous at $x = a$ (no plastic hinge), Eq. (7.57) can be written

$\lambda'(a)\{M'_{(2)}(a) - M'_{(1)}(a)\} = 0$, or

$$R\lambda'(a) = 0, \tag{7.58}$$

where R is the support reaction. From Eq. (7.58) we conclude that the optimal location of a support with nonvanishing reaction is the abscissa of a section with vanishing rotation rate.

7.4.3. Assigned Elastic Compliance—Optimal Segmentation

When a maximum value $\mathscr{E}_0$ is assigned to the elastic compliance of a beam in bending, the minimum-cost problem can be formulated in a most compact manner by using the functional

$$\mathscr{L} = \int_0^l \gamma(t)\,dx + \beta\left\{\int_0^l (pv - \tfrac{1}{2}Bv''^2)\,dx - \tfrac{1}{2}\mathscr{E}_0\right\}. \tag{7.59}$$

Indeed, the constraint $\int_0^l (pv - \frac{1}{2}Bv''^2)\,dx = \frac{1}{2}\mathscr{E}_0$ contains both the assignment of a given compliance in the form $\int_0^l pv\,dx = \mathscr{E}_0$ and Hooke's law in the form

$$\frac{1}{2}\int_0^l pv\,dx = \frac{1}{2}\int_0^l Bv''^2\,dx.$$

In Eq. (7.59) the Lagrange multiplier β is a constant. The stationary conditions with respect to t and v furnish the optimality condition

$$\gamma_t - \beta B_t v''^2/2 = 0 \tag{7.60}$$

and the equation of elastic equilibrium

$$p - (Bv'')'' = 0. \tag{7.61}$$

We can, as in Section 7.4.1, divide the integrals of Eq. (7.59) and write

$$\mathscr{L} = \int_0^a \gamma_{(1)}\,dx + \int_a^l \gamma_{(2)}\,dx + \beta\left\{\int_0^a (p_{(1)}v_{(1)} - \tfrac{1}{2}B_{(1)}v''^2_{(1)})\,dx \right.$$
$$\left. + \int_a^l (p_{(2)}v_{(2)} - \tfrac{1}{2}B_{(2)}v''^2_{(2)})\,dx - \tfrac{1}{2}\mathscr{E}_0\right\}. \tag{7.62}$$

Because p and v are continuous at a, $\partial\mathscr{L}/\partial a = 0$ gives

$$\gamma_{(1)}(a) - \gamma_{(2)}(a) = (\beta/2)\{B_{(1)}(a)v''^2_{(1)}(a) - B_{(2)}(a)v''^2_{(2)}(a)\}, \tag{7.63}$$

a formula already used in Eq. (3.230). Note that the constant β depends on $\mathscr{E}_0$.

7.4.4. Assigned Elastic Compliance—Optimal Support Location

We first transform the relation in Eq. (7.62), using integration by parts:

$$\int_0^a Bv''^2\,dx = \int_0^a (Bv'v'')'\,dx - \int_0^a (Bv'')'v'\,dx$$
$$= [Bv'v'']_0^a - \int_0^a (Bv'')'v'\,dx.$$

Obviously, the same property holds for the interval $\{a, l\}$. But the boundary conditions of the beam are such that $Bv'v''$ vanishes at $x = 0$ and $x = l$. Also, continuity of moment Bv'' and slope v' give

$$[B(a)v'(a)v''(a)]_{(1)} = [B(a)v'(a)v''(a)]_{(2)}.$$

Hence Eq. (7.62) becomes

$$\mathscr{L} = \int_0^a \gamma_{(1)}\,dx + \int_a^l \gamma_{(2)}\,dx + \beta\left\{\int_0^a [p_{(1)}v_{(1)} + \tfrac{1}{2}(B_{(1)}v''_{(1)})' \times v'_{(1)}]\,dx + \int_a^l [p_{(2)}v_{(2)} + \tfrac{1}{2}(B_{(2)}v''_{(2)})'v'_{(2)}]\,dx - \tfrac{1}{2}\mathscr{E}_0\right\}. \tag{7.64}$$

Because p, v, v', and γ are continuous at the support location, $\partial\mathscr{L}/\partial a$ gives

$$v'(a)[(Bv'')'_{(2)} - (Bv'')'_{(1)}]|_{x=a} = 0. \tag{7.65}$$

The factor in the braces is the discontinuity in shear force at $x = a$ and equals the support reaction R. Hence, Eq. (7.65) can be written

$$Rv'(a) - 0. \tag{7.66}$$

The reaction R is, in general, nonvanishing. The optimal support location then corresponds to vanishing slope of the deflection curve.

7.4.5. Assigned Elastic Buckling Load—Optimal Support Location

Equation (7.66) remains valid if $v(x)$ is regarded as the buckled shape. But any intermediate support (with unknown location) will impose a new type of buckled shape for which, in general, $v'(a) \neq 0$. Hence, optimality will imply $R = 0$; that is, the support must be located at a "natural node" of the buckled configuration.

7.5. General Discussion

In Sections 7.1–7.4 it has been shown that the variational formulation using Lagrange multipliers for constraints furnishes a unified and systematic procedure for establishing optimality criteria in various problems of structural optimization. Besides those explicitly derived, all the other criteria obtained in Chapters 3–6 can be recovered without special difficulties [see, for example, Cinquini and Sacchi (1980)]. Moreover, we can easily derive criteria for which a mechanical basis does not yet seem to exist, as in Section 7.4. Hence, the method can be regarded as a very (if not the most) powerful way of obtaining optimality criteria. Many authors have used it widely [see, for example, Rozvany (1976)]. It must be noted that the resulting optimality conditions are, of course, only necessary. Sufficiency as well as the relative or absolute character of the optimum, which are often easily dealt with on a mechanical basis, are here connected to a sometimes intricate discussion of convexity of the functions and sets involved. Also, determining the physical significance of the Lagrange multipliers is not always straightforward.

The solution methods employed (analytical or numerical, discretized or nondiscretized) are called "optimality criteria methods" since satisfactory progress in the calculation is not judged on evaluation of cost or cost variation, but on satisfaction of optimality criteria either at every stage of a possible successive approximation procedure or in a closed form solution. Cost evaluation is done only at the very end after optimality has been obtained. This solution method has already been used extensively in Chapters 3–6.

It will be seen in Volume 2 of this treatise that for discretized systems it is, as a rule, more convenient to rewrite the functional and its stationary conditions completely in matrix form. Knowledge of the procedure and of the general results of this chapter will help in understanding those other forms of the optimality calculation.

8

Review of Some Extensions of Design Procedures

8.1. Introduction

We have so far derived optimality criteria for different types of single-purpose or multipurpose structures subject to a variety of behavioral constraints and constructed the optimal design for a number of simple problems. Besides the intrinsic interest of the examples treated, at least as a reference basis for more practical designs, they enabled us to gain deep mechanical insight into the procedure of optimal design, as well as a fairly unified view of the various facets of that problem. When one is faced with large structures subjected to numerous constraints and loading cases, a numerical approach, preceded by proper discretization, is, as a rule, necessary. Application of a criterion to guide the design process toward optimality may very often be recommended [see, for example, Venkayya (1978)] independently of the derivation of the criterion (by mechanical arguments, by the Lagrange multiplier technique, or even as a Kuhn–Tucker condition of the optimal solution of a mathematical programming problem).

This question will be discussed at length in Volume 2 of this treatise. Hence, we shall devote ourselves in this chapter to a broad, though not exhaustive, overview of many aspects that have not been considered in the preceding chapters; namely, the extensions to other types of structures, to other behavioral constraints, and to other cost functions.

8.2. Extensions to Other Structures

The first natural extension from beams in bending is to arches or rings in which bending moment and axial force are, in general, acting simultaneously on the cross section. Minimum-weight plastic design of a circular sandwich ring was considered by Prager (1969), who treated the case of two diametrically opposed outward forces with a lower bound set on the yield moment.

This is a very simple example of a *beam subjected to several simultaneous generalized stresses*, a problem that can be solved, as a rule, along the same lines as plate and shell problems. The ring studied by Prager (1969) was reconsidered by Mroz and Gawecki (1975) in the absence of a lower bound on the yield moment. They showed that *the post-yield behavior* of the optimized ring is always unstable, in the sense that increasing displacements occur for a decreasing load parameter, even when the applied forces are directed outward. It might then be feared that, due to the change of geometry in the elastic range (possibly amplified by shape imperfections), the structure might "jump" to the descending branch of the load-displacement curve (Fig. 8.1) with the corresponding lowering of the limit load. A similar situation was pointed out by Thompson and Supple (1973) in connection with minimum-weight design for assigned elastic buckling load. This phenomenon of *erosion of optimum designs*, as it is called by these authors, must be kept in mind and properly accounted for in setting the required safety factors, but it should also be tested experimentally in order to take

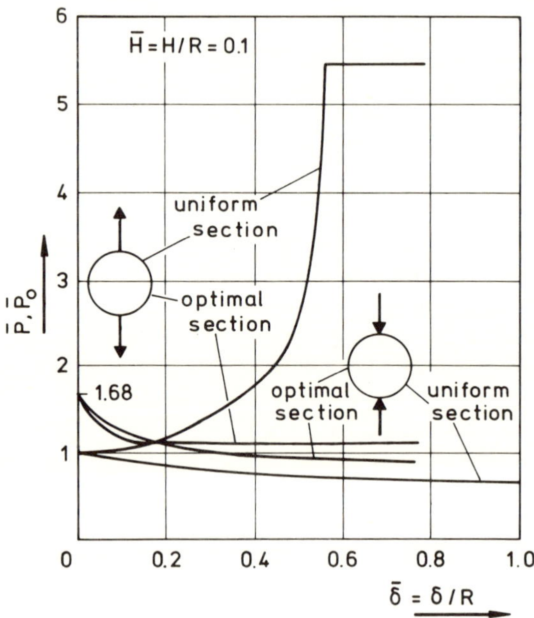

Figure 8.1. Post-yield behavior of rigid plastic ring.

into account the differences between real structures and the models used in the very evaluation of this "erosion" phenomenon. Indeed, experiments by Garstecki and Gawecki (1978) for the circular ring problem of Fig. 8.1 have shown that theoretical predictions were overly pessimistic.

Elastic sandwich circular rings and semicircular arches were studied by Huang and Sheu (1970) to achieve minimum compliance for given weight. Batterman and Felton (1971) considered the minimum-volume plastic design of doubly symmetric sandwich rings under uniform internal pressure, with detailed solutions for elliptic and rectangular rings.

When the shape of the arch axis is at the discretion of the designer, it may be identified with the funicular curve of the applied forces to avoid any bending in the arch. By setting the axial stress at all sections equal to the yield (or limit) stress, we complete the design, except for the value of the thrust, which is chosen to give minimum volume. More generally, loads can be transmitted to rigid supports along the boundaries of a plane domain by two such arches at every loading point. We so arrive at the construction of a new class of optimal structures: namely, an arch grid (for concentrated forces) or an arch-grid-like continuum (for distributed loads), as introduced by Rozvany and Prager (1979) and later called Prager structures (in analogy with Michell structures) by Rozvany (1981) in a paper in which he gives various examples of such structures. Related papers are Rozvany *et al.* (1980), Rozvany *et al.* (1982), and Rozvany and Wang (1982). In this latter paper Rozvany says

> Despite extensive research on Michell frames for the best part of this century, they remained restricted to a few simple load conditions. It is the more remarkable therefore that, within a few years, the proposed layout theory has furnished both optimal grillages and Prager structures in a closed form for almost any loading condition.

From what we know from the preceding chapters, we easily understand that the minimum-volume condition for the Prager structure is that the absolute value of the strain rate along the members is a positive constant, and is smaller than or equal to that constant along all vanishing members. Moreover, because the loads are allowed to be moved along their lines of action in order to enable optimal shaping of the arches, they are regarded as connected by cost-free rods or ties to the structure. Hence, the strain rate must vanish along their lines of action. These kinematic conditions are used, together with the equations of equilibrium in the undeformed geometry and the yield equation in each section, to derive the optimal designs.

Plates, shells, and disks may be constructed with anisotropic materials. As a matter of fact, the appearance of ribs in solid plates is a natural tendency of the structure itself toward anisotropy. An optimality criterion for the assigned load factor at collapse of perfectly plastic structures with several design parameters t_i $(i = 1, \ldots, m)$ at every point was given by

Mroz (1974), using the Lagrange multiplier technique. The result is simply

$$\frac{\partial D}{\partial t_i} \left\{ \begin{matrix} \leq \\ = \\ \geq \end{matrix} \right\} c^2 \frac{\partial \gamma}{\partial t_i} \qquad \text{when} \left\{ \begin{matrix} t_i = t_i^-, \\ t_i^- < t_i < t_i^+, \\ t_i = t_i^+, \end{matrix} \right. \tag{8.1}$$

which we could expect from the criteria already obtained. We leave it to the reader (Exercise 8.1) to show that Eq. (8.1) implies that the directions of the bars of a reinforced concrete plate coincide with the principal directions of the curvature rates, as already shown in Section 7.3. We recall that reinforced concrete plates for which minimum volume of reinforcement is desired can be regarded as gridlike continua (Section 6.2.1).

Shape optimization of the cross section of a bar subjected to given generalized stresses has been studied by several authors [for example, Krzys and Zyczkowski (1963), Krzys (1964), Banichuk (1976), Banichuk and Karihaloo (1976), and Dems (1980)].

Exercise 8.1. Prove that the optimality condition in Eq. (8.1) imposes coincidence of directions of reinforcement layers with principal directions of curvature rates in a reinforced concrete plate with orthogonal nets of reinforcing bars.

8.3. Extensions to Other Behavioral Constraints

8.3.1. Elastic–Plastic Material

For structures in the linear elastic range subjected to dynamic loading, a simple analytic expression for the maximum displacement at a given point does not, in general, exist. For certain types of dynamic loading, however, it is possible to formulate a comparatively simple expression for an *upper bound on the maximum deflection at a given point* [Martin (1964, 1968, 1972)]. Hence, one may think of distributing the structural material optimally in the sense of minimizing the total volume for fixed maximum value of this upper bound, or of minimizing this upper bound for a given amount of material, which was done by Plaut (1970) and by Chern *et al.* (1973).

A necessary and sufficient condition for stationarity of the displacement bound with respect to small variations of the distribution of material is derived, from which analytical solutions can be obtained for statically determinate beams and simple indeterminate problems. An iterative numerical procedure may be used for more complex cases.

Minimum-volume design of beams for given deflection at a specified cross section when subjected to given loads and temperature changes was treated by Prager (1970). Masur (1978) has considered minimum-volume design against post-buckling collapse, which is defined as indefinitely increasing elastic deflection under a finite elastic stress rate and a finite load parameter.

For elastic structures subject to the constraint of bounded stress intensity, it may be thought that *prestressing* would help in volume minimization. This problem was considered by Nagtegaal (1972), who gave relations between the minimum-volume prestressed elastic design and the minimum volume for assigned plastic limit load and assigned shakedown load (see Section 1.4). His results hold under the limiting assumption of costless prestressing. No application seems to have been given to date.

The influence of moderately large deflections and shear deformation on optimal design for minimum elastic compliance with assigned upper bound on the total cost has been studied by Gierlinski and Mroz (1981).

Rigid, perfectly plastic structures subjected to dynamic loading have been successfully studied by using the *mode-approximation technique*, which assumes the dynamic velocity field to be the product of a function of time only by a function of space coordinates only. Using such a velocity field Lepik and Mroz (1977) have considered beams with piecewise-constant thickness for which they studied the variation of the final central deflection with respect to the design parameters, assuming given total volume.

Optimal solutions corresponding to minimum final deflection were found to collapse as a single degree-of-freedom mechanism in opposition to the static case. An approximate kinematic approach to the same type of problem has been given by Kaliszky (1981).

Despite the existence of a static theorem (see Section 1.4) and its kinematic counterpart, given by Koiter (1956), very little seems to have been done in *optimal design for assigned shakedown load* [König (1971, 1975)]. The basic difficulty lies in the coupling of the elastic and plastic properties of the design. Uncoupling may be achieved, resulting, however, in a somewhat artificial problem, by considering a given structure in which only the yield limit is allowed to vary, and by minimizing some average value of this yield limit over the structure. After proper discretization, the resulting convex programming problem can then be solved, which was done for plates by Borkauskas and Atkociunas (1975).

For rigid-plastic structures exhibiting *isotropic work-hardening*, Prager (1969b) showed that an optimal design satisfying the criterion of constant specific dissipation per unit of strength remains optimal under proportionally increasing loads beyond the yield-point load. If the hardening law is of the so called "kinematic" type—that is, with a fixed magnitude of the domain of rigidity—minimum-volume plastic design for Bauschinger adaptation has been considered by Prager (1974) in the case of trusses.

8.3.2. Viscous Materials

Suppose a disk in plane stress satisfies the generalized Norton law for stationary creep [see Odqvist (1966)]. The compliance is defined by the power of the loads on the creep velocities. Prager (1968) has given the optimality criterion for minimum volume and applied it to the example of the annular disk. Because of the similarity of Norton's law to the constitutive equations of perfect plasticity, it is not surprising that the optimality criterion is also one of constant dissipation per unit volume.

For beams in bending, Martin (1971) has generalized the optimality criterion to one that holds for any convex cost function. Because the behavioral constraint is that of an assigned upper bound on the compliance, optimality criteria for elastic and for rigid, perfectly plastic beams are recovered as limiting cases.

Minimum-volume designs of rotating disks in stationary creep for a prescribed value of the creep velocity at the edge have been treated by Nemirovskii (1971) and Gunneskov (1975), with Norton's law generalized on the basis of the Tresca and the von Mises criteria, respectively.

One-dimensional viscoelastic structures subjected to forced steady-state vibration are considered in the paper by Lekszycki and Olhoff (1980). They derive the stationarity criterion for some functional response of the structure when the total volume of the viscoelastic material is given. Examples of optimal thickness distribution of cover sheets of sandwich beams are then given.

8.4. Extensions to Other Cost Functions

To begin with, let us point out that, although our basic problem is to minimize some cost functional while satisfying a given behavioral constraint, we have so far considered that it is equivalent to finding the distribution of the design variable that maximizes (or minimizes) the scalar behavioral parameter for assigned cost. It may, however, turn out that, in special cases, the solution of the second problem exists, whereas that of the first does not [see Olhoff (1976)]. For example, when the fundamental frequency of free elastic vibrations is assigned, the minimum-cost design of a beam does not exist if there are no "dead" masses (distributed or concentrated) and if the cross-sectional stiffness is proportional to the unbounded design parameter regarded as the specific cost. A solution of the dual problem (given cost, maximum frequency) exists only for simple support conditions, according to Brach (1968, 1973). In all other cases, however, the equivalence of the two problems holds.

Another type of structural optimization problem consists of finding the *optimal load distribution* on a given structure. It is assumed that, whereas some part S_1 of the structure is subjected to assigned loads or kinematic constraints, the remaining part S_2 is to be loaded by an unknown field of vector forces $\bar{F}$ that maximizes the merit functional

$$\mathcal{M}(\bar{m}) = \int_{S_2} \bar{F}\bar{m}\, dx \tag{8.2}$$

when $\bar{m}$ is a *given* field of unit vectors applied on S_2. Obviously, the relevant behavioral constraint must simultaneously be satisfied. This problem has been considered by Collins (1968) for the yield-point loading of rigid, perfectly plastic structures. He found that a sufficient condition for an absolute maximum of $\mathcal{M}$ is that the velocities of the collapse mechanism on S_2 are proportional (with positive factor) to the distribution of vectors $\bar{m}$. The same problem has been studied independently by Cyraz (1975) for elastic, perfectly plastic, as well as for rigid, perfectly plastic, bodies, on the basis of the theory of convex programming. A practical example of this optimization problem arises when one desires to erect a series of buildings on a layer of soil.

Finally, it is worth pointing out that, in practice, the real cost function of the structure will contain not only the manufacturing cost but also the costs of erection and maintenance. These questions will be discussed in Volume 3 of this series, dealing with practical optimization of metal and concrete structures.

9

Solutions to Exercises

Solutions for Chapter 1

1.1. The two first equations of equilibrium Eq. (1.3), remain valid, but the third must be replaced by

$$Q_3' - Q_1 \sin \phi - Q_2 \cos \phi + P_3 = 0.$$

Use of the principle of virtual work then yields

$$q_1 = p_1' + p_3 \sin \phi, \qquad q_2 = p_2' + p_3 \cos \phi, \qquad q_3 = p_3'.$$

1.2. The expression on the left of the first inequality in Eq. (1.45) is twice the negative potential energy for the fields p_α^* and q_j^*. In view of Eq. (1.14), twice the negative potential energy of the solution fields p_α and q_j has the value $\int \sum_\alpha (P_\alpha p_\alpha)\, dx$, which equals twice the complementary energy of the solution field Q_j. The left-hand inequality in Eq. (1.45) is thus equivalent to the principle of minimum potential energy, and the right-hand inequality is equivalent to the principle of minimum complementary energy.

1.3. Maximization of the expression $2cW(p^*) - 2c^2\Phi(q^*)$, where $W(p^*) = \int \Sigma_\alpha (P_\alpha p_\alpha^*)\, dx$, furnishes $c = W(p^*)/2\Phi(q^*)$. Substituting this value of c into the expression on the left of the first inequality sign in Eq. (1.45) yields the highest lower bound

$$W^2(p^*)/2\Phi(q^*) \leq W.$$

1.4. Minimization of $2\Psi(Q^{**}) + 4c\Psi(Q^{**}, Q^0) + 2c^2\Psi(Q^0)$, where $\Psi(Q^{**}, Q^0) = \frac{1}{2} \int \Sigma_j \Sigma_k (C_{jk} Q_j^{**} Q_k^0)\, dx$, furnishes $c = -\Psi(Q^{**}, Q^0)/\Psi(Q^0)$, and substituting this value of c into the rightmost term of Eq. (1.45) yields the lowest upper bound

$$2\left[\Psi(Q^{**}) - \frac{\Psi^2(Q^{**}, Q^0)}{\Psi(Q^0)}\right] \geq W.$$

1.5. Because of the orthogonality condition, the coefficient c_1 of $p_\alpha^{(1)}$ in the development of $\bar{p}_\alpha$ into the natural modes vanishes. The arguments used in Section 1.2.8 thus remain valid, provided ω_1^2 is replaced by ω_2^2.

1.6. According to Eq. (1.47), we have $-Q^* \cdot q \geq -D(q)$, while $Q \cdot q = D(q)$ because the strain rate q can be produced by the stress Q. Adding the two relations furnishes Eq. (1.58).

1.7. For a strictly convex yield locus, the first relation in Solution 1.6 and, hence, Eq. (1.58) become strong inequalities.

1.8. Applied to the static variables P_α, Q_j, or Q_j^* and the normalized kinematic variables p_α, q_j or p_α^*, q_j^*, the principle of virtual work furnishes the equations

$$1 = \int Q \cdot q \, dx = \int Q^* \cdot q \, dx = \int Q \cdot q^* \, dx = \int Q^* \cdot q^* \, dx,$$

which establish the validity of Eq. (1.59). For the kind of structure considered here, $(Q - Q^*) \cdot q > 0$ unless $Q = Q^*$ or $q = 0$ (see Exercise 1.7). Similarly, $(Q^* - Q) \cdot q^* > 0$ unless $q = Q^*$ or $q^* = 0$. The integrand in Eq. (1.59) is thus nonnegative and must vanish identically, since the integral vanishes. Accordingly, $Q = Q^*$ unless $q = q^* = 0$.

1.9. According to Exercise 1.8, a part of the structure in which the two stress fields differ performs a rigid-body motion, which is compatible with either one of the two stress fields.

1.10. If the vertical bar has elastic coefficient $\tilde{C}$, the oblique bars have coefficients $\tilde{C}\sqrt{2}$. The bar forces and the common yield force Y of the bars will be rendered dimensionless by division by P. The first two lines of Table 1 show the dimensionless elastic bar forces $Q_i'^{(1)}$ and $Q_i'^{(2)}$ corresponding to the loads $3P$ and $2P$, respectively. The next two lines of the table give the extreme values $Q_i'^{+}$ and $Q_i'^{-}$ of nonnegative and nonpositive elastic bar forces. The last line contains the bar forces of the typical state of self-stress. According to the shakedown theorem, P_0 is the greatest value of P for which the following inequalities are satisfied:

$$\begin{aligned} 1.4142 + c &\leq Y/P, \\ 0.8787 + c &\leq Y/P, \\ 1.7574 - 1.4142c &\leq Y/P, \\ c &\geq -Y/P, \\ -1.4142 + c &\geq -Y/P, \\ -1.4142c &\geq -Y/P. \end{aligned} \tag{S.1}$$

With equality signs, the first and third relations yield $P = 0.6425\ Y$ and

Table 1. Dimensionless Bar Forces

i	1	2	3
$Q_1'^{(1)}$	0.8787	0.8787	1.7574
$Q'^{(2)}$	1.4142	−1.4142	0
$Q_i'^{+}$	1.4142	0.8787	1.7574
$Q_i'^{-}$	0	−1.4142	0
$\bar{Q}_i''$	c	c	$-1.4142c$

$c = 0.1421$; with these values the remaining four relations are satisfied as strong inequalities. Accordingly, $P_0 = 0.6425\ Y$.

Solutions for Chapter 3

3.1. If the reaction at $x = l$ is denoted by $\rho p l$, the conditions of vanishing bending moment at $x = x_1$ and $x = x_2$ are

$$(0.5 + \xi + \eta)^2 - 2\rho(\xi + \eta) = 0, \qquad (0.5 + \xi)^2 - 2\rho\xi = 0.$$

Elimination of ρ between these equations furnishes

$$\eta = \frac{(0.5 - \xi)^2}{\xi}. \tag{S.2}$$

The *conjugate* beam is free at $x = 0$, built in at $x = 1.5l$, and has a hinge at $x = l$. When it is loaded with the rates of curvature $-\kappa_0$ in $0 \le x < (1 - \xi - \eta)l$ and $(1 - \xi)l < x \le 1.5l$, and κ_0 in $(1 - \xi - \eta)l < x < (1 - \xi)l$, the condition of vanishing bending moment at the hinge yields

$$\eta = -\xi + (\xi^2 + 0.5)^{1/2}. \tag{S.3}$$

From Eqs. (S.2) and (S.3) there follows $\xi = 0.3218l$, $\eta = 0.4551l$, and, hence,

$$x_1 = 0.2231l, \qquad x_2 = 0.6782l.$$

3.2. If the bending moment vanishes at $x = \xi l$, the rates of curvature of the collapse mechanism of the optimal design are κ_0 for $0 \le x < \xi l$ and $-\kappa_0$ for $\zeta l < x \le 1.5l$. The conjugate beam is free at $x = 0$ and built in at $x = 1.5l$; it has a hinge at $x = l$. When it is loaded with the rates of curvature, the condition that its bending moment must vanish at the hinge yields $\xi = l(2 - \sqrt{2})/2$.

3.3. Consider the propped cantilever with

$$h = \left(\frac{1}{h_0} - \left(\frac{1}{h_0} - \frac{1}{h_1}\right)\frac{x}{l}\right)^{-1}.$$

The optimality condition $c^2\gamma_Y = |\kappa|$, with $\gamma = bh$, gives $|\kappa h| = \alpha^2$ or

$$|\kappa| = \frac{\alpha^2}{h} = \alpha^2\left[\frac{1}{h_0} - \left(\frac{1}{h_0} - \frac{1}{h_1}\right)\right]\frac{x}{l}.$$

Let $|\kappa_0| = \alpha^2/h_0$, $|\kappa_1| = \alpha^2/h_1$; then

$$|\kappa| = \kappa_0 - (\kappa_0 - \kappa_1)x/l. \tag{S.4}$$

We load the conjugate beam. The abscissa x^* of the counterflexure point is such that moment equilibrium is satisfied. Resultants and arms are, respectively,

$$\frac{\kappa_0 + \kappa^*}{2}x^* \text{ and } \frac{x^*}{3}\frac{\kappa_0 + 2\kappa^*}{\kappa_0 + \kappa^*},$$

$$\frac{\kappa^* + \kappa_1}{2}\cdot(l - x^*) \text{ and } x^* + \frac{l - x^*}{3}\cdot\frac{\kappa^* + 2\kappa_1}{\kappa^* + \kappa_1} = \frac{x^*(2\kappa + \kappa_1) + l(\kappa^* + 2\kappa_1)}{3(\kappa^* + \kappa_1)}.$$

Equilibrium requires

$$x^{*2}(\kappa_0 + 2\kappa^*) = (l - x^*)[x^*(2\kappa^* + \kappa_1) + l(\kappa^* + 2\kappa_1)]. \tag{S.5}$$

We thus have the following equations for x^* and κ^*:

$$l^2(\kappa^* + 2\kappa_1) + x^*l(\kappa^* - \kappa_1) - x^{*2}(\kappa_0 + 4\kappa^* + \kappa_1) = 0 \tag{S.5'}$$

and

$$\kappa^* = \kappa_0 - (\kappa_0 - \kappa_1)x^*/l. \tag{S.4'}$$

If, for example, we assume $l = 10$, $h_0 = 0.5$, and $h_1 = 1$, and take $\alpha^2 = 1$, we obtain $x^* = 6.527$, and the design is readily obtained for all downward acting loads.

3.4. The total cost is $\Gamma = a\int Y\,dx + \beta\bar{Y} = \int pv\,dx$ according to the final remark of Section 3.1. From Fig. 3.16 we have

$$v = \beta + \alpha(2x_1 - l)x - \alpha x^2/2 \qquad \text{for } 0 \le x \le x_1,$$

$$v = \beta + \alpha(2x_1 - l)x - 2\alpha x x_1 + \alpha x_1^2 + \alpha x^2/2 \qquad \text{for } x_1 \le x \le l$$

because the reaction of the conjugate beam is $2\alpha x_1 - \alpha l$. Hence,

$$\int_0^l pv\,dx = \int_0^{x_1} pv\,dx + \int_{x_1}^l pv\,dx = p\left(\beta l - \frac{\alpha l^3}{3} - \frac{\alpha x_1^3}{3} + \alpha x_1^2 l\right)$$

and

$$\Gamma/\alpha p l^3 = k - 1/3 - x_1^3/3l^3 + x_1^2/l^2.$$

As the bending moment vanishes at $x = x_1$, the reaction of the given beam is $R = px_1/2$, and its contribution to the cost is $\beta R = \alpha k l^2 R = \alpha k l^2 p x_1/2$.

The last value is rendered nondimensional by dividing it by αpl^3. The percent contribution of the cost of the reaction is then given by

$$A = \frac{\beta R}{\Gamma + \beta R} = \frac{kx_1/2l}{k} - \frac{1}{3} - \frac{x_1^3}{3l^3} + \frac{x_1^2}{l^2} + \frac{kx_1}{2l}.$$

A results are summarized in the following table.

k	0	0.1	0.2	0.3	0.4	0.5
x_1/l	0.70711	0.63246	0.54772	0.44721	0.31623	0.0000
$\Gamma/\alpha pl^3$	0.04882	0.08234	0.11189	0.13685	0.15613	0.16667
A	0	38.41%	48.95%	49.02%	40.51%	0

The maximum of A is 50% for $k = 0.25$, $x_1/l = 0.5$, and $\Gamma/\alpha pl^3 = 0.125$. For $k = 0.5$, $x_1 = 0$, and the simple support becomes superfluous because the optimal cantilever beam is less expensive than the optimal propped cantilever.

3.5. The problem is equivalent to one of optimal design of a doubly built-in beam for central collapse load $12Y_0/L$, when the lower bound $Y^-(x)$ on the yield moment varies as shown in Fig. 3.17b. On account of the symmetry with respect to the center of the span, we need only discuss the optimal variation of the yield moment in the left half of the span, where

$$M(x) = M(0) + 6Y_0x/L, \qquad Y^-(x) = Y_0(1 + x/L). \tag{S.6}$$

As for the beam considered in Example b of Section 3.2.2, the yield moment will equal Y^- in some interval $(a, L - a)$. In view of Eqs. (S.6), the conditions $M(a) = -Y^-(a)$ and $M(L - a) = Y^-(L - a)$ furnish $a = L/4$ and $M(0) = -11Y_0/4$, respectively. Accordingly, since $Y(x) = |M(x)|$ if $M(x) > Y^-(x)$,

$$Y(x) = \begin{cases} Y_0(11/4 - 6x/L) & \text{for } 0 \le x \le a, \\ Y_0(1 + x/L) & \text{for } a \le x \le L - a, \\ Y_0(6x/L - 11/4) & \text{for } L - a \le x \le L. \end{cases}$$

3.6. With $\xi = x/l$ the yield moment is given by

$$Y(\xi) = (1 - \xi)Y_1 + \xi Y_2. \tag{S.7}$$

If we denote the unknown reaction at $x = 0$ by ρpl, the bending moment is

$$M(\xi) = pl^2\xi(2\rho - \xi)/2. \tag{S.8}$$

For yield hinges to occur at some intermediate section ξ_0 and at the built-in end, we must have

$$Y(\xi_0) = M(\xi_0) \qquad \text{and} \qquad Y_2 = -M(l). \tag{S.9}$$

Moreover, the bending moment must not exceed the yield moment in the vicinity of $\xi = \xi_0$, and this requires that

$$M'(\xi_0) = Y'(\xi_0), \tag{S.10}$$

where $'$ denotes differentiation with respect to ξ. If the rotation rate of the mechanism at $\xi = \xi_0$ is θ, it will be $-\xi_0\theta$ at $\xi = 1$. The internal power of dissipation is

$$D = \{Y(\xi_0) + \xi_0 Y_2\}\theta. \tag{S.11}$$

On the other hand, the volume of the beam is proportional to

$$\Gamma = Y_1 + Y_2. \tag{S.12}$$

The optimality condition requires that the derivatives of D with respect to Y_1 and Y_2 have the same ratio to each other as the derivatives of Γ. Thus, $1 - \xi_0 = 2\xi_0$ or $\xi_0 = 1/3$. Using this value of ξ_0 in Eqs. (S.8)-(S.10), we finally obtain $p = 7/18$, $Y_1 = pl^2/18$, and $Y_2 = pl^2/9$.

3.7. The conjugate beam must be loaded as shown in Fig. S.1 (top). The value of x_0 is given by rotational equilibrium:

$$\alpha a^2 + \beta(b^2 - a^2) + \alpha(x_0^2 - b^2) = \alpha(l^2 - x_0^2),$$

or

$$x_0 = (1/\sqrt{2})[l^2 - (b^2 - a^2)(1 - \beta/\alpha)]^{1/2}.$$

When

$$\beta > \alpha, \qquad x_0 < l/\sqrt{2},$$
$$\beta < \alpha, \qquad x_0 > l/\sqrt{2}.$$

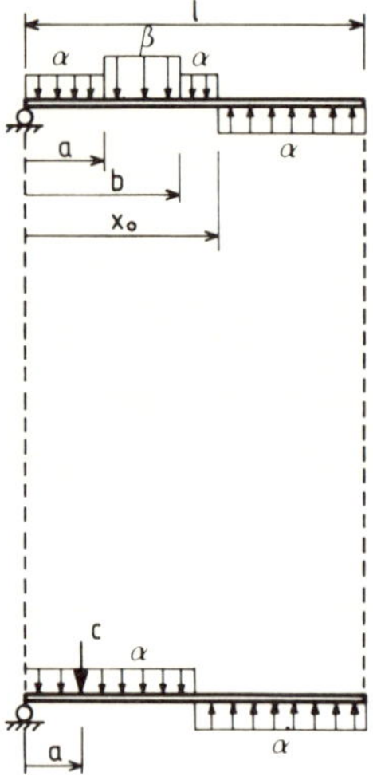

Figure S.1. Conjugate beam loaded with rate of curvature.

3.8. In Exercise 3.7, x_0 tends to $l/\sqrt{2}$ when β tends to α or $(b-a)$ tends to zero, except when the product $(b-a)\beta$ has a finite limit C with vanishing $b-a$. This is the case of concentrated cost C at abscissa b. Indeed, if $\lim(b-a)\beta = C$ when $\lim(b-a)=0$, then, because

$$(b^2-a^2)(1-\beta/\alpha) = (b^2-a^2) - (b+a)(b-a)\beta/\alpha,$$

we find

$$x_0 = \left[\frac{l^2}{2} - \frac{b+a}{2}\frac{c}{\alpha}\right]^{1/2} = \left(\frac{l^2}{2} - a\frac{c}{\alpha}\right)^{1/2} \qquad \text{for } \lim(b-a)=0.$$

This result is obtained by loading the conjugate beam as shown in Fig. S.1 (bottom). The concentrated force C gives rise to a slope discontinuity of the collapse mechanism at the location of the concentrated cost (due, for example, to the presence of a joint).

3.9. Let B^* and B be the values of the bending stiffness when the variable at the choice of the designer has the values t^* and t, respectively. The convexity of $\gamma(t)$ and the convexity of $B(t)$ are expressed by

$$\gamma(t^*) - \gamma(t) \geq (t^*-t)\gamma_t(t), \tag{S.13}$$

$$B(t^*) - B(t) \leq (t^*-t)B_t(t), \tag{S.14}$$

where the subscript t denotes differentiation with respect to t. Since $B(t)$ is monotonically increasing, $B_t(t) > 0$, and Eq. (S.14) furnishes

$$t^* - t \geq \frac{B^*-B}{B_t(t)}. \tag{S.15}$$

Now $\gamma(t)$ may be written as $\bar{\gamma}(B)$, where $B = B(t)$. Accordingly,

$$\gamma_t(t) = \bar{\gamma}_B[B(t)]B_t(t). \tag{S.16}$$

Using Eqs. (S.14)–(S.16) in Eq. (S.13) furnishes the relation

$$\bar{\gamma}(B^*) - \bar{\gamma}(B) \geq \frac{(B^*-B)\gamma_t(t)}{B_t(t)} = (B^*-B)\bar{\gamma}_B(B),$$

which establishes the convexity of $\bar{\gamma}(B)$.

In the special case where $B = (pt+q)^n$, with $p > 0$, $q \geq 0$, the bending stiffness B is a concave function of t if $0 \leq n \leq 1$, and our optimality condition assures global optimality if $\gamma(t)$ is convex. If, on the other hand, $n > 1$, global optimality is not assured even if $\gamma(t)$ is convex.

3.10. For assigned limit load the optimality criterion is $|\dot{\kappa}| = c^2\gamma_Y$. But $Y = bh^2\sigma_0/4$ and $\gamma = bh = 4Y/h\sigma_0$. Hence, $\gamma_Y = 4/h\sigma_0$ and the optimality condition is

$$|\dot{\kappa}| = 4c^2/h\sigma_0, \qquad \text{or} \qquad |\dot{\kappa}h| = \alpha^2. \tag{S.17}$$

In simple cases where regions of positive and negative rates of curvature can be predicted approximately, they furnish the locations of the counterflexure points, using, for example, the conjugate beam loaded with α^2/h. In any case the design is given by

$$b(x) = 4|M(x)|/\sigma_0 h^2(x) = k_1|M|/h^2 \tag{S.18}$$

with $k_1 = 4/\sigma_0$.

For assigned elastic compliance the optimality condition is $\kappa^2 = c^2\gamma_B$. But $B = bh^3E/12$ and $\gamma = bh = 12B/h^2E$. Then $\gamma_B = 12/h^2E$, and the optimality condition is

$$\kappa^2 h^2 = 12c^2/E. \tag{S.19}$$

Equation (S.19) yields the same counterflexure points as Eq. (S.17) and, hence, the same bending moment diagram $M(x)$. The corresponding design is given by $B = |M/\kappa|$, or

$$b(x) = \frac{1}{c}\sqrt{\frac{12}{E}} \cdot \frac{|M(x)|}{h^2(x)} = k_2\frac{|M|}{h^2} \tag{S.20}$$

with $k_2 = \sqrt{12}/c\sqrt{E}$.

Finally, when the behavioral constraint is that all bending stresses must lie in the given allowable interval $\{-\sigma_0, \sigma_0\}$, the optimality condition is $|\kappa| = c^2\gamma_{B/h}(B/h)$. But $B = bh^2/12E$, $\gamma = 12B/h^2E$, and, hence, $\gamma_{B/h} = 12/hE$. The optimality condition thus becomes

$$|\kappa| = 12c^2/hE \qquad \text{or} \qquad |\kappa h| = \alpha^2 \tag{S.21}$$

identical to Eqs. (S.17) and (S.19). The corresponding design is given by $|\sigma_{\max}(x)| = \sigma_0$, with $|\sigma_{\max}(x)| = 6|M|/bh^2$. It yields

$$b(x) = \frac{6|M(x)|}{\sigma_0 h^2(x)} = k_3\frac{|M|}{h^2}. \tag{S.22}$$

Comparing Eqs. (S.18), (S.20), and (S.22) shows that, for the same loading, the various designs differ only by coefficients k_1, k_2, and k_3, such that $k_1 = 1.5k_3 = (2c\sqrt{E}/\sigma_0\sqrt{3})k_2$.

Solutions for Chapter 4

4.1. Because the point R with radius vector $\mathbf{V}_R$ is on the strip boundaries 1^+, 3^-, and 4^-, it follows from Eqs. (4.10) and (4.13) that $\mathbf{V}_R \cdot \mathbf{U}_1 = l_1$, $\mathbf{V}_R \cdot \mathbf{U}_3 = -l_3$, and

$$-l_4 = \mathbf{V}_R \cdot \mathbf{U}_4 = \mathbf{V}_R \cdot (\alpha\mathbf{U}_1 - \beta\mathbf{U}_2) = \alpha l_1 + \beta l_3.$$

Substituting these expressions into $\mathbf{P} \cdot \mathbf{V}_R$, computed from Eqs. (4.12) and (4.14), and comparing the results furnish Eq. (4.15).

4.2. Because the yield force of bar i is now $Y_i = \sigma_0 A_i$ or $Y'_i = \sigma'_0 A_i$, depending on whether the axial strain rate ε_i is positive or negative, the cross-sectional areas A_i are more convenient design variables than the yield forces. The total power of dissipation is

$$D = \sigma_0 \sum_{\varepsilon_i>0} \varepsilon_i A_i l_i - \sigma'_0 \sum_{\varepsilon_i<0} \varepsilon_i A_i l_i,$$

and the total volume of the bars is $V = \Sigma_i A_i l_i$. When A_i is bounded by 0 and ∞, the optimality condition, Eq. (4.9), must therefore be replaced by

$$\mathbf{v} \cdot \mathbf{U}_i \begin{Bmatrix} \leq \\ = \end{Bmatrix} c^2 l_i \qquad \text{if } \mathbf{v} \cdot \mathbf{U}_i > 0 \text{ and} \begin{cases} A_i = 0 \\ A_i > 0 \end{cases}$$

$$\mathbf{v} \cdot \mathbf{U}_i \begin{Bmatrix} \geq \\ = \end{Bmatrix} \frac{-c^2 l_i \sigma_0}{\sigma'_0} \qquad \text{if } \mathbf{v} \cdot \mathbf{U}_i < 0 \text{ and} \begin{cases} A_i = 0 \\ A_i > 0. \end{cases}$$

These conditions restrict the point with radius vector $\mathbf{V}$ to a convex polygon Π that is no longer symmetric with respect to the point O (Fig. 4.1b), but this lack of symmetry neither invalidates the conclusion that there exists, for any direction of the load $\mathbf{P}$, an optimal truss with at most two bars, nor affects the choice of bars that will most economically transmit a load of given direction to the wall.

4.3. Since the layout in Fig. 4.3 is based on the velocity field of Eq. (4.16), in which the velocity of point A is horizontal, using the principle of virtual power, as in the derivation of Eq. (4.20), shows that the required volume of material does not depend on the vertical load. The magnitude of this load, however, must not exceed P, because otherwise the resultant of the two loads would form an angle smaller than 45° with the line OA, and the forces in the two bars at A would have the same sign contrary to the fact that, in the velocity field in Eq. (4.16), these bars experience axial strains of opposite signs.

4.4. The principle of virtual work yields Eq. (4.22) and

$$W = \sum_{i^*} S^*_i \int \varepsilon^*_i ds^*_i = \sum_{s^*_i>0} \sigma_0 A^*_i \int \varepsilon^*_i ds^*_i - \sum_{s^*_i<0} \sigma'_0 A^*_i \int \varepsilon^*_i ds^*_i.$$

Using $\varepsilon^*_i \leq \varepsilon_0$ and $\varepsilon^*_i \geq -\varepsilon'_0 = -\varepsilon_0\sigma_0/\sigma'_0$ in the last two integrals, we again obtain the inequality of Eq. (4.24), whose comparison with Eq. (4.22) establishes the global optimality of the considered truss-like continuum.

4.5. In view of the symmetry of the truss with respect to the center of the square, the bars of the truss form three groups, each of which consists of identical bars that transmit forces of the same magnitude. Representative members of these groups have been labelled 1, 2, 3 in Fig. 4.8; their lengths will be denoted by l_1, l_2, l_3. If the bars of group 1 are to carry the loads α, where $0 \leq \alpha \leq 1$, the forces in the bars of the three groups have magnitudes

$$S_1 = \alpha \Lambda P/\sqrt{2},$$

$$S_2 = (1-\alpha)\Lambda P l_2/\{(a+b)\sqrt{2}\},$$

$$S_3 = S_2 b\sqrt{2}/l_2.$$

Since $l_1 = 2a$, $l_2 = (a^2 + b^2)^{1/2}$, $l_3 = (a-b)\sqrt{2}$, the necessary volume of structural material is

$$V = 4(S_1 l_1 + 2S_2 l_2 + S_3 l_3)/\sigma_0 = 4\Lambda Pa\sqrt{2}/\sigma_0,$$

which agrees with Eq. (4.25).

4.6. With the values of the angles indicated in Fig. S.2(a), one finds

$$l_1/R = \sin 20°/\cos 55° = 0.59629 = l_2/R,$$

$$l_3/R = (l_1/R)\tan 55° = 0.85160.$$

The Maxwell diagram in Fig. S.2(b) furnishes

$$S_3/P = 0.5/\cos 55° = 0.87172,$$

$$S_2/P = S_3/(P\cos 20°) = 0.92767,$$

$$S_1/P = -(S_3/P)\tan 20° = -0.31728.$$

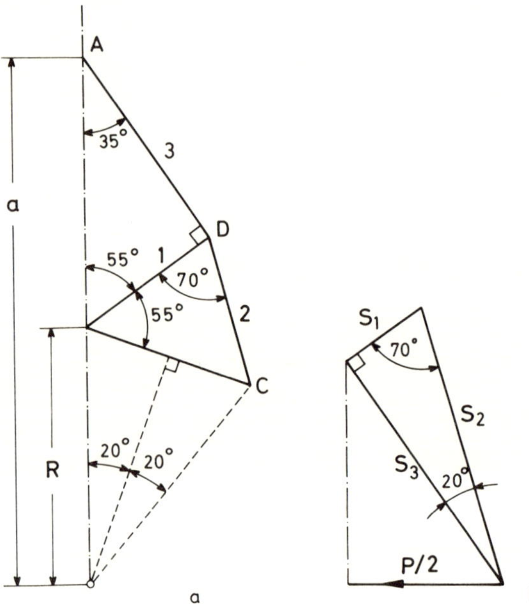

Figure S.2. (a) Truss. (b) Maxwell diagram.

Accordingly,

$$V^* = \frac{2\Sigma|S_i|l_i}{\sigma_0} = 2.96942\,\frac{PR}{\sigma_0}.$$

The distance OA is

$$a = R + l_1/\sin 35° = 2.03960\ R.$$

Thus,

$$V = (2P/\sigma_0)a\ \ln(a/R) = 2.90746\ PR/\sigma_0,$$

$$\eta = 100\ v/v^* = 97.9\%.$$

4.7. The lengths of the bars of the modified truss are

$$l_1^* = l_1/\cos 5° = 0.59857\ R,$$

$$l_2^* = \{l_1^{*2} + (2R\sin 20°)^2 - 4Rl_1'\sin 20°\cos 60°\}^{1/2} = 0.65456\ R,$$

$$l_3^* = l_3 - l_1^*\sin 5° = 0.79943\ R,$$

(l_1 and l_3 are the values in Solution 4.6). The angles $\alpha = \measuredangle\ CD^*B$ and $\beta = \measuredangle\ BD^*A$ are

$$\alpha = \arcsin\{(2R/l_2^*)\sin 20°\sin 60°\} = 66.5838°,$$

$$\beta = 95°,$$

and the forces in the bars are

$$S_1^* = -S_3\,\frac{\sin(\alpha+\beta)}{\sin\alpha} = -0.30011\ P,$$

$$S_2^* = S_3\,\frac{\sin\beta}{\sin\alpha} = 0.94634\ P,$$

$$S_3^* = S_3 = 0.87172\ P.$$

(S_3 is the value in Solution 4.7.) The necessary volume V^* of structural material and the efficiency η^* are thus

$$V^* = 2\sum_i\frac{|S_i^*|l_i^*}{\sigma_0} = 2.9749\,\frac{PR}{\sigma_0},$$

$$\eta^* = 100\ V/V^* = 97.7\%.$$

(V is the value in Solution 4.7.)

4.8. The bars of the modified truss have lengths

$$l_1^{**} = l_1 - l_3\tan 5° = 0.52179\ R,$$

$$l_2^{**} = [l_1^{**2} + (2R\sin 20°)^2 - 4l_1^{**}R\sin 20°\cos 55°]^{1/2} = 0.57509\ R,$$

$$l_3^{**} = l_3/\cos 5° = 0.85489.$$

(l_1 and l_3 are the values in Solution 4.7.) The angles $CD^{**}B$ and $BD^{**}A$ are

$$\gamma = \arcsin[(2R/l_2^{**})\sin 20^\circ \sin 55^\circ] = 76.993^\circ, \qquad \delta = 95^\circ,$$

and the bar forces are

$$S_1^{**} = -S_3^{**}\frac{\sin(\gamma+\delta)}{\sin\gamma} = 0.14297\ P,$$

$$S_2^{**} = \frac{S_3^{**}\sin\delta}{\sin\gamma} = 1.02243\ P,$$

$$S_3^{**} = \frac{P}{2\sin 30^\circ} = P.$$

The volume V^{**} of the bars and the efficiency η^{**} of the truss are thus

$$V^{**} = 2\sum_i \frac{|S_i^{**}|l_i^{**}}{\sigma_0} = 3.0349\frac{PR}{\sigma_0},$$

$$\eta^{**} = 100\ V/V^{**} = 95.8\%.$$

(V is the value in Solution 4.7.)

4.9. To the horizontal displacement of joint J correspond bar elongations with absolute value $v\cos\alpha$. Denoting the common cross-sectional area of the bars by A, and writing the bar lengths as $l = h/\sin\alpha$ and the magnitudes of the bar forces as $S = P/(2\cos\alpha)$, we have

$$V\cos\alpha = Sl/EA = Ph(EA\sin 2\alpha).$$

Solving for A and multiplying the result by $2l$, we obtain the following expression for the total volume V of the bars:

$$V = \frac{4Ph^2}{Ev\sin^2 2\alpha}.$$

To minimize V we must maximize $\sin 2\alpha$—that is, choose $\alpha = 45^\circ$.

Solutions for Chapter 5

5.1. Figure S.3(a)-(e) correspond to Figs. 5.1(a)-(e). In place of Eq. (5.2) we now have

$$\omega_x\gamma = \omega_y(2\alpha-\beta) = \omega'_y\beta. \tag{S.23}$$

and the optimality condition again takes the form of Eq. (5.3). Eliminating a between Eqs. (5.3) and using Eq. (S.23), we find

$$\frac{2\gamma}{2\alpha-\beta}+\frac{\gamma}{\beta} = 2\alpha. \tag{S.24}$$

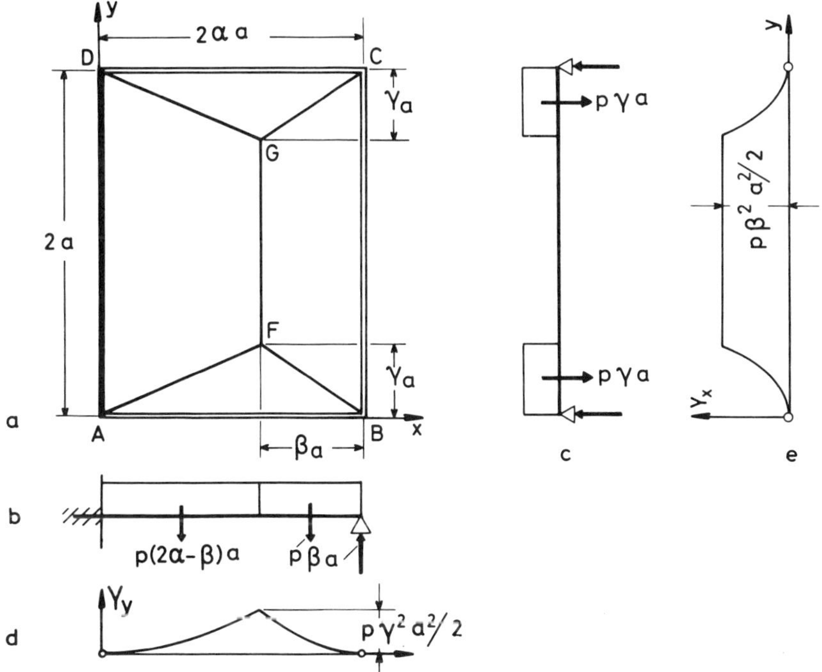

Figure S.3. (a) Velocity field. (b) Loading in x direction. (c) Loading in y direction. (d) Yield moment for beam in x direction. (e) Yield moment for beam in y direction.

The fact that the bending moments at $x = (2\alpha - \beta)a$ and $x = 0$ are $Y_x(a) = p\beta^2a^2/2$ and $-Y_x(a)$, respectively, furnishes

$$\beta = 2\alpha(\sqrt{2} - 1), \tag{S.25}$$

and using this relation in Eq. (S.24) yields

$$\gamma = [2\alpha(\sqrt{2} - 1)]^2 = \beta^2. \tag{S.26}$$

Note that the present discussion is only valid when $\gamma \leq 1$, and that, for $\gamma = 1$, the relation in Eq. (S.26) gives an α value that agrees with the right-hand side of Eq. (5.11).

The cost of the grillage is found by multiplying the areas of the diagrams in Fig. S.3(d), (e) by $2a$ and $2\alpha a$, respectively, and adding the products. Thus,

$$\Gamma = 2\alpha pa^4\{\gamma^2 + (3 - 2\gamma)\beta^2\}/3 = 2\alpha\gamma pa^4(3 - \gamma)/3. \tag{S.27}$$

For $\gamma = 1$, α equals the right-hand side of Eq. (5.11), and Eq. (5.27) furnishes the same value of Γ as Eq. (5.10): namely, $\Gamma = 1.6094pa^4$.

5.2. Figure S.4 shows the assumed division into four regions with linear rates of deflection. If the angular velocity of region $CDFG$ about the edge

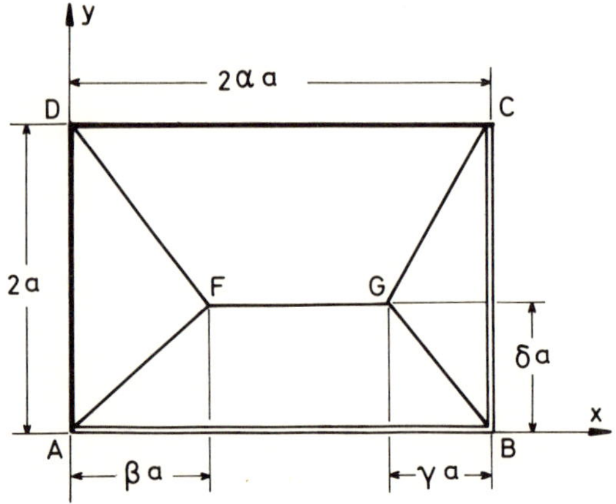

Figure S.4. Velocity field.

CD has magnitude ω'_x, and if ω_x, ω_y, ω'_y have the same meaning as in the discussion of Fig. S.3, we have

$$\omega_x\delta = \omega'_x(2-\delta) = \omega_y\beta = \omega'_y\gamma, \tag{S.28}$$

and optimality requires that

$$\omega_x + \omega'_x = a, \qquad 2\omega_y + \omega'_y = 2\alpha a. \tag{S.29}$$

Static considerations for the x-beam through F and for any y-beam between F and G yield Eq. (S.28) and

$$\delta - 2(\sqrt{2}-1). \tag{S.30}$$

Using Eq. (S.28) to express ω'_x, ω_y, ω'_y as multiples of ω_x, and substituting the resulting expressions as well as Eqs. (S.28) and (S.30) into Eq. (S.29), we obtain

$$\gamma = 2(\sqrt{2}-1)/\alpha. \tag{S.31}$$

The cost of the structure may be evaluated as the product of p by the integral of the rate of deflection over the area of the grillage-like continuum. One finds

$$\Gamma = \frac{8pa^4(\sqrt{2}-1)^2(3\alpha - 1/\alpha)}{3}. \tag{S.32}$$

Note that, strictly speaking, the preceding discussion applies only if the sum of β and γ does not exceed 2α, that is, for $\alpha \geq 1$. This, however, only means that, in the case of built-in edges of unequal length, the shorter one should be denoted by $2a$.

5.3. The beam BB' transmits the loads $Q = Py'/(y + y')$ at B and $Q' = Py/(y + y')$ at B' to the beams BC and $B'C'$, respectively; and these in turn transmit the loads $R = Q(3a - y)/(x - y)$ at E and $R' = Q'(3a - y')/(x - y')$ at E' to the beam DD'. The moment areas thus have the following values:

Beam BB': $\frac{1}{2}(Pyy')$,

Beam BC: $\frac{1}{2}[Q(3a - x)(3a - y)]$,

Beam $B'C'$: $\frac{1}{2}[Q'(3a - x)(3a - y')]$,

Beam DD': $\frac{1}{2}[R(x - y)(x + y)] + \frac{1}{2}[R'(x - y')(x + y')]$.

With the above values of Q, Q', R, and R', the sum of these moment areas again has the value $\Gamma = 9Pa^2/2$.

5.4. The beam AB transmits the loads P and $2P$ at B and C to the beams EE' and DD', respectively. Thus the moment areas are as follows:

Beam AB: Pa^2; Beam DD': $4Pa^2$; Beam EE': $Pa^2/2$.

Accordingly, $\Gamma^* = 5.5Pa^2$. Since the cost of the optimal grillages in Fig. 5.3(c)–(e) is $\Gamma = 4.5\ Pa^2$, the efficiency of the grillage in Fig. 5.3(f) is only $\eta^* = 100\Gamma/\Gamma^* = 81.8\%$.

Note that the value of Γ^* may also be obtained by applying, on the one hand, the principle of virtual power to the rate of deflection in Eq. (5.26) and, on the other hand, the load P and the bending moments M^* caused by it in the beams of the grillage of Fig. 5.3. Since the yield moment is $Y^* = M^*$ for beams AB and DD', but $Y^* = -M^*$ for beam EE', we have

$$Pv(3a, 0) = \int_{AB} Y^* ds^* - \int_{DD'} Y^* ds^* \int_{EE'} Y^* ds^*.$$

Here, the last integral equals the moment area $Pa^2/2$ of the beam EE'. The sum Γ^* of the moment areas of the three beams is thus

$$\Gamma^* = Pv(3a, 0) + 2(Pa^2/2) = 11\ Pa^2/2.$$

5.5. Since $\rho + \rho_0$ is the radius of curvature ρ_2 of the second principal line through G, the first equation in Eq. (5.35) may be written in the form

$$\frac{d\kappa_2}{d\rho} = \frac{\kappa - \kappa_2}{\rho + \rho_0}.$$

Integrating this differential equation, with initial condition $\kappa_2 = -\kappa$ for $\rho = 0$, yields the desired result.

5.6. If the coordinates of the point W in Fig. 5.15 are denoted by ξ, η, the equality of WU and WU' furnishes

$$\xi/\sin\alpha - \eta = h - \delta\cot\alpha$$

or

$$\xi = \frac{h \sin \alpha}{1 + \cos \alpha}, \qquad \eta = \frac{h}{1 + \cos \alpha}.$$

The coordinates of Q are $x = \xi/2$ and $y = (h + \eta)/2$. Substituting the expressions for ξ and η and eliminating α between the resulting equations show that the curve c in Fig. 5.15 is the parabola

$$y = x^2/h + 3h/4.$$

In a similar manner, eliminating α between the expressions $x = \xi$ and $y = \eta/2$ for the coordinates of point Q' shows that the curve c' in Fig. 5.15 is the parabola

$$y = x^2/4h + h/4.$$

5.7. The coordinates ξ, η of point W in Fig. 5.16 follow from the condition that $WU = WV/2$. One finds

$$\xi = \frac{h\sqrt{2} \sin \alpha}{1 + \sqrt{2} \cos \alpha}, \qquad \eta = \frac{h}{1 + \sqrt{2} \cos \alpha}.$$

The coordinates of Q are $x = \xi/2$, $y = (h + \eta)/2$. Substituting the expressions for ξ and η and eliminating α show that the curve c in Fig. 5.16 is the hyperbola

$$y^2 = x^2 + h^2/2.$$

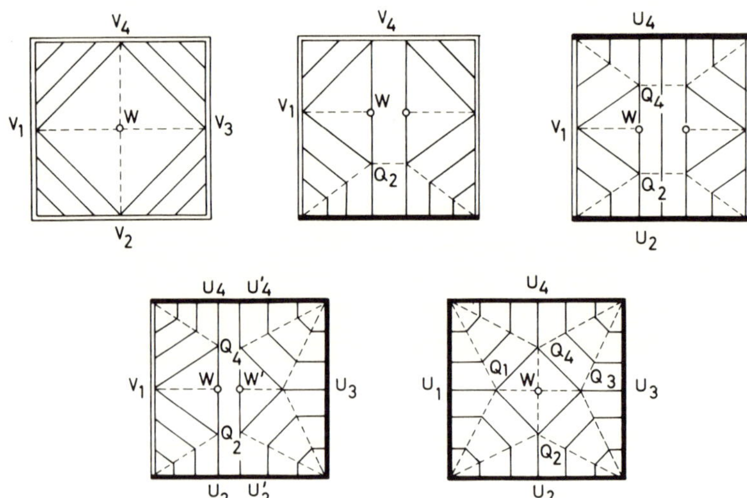

Figure S.5. Optimal grids for various loadings and support conditions.

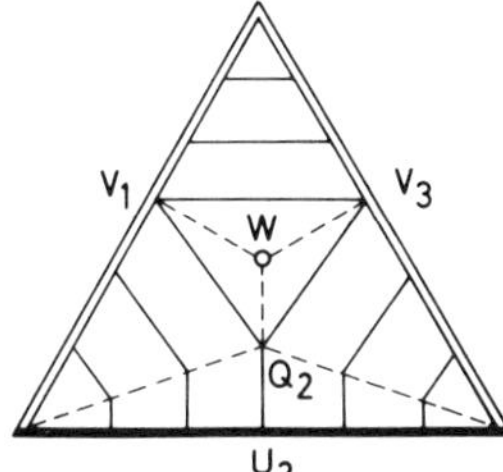

Figure S.6. Optimal triangular grid.

5.8 and **5.9.** Figures S.5 and S.6 show the optimal layouts, which are determined by their extremal points. In regions of type S^+, in which no beams are indicated, the beams may be given any direction. Only beams with positive rate of curvature are shown in regions of type T.

5.10. The grillage must be regarded as clamped at the obtuse corners. The optimal layout is shown in Fig. S.7. Regions UV_1Q and $UQ'V_4'$ are of type T. The hyperbolic boundary QQ' of the region $QV_2V_2'Q'$, which is of type R^+, and the beams of this region are determined as in Fig. 5.16.

5.11. Regions ABV_2 and V_2CD (Fig. S.8) are of type R^+ with beams normal to the bisectors of the angles at B and C. Region AV_2D is of type S^+ and has a beam along the free edge AD. The other beams of this region may be given any direction.

5.12. The numbers in the upper half of the plan form in Fig. S.9 give the multiples of P that are acting on the short beams, positive numbers indicating downward forces. The diagrams below and to the right of the plan form show the variation of the yield moment along the beams as multiples of $Pa/4$. The areas of these diagrams are given as multiples of $Pa^2/16$. The sum of the moment areas is thus

$$(Pa/16)[1 + 2(1 + 1.5 + 2) + 1 + 4 + 7 + 10] = 2Pa^2.$$

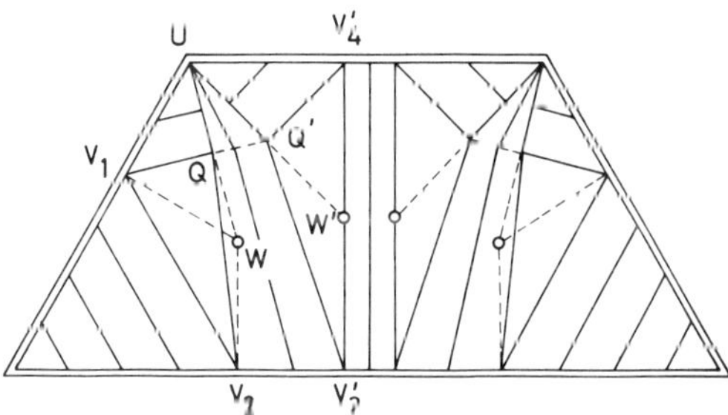

Figure S.7. Optimal trapezoidal grid, all edges supported.

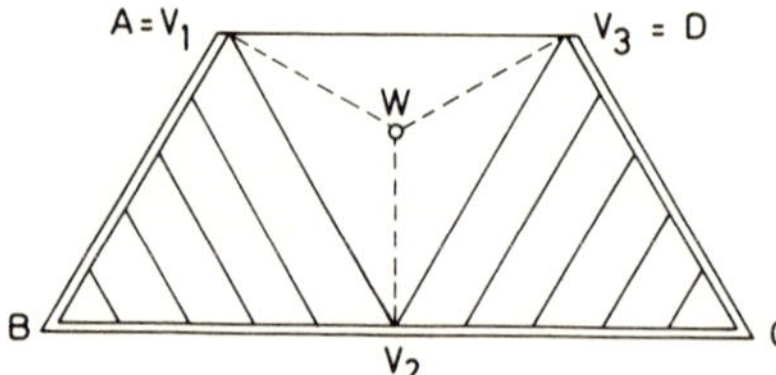

Figure S.8. Optimal trapezoidal grid, one edge free.

5.13. In Fig. S.10, the point E is the center of AC, B is the origin, and BE is the x-axis of the rectangular coordinates x, y. The rate of deflection $v_1(x, y)$ in ABC vanishes along BA and BC—that is, for $y = \pm x/\sqrt{3}$—and has positive unit rate of curvature in the y-direction. Accordingly,

$$v_1(x, y) = \frac{x^2 - 3y^2}{6}.$$

At E we have $v_1(a\sqrt{3}/2, 0) = a^2/8$ and $\partial_x v_1(a\sqrt{3}/2, 0) = a\sqrt{3}/6$. The rate of deflection v_2 in ACD will be referred to coordinates ξ, η with origin E and ξ-axis along BE. Since the rates of curvature of this T-field are 1 and

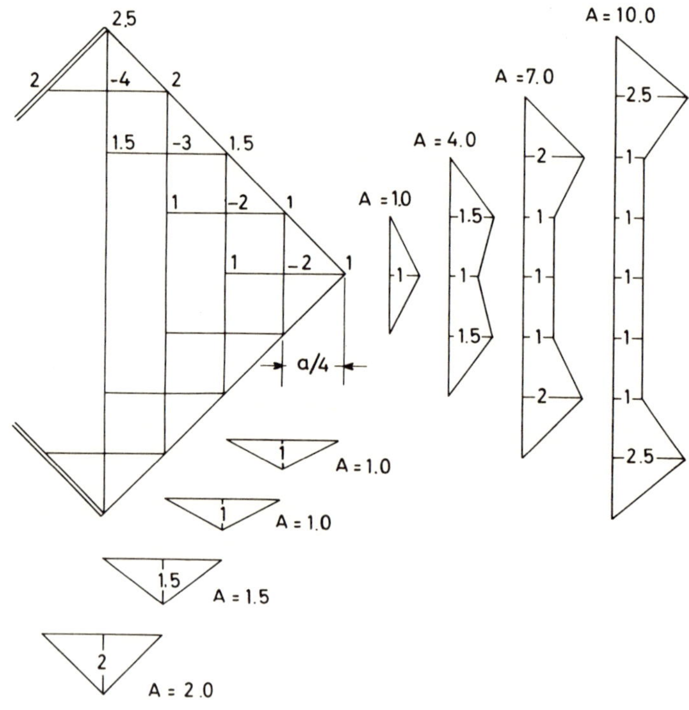

Figure S.9. Optimal grid arrangement and yield moments.

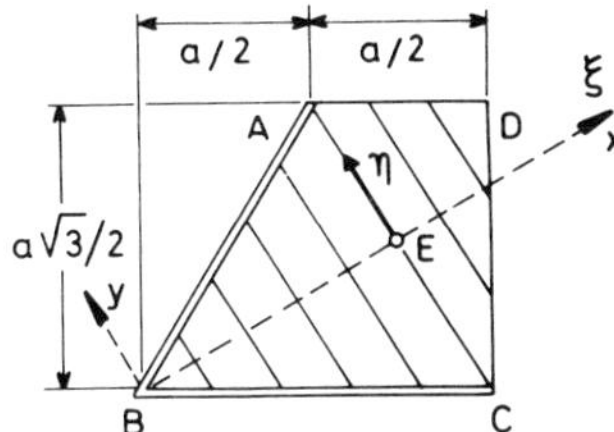

Figure S.10. Optimal trapezoidal grid, two edges free.

-1 in the η- and ξ-directions, we have

$$v_2(\xi, \eta) = \tfrac{1}{2}(\xi^2 - \eta^2) + q\xi + r,$$

where q and r must be determined from the continuity requirements that $v_1 = v_2$ and $\partial_x v_1 = \partial_\xi v_2$ at the point E. Thus,

$$v_2(\xi, \eta) = \tfrac{1}{2}(\xi^2 - \eta^2) + a\xi\sqrt{3}/6 + a^2/8.$$

The coordinates of D are readily found to be $\xi_D = a\sqrt{3}/4$ and $\eta_D = a/4$. The power of the load P at D is thus $Pv_2(\xi_D, \eta_D) = 5Pa^2/16$, and, since $\gamma = Y$, the cost equals this power.

5.14. To verify the formula introduce coordinate axes ξ, η along AB and AC and check that the formula is valid for each of the functions l, ξ, ξ^2, and $\xi\eta$.

5.15. *Simply supported edges.* Figure S.11(a) shows a quarter of the grillage. Because RS and RQ have unit rates of curvature and the derivatives of the rates of deflection vanish at Q, we have

$$v_Q = a^2/2, \qquad v_A = v_B = v_Q - \tfrac{1}{2}(a/2)^2 = 3a^2/8,$$
$$V_C = \tfrac{1}{2}(a\sqrt{2}/2)^2 = a^2/4.$$

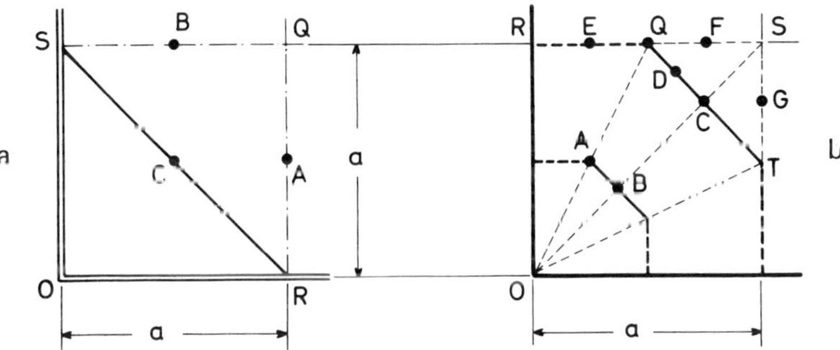

Figure S.11. (a) Simply supported edges, quarter of grillage. (b) Built-in edges, quarter of grillage.

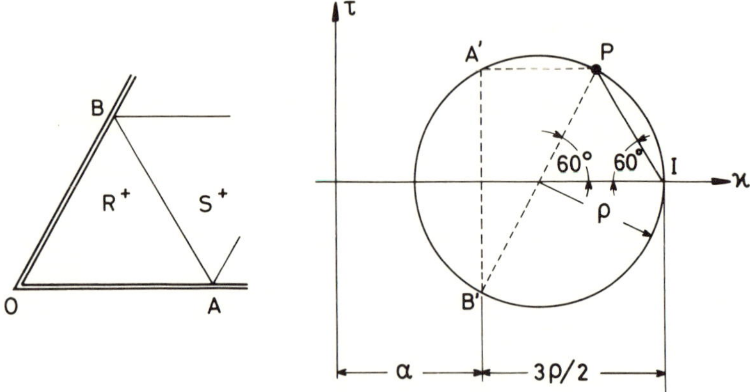

Figure S.12. Corner of grid and Mohr's circle.

The sum of the integrals over the triangles *ORS* and *RQS* is thus

$$(a^2/6)[v_C + (v_A + v_B + v_C)] = 5a^4/24.$$

According to Eq. (5.23) the cost Γ of the optimal grillage for uniform loading p is $4p$ times this amount: $\Gamma = 5pa^4/6$.

Built-in edges. Figure S.11(b) shows a quarter of the grillage. Because the beams shown by full or dashed lines have positive or negative unit rates of curvature, the relevant rates of deflection are given by

$$\begin{aligned}
2v_Q/a^2 &= (\tfrac{1}{2})^2 = \tfrac{1}{4},\\
2v_A/a^2 &= (\tfrac{1}{4})^2 = \tfrac{1}{16} = 2v_E/a^2,\\
2v_B/a^2 &= 2v_A/a^2 + (\sqrt{2}/8)^2 = \tfrac{3}{32},\\
2v_C/a^2 &= 2v_Q/a^2 + (\sqrt{2}/4)^2 = \tfrac{3}{8},\\
2v_D/a^2 &= 2v_C/a^2 - (\sqrt{2}/8)^2 = \tfrac{11}{32},\\
2v_S/a^2 &= 2v_Q/a^2 + (\tfrac{1}{2})^2 = \tfrac{1}{2},\\
2v_F/a^2 &= 2v_S/a^2 - (\tfrac{1}{4})^2 = \tfrac{7}{16}.
\end{aligned}$$

The contributions of the triangles *OCQ*, *OQR*, and *QTS* to the cost are thus

$$\begin{aligned}
\Gamma_1 &= (\tfrac{1}{6})(3a\sqrt{2}/4)(a\sqrt{2}/4)(v_A + v_B + v_D)p = pa^4/64,\\
\Gamma_2 &= (\tfrac{1}{6})a(a/2)(v_A + v_E)p = pa^4/192,\\
\Gamma_3 &= (\tfrac{1}{6})(a/2)(a/2)(v_C + v_F + v_G)p = 5pa^4/192.
\end{aligned}$$

Since there are, respectively, eight, eight, and four triangles of each kind, the total cost of the grillage is $\Gamma = 8(\Gamma_1 + \Gamma_2) + 4\Gamma_3 = 13pa^4/48$.

5.16. The points A and B in Fig. S.12 are the centers of the edges through O. In OAB we have a regime R^+ with unit rate of curvature in the direction of AB. The rightmost point I of the Mohr circle has unit abscissa, and the line joining it to the pole P is parallel to AB. Lines parallel to the edges OA and OB through P intersect the circle at points A' and B' with common abscissa α. Accordingly, $\alpha = 1 - 3\rho/2$ or

$$\rho = \frac{2(1-\alpha)}{3}, \qquad \kappa = 1 - \rho = \frac{1+2\alpha}{3}.$$

Solutions for Chapter 6

6.1. We use the superposition method of Save and Shield to obtain the design by adding the design under uniform pressure $p = a$ and the one subjected to hydrostatic pressure with $p(0) = b$.

6.2. Denote by Γ_1 and Γ_2 the volumes without and with reinforcing ring, respectively. We have

$$\Gamma_1 = \frac{8\pi R^2 p}{\omega\sigma_0}\left(\frac{1 - \cosh 2\omega L^*}{\sinh 2\omega L^*} + I^*\omega\right),$$

or

$$\Gamma_1 = \frac{16\pi R^2 p}{\omega\sigma_0}\left(\frac{1 - 2\cosh^2 \omega L^* + 1}{(4\sinh\omega L^* \cosh\omega L^*)} + L^*\omega/2\right),$$

and

$$\Gamma_2 = \frac{16\pi R^2 p}{\sigma_0\omega}\left\{\left[\frac{1-\cosh\omega L^*}{\sinh\omega L^*} + \frac{\omega L^*}{2} - \frac{(\cosh\omega L^* - 1)(\cosh\omega L^*)}{4}\sinh\omega L^*\right] + \frac{\sinh\omega L^*}{4}\right\}.$$

Taking into account that $\cosh\omega L^* = 2$, we finally obtain

$$\Gamma_1 = \Gamma_2 = \frac{16\pi R^2 p}{\sigma_0\omega}\left\{(L^*\omega/2) - \left[\frac{3}{4\sinh\omega L^*}\right]\right\}.$$

6.3. The design is obtained, on the basis of the Save–Shield superposition method, by adding the minimum-volume designs of the given shell subjected to $p(x) = p_0(1 - x/L)$ and $p(x) = p_0 x/L$, respectively, where p_0

is the given uniformly distributed load. The explicit solution is

$$t(x) = \frac{2pR}{\sigma_0}\left\{1 - \cos(\omega x - l_1) + \frac{[1 - \cosh(l_2 - l_1)]\sin(\omega x - l_1)}{\sinh(l_2 - l_1)}\right\},$$
$$0 \le x \le l_1/\omega,$$

$$t(x) = \frac{2pR}{\sigma_0}\left\{1 + [\sinh(\omega x - l_2) + \frac{\sinh(l_1 - \omega x)]}{\sinh(l_2 - l_1)}\right\}, \qquad l_1/\omega \le x \le l_2/\omega,$$

$$t(x) = pR/\sigma_0, \qquad l_2/\omega \le x \le L/2,$$

$$l_1 = 0.65907, \qquad l_2 = 1.69078,$$

$$\omega L > 3.38156.$$

At $x = l_2/\omega$ there is a reinforcing ring of cross-sectional area

$$A = (pR/\sigma_0\omega)\left[\frac{\cosh(l_2 - l_1) - 1}{\sinh(l_2 - l_1)}\right].$$

The total volume is

$$\Gamma = (4\pi R^2 p/\omega\sigma_0)(\omega L/2 + l_2 - 4\sin l_1).$$

6.4. In the vicinity of the built-in (lower) end, regime B, Fig. 6.18, applies in the optimality condition in Eq. (6.93) to give Eq. (6.111): $\dot{w} = 2c^2R(1 - \cos\omega x)/\sigma_0$. This velocity field can be used alone as long as the normality law at B is satisfied; that is,

$$\frac{d^2\dot{w}/dx^2}{\dot{w}/R} \ge \frac{N_p}{2M_p}. \tag{S.33}$$

Equation (S.33) implies

$$\omega L \ge \pi/3 = 1.047. \tag{S.34}$$

When ωL becomes larger than $\pi/3$, regime B continues to apply near the built-in end, but it is followed by regime F. With these two regimes the optimality condition gives

$$\dot{w}_1(x) = \frac{2c^2R(1 - \cos\omega x)}{\sigma_0}, \qquad 0 \le x \le x_1,$$

$$\dot{w}_2(x) = \frac{2c^2R}{\sigma_0}\{\sin\omega x_1 \sinh\omega(x - x_1) - \cos\omega x_1 \cosh\omega(x - x_1) + 1\},$$

$$x_1 \le x \le l,$$

when the boundary conditions $\dot{w}_1(0) = (d\dot{w}_1/dx)_{x=0} = 0$ and $\dot{w}_2(x_1) = \dot{w}_1(x_1)$, $(d\dot{w}_2/dx)_{x=x_1} = (d\dot{w}_1/dx)_{x=x_1}$ have been used. As a rule, $(d\dot{w}_2/dx)_{x=l} \ne 0$. When $d\dot{w}_2/dx$ vanishes at $x = l$, we reach the limiting case where the shell behaves as half of a long, doubly built-in shell with *no* central region in a membrane state. Hence, from the solution of Exercise 6.3 we know that this limiting case occurs for $\omega l = 1.69078$ and that regime B then extends from $x = 0$ to x_1 such that $\omega x_1 = 0.65907$. For $\omega l > 1.69078$ the solution is given by half of the doubly built-in shell and exhibits a central region in a pure membrane state. For intermediate values of ωl such that $1.047 \le \omega l \le 1.69078$, the normality law in regime F implies that

$$\frac{d^2\dot{w}_2/dx^2}{\dot{w}_2/R} \le -\frac{N_p}{2M_p},$$

or

$$2\sin\omega x_1 \sinh\omega(x - x_1) - 2\cos\omega x_1 \cosh\omega(x - x_1) + 1 \le 0.$$

Denoting by f the function on the left-hand side of this last relation, it can be verified that f will remain negative if we choose x_1 such that $f'(l) = 0$. Indeed, for ωl varying from 1.047 to 1.69, the corresponding values of ωx_1 varying uniformly from 0.462 to 0.659, and f varies uniformly from -0.551 to zero.

6.5. A long cylinder will exhibit, in the vicinity of the free end, a region in a pure membrane state ($N_\theta \neq 0$, $\dot{\varepsilon}_\theta =$ constant, $M_x = \dot{\kappa}_x = 0$). Hence, the boundary conditions at this free end are those encountered in the central section of a cylinder built-in at both ends with the same radius and twice the length. The optimal design of one-half of this latter cylinder is the desired solution, as found solving Exercise 6.3.

6.6. With regime B with $\dot{w}(0) = (d\dot{w}/dx)_{x=0} = 0$, the optimality condition in Eq. (6.93) gives $\dot{w} = (2c^2R/\sigma_0)(1 - \cos\omega x)$. The normality law, applied at point B, implies $0 \le \dot{w}/R \le Hd^2\dot{w}/dx^2$, a condition satisfied if $L \le \pi(RH)^{1/2}/3$, or $\omega \le 1.047$.

6.7. The relevant plastic regimes are given in Section 6.4.3.2. The corresponding expressions for M_x and N_θ are introduced in the equation of equilibrium, which we integrate with the following boundary conditions: $M_x(0) = M_x(L) = M_x(L^*) = M_x(L - L^*)$. We obtain

$$A_x = \left[\frac{p(0)x}{2\sigma_0 H}\right]\left[\frac{x - L^* - (x^2 - L^{*2})}{3L}\right], \quad A_\theta = 0, \qquad 0 \le x \le L^*;$$

$$A_x = 0, \quad A_\theta = \frac{p(0)R[1 - x/L]}{\sigma_0}, \qquad L^* \le x \le L - L^*;$$

$$A_\theta = 0, \quad A_x = \frac{p(0)}{2\sigma_0 H}\left[\left(1 - \frac{x}{3L}\right)x^2 + \frac{(L - L^*)^2(2L + L^*)(x - L)}{3LL^*} - \frac{2L^2(x - L + L^*)}{3L}\right], \qquad L - L^* \le x \le L.$$

We recall that $L^* = 1.4142\,(RH)^{1/2}$. Taking into account the relation $M_x = A_x H\sigma_0$, discontinuities of the shear V_x are found to be

$$V_{x1} = [p(0)L^*/2][1 - 2L^*/3L] \qquad \text{at } x = L^*,$$

$$V_{x2} = p(0)\left[\frac{(L - L^*)^3 - L}{3LL^*} + L - L^*\right] \qquad \text{at } x = L - L^*,$$

and the corresponding, concentrated, circumferential, reinforcement areas (reinforcing rings) are

$$A_{\theta 1} = V_{x1}R/\sigma_0 \qquad \text{and} \qquad A_{\theta 2} = V_{x2}R/\sigma_0.$$

6.8. With collapse in bending $(\dot{\kappa}_x < 0)$, integration of the equilibrium equation $(d^2M_x/dx^2) = p$, with $M_x = A_x\sigma_0 H$ and $M_x(0) = M_x(L) = 0$, gives $A_x = px(x - L)/2\sigma_0 H$. Adding the designs given by Eq. (6.132) and by (6.133) with $x - L$ substituted for x gives the expression of A_x obtained in Solution 6.7.

For a *long* shell $\{L \geq 2L^* = 2.8184(RH)^{1/2}\}$, we proceed to the same superposition of solutions of Exercise 6.7 and obtain

$$A_\theta = 0, \quad A_x = (px/2\sigma_0 H)(x - L^*), \qquad 0 \leq x \leq L^*,$$

$$A_x = 0, \quad A_\theta = pR/\sigma_0, \qquad L^* \leq x \leq L/2.$$

At $x = L^*$ the discontinuity of the shear V_x necessitates a concentrated reinforcement of area

$$A_\theta = pL^*R/2\sigma_0.$$

The design is symmetric with respect to the central cross section.

Solutions for Chapter 8

8.1. We assume that the principal curvature rates are different: $\dot{\kappa}_1 \neq \dot{\kappa}_2$.

Let $\dot{\kappa}_1 > \dot{\kappa}_0 > 0$. Then, if α is the positive clockwise angle of direction 1 with direction x of the reinforcing bars, we have [see Save and Massonnet (1972), for example] that the yield moments in directions 1 and 2 are given by

$$Y_1 = Y_x \cos^2\alpha + Y_y \sin^2\alpha, \tag{S.35}$$

$$Y_2 = Y_x \sin^2\alpha + Y_y \cos^2\alpha. \tag{S.36}$$

The specific dissipation $D = Y_1\dot{\kappa}_1 + Y_2\dot{\kappa}_2$ is then

$$D = Y_x(\dot{\kappa}_1 \text{cops}^2\alpha + \dot{\kappa}_2 \sin^2\alpha) + Y_y(\dot{\kappa}_1 \sin^2\alpha + \dot{\kappa}_2 \cos^2\alpha).$$

We apply the condition in Eq. (8.1), expressing that the specific cost does *not* depend on α; that is, $\partial\gamma/\partial\alpha = 0$. Hence, $\partial D/\partial\alpha = (\dot{\kappa}_2 - \dot{\kappa}_1)(Y_x - Y_y)2\sin\alpha\cos\alpha = 0$, from which we conclude that, because $\dot{\kappa}_2 \neq \dot{\kappa}_1$, the nonisotropic reinforcement ($Y_x \neq Y_y$) must be such that $\sin 2\alpha = 0$; that is, it must be placed in the principal directions 1 and 2. The case $0 > \dot{\kappa}_2 > \dot{\kappa}_1$ is treated in a simlar manner.

Now let $\dot{\kappa}_1 > 0 > \dot{\kappa}_2$. Equation (S.35) still holds, whereas Eq. (S.36) becomes $Y_2 = Y'_x \sin^2\alpha + Y'_y \cos^2\alpha$, and $D = Y_1\dot{\kappa}_1 - Y_2\dot{\kappa}_2$ becomes

$$D = (\dot{\kappa}_1 Y_x - \dot{\kappa}_2 Y'_y)\cos^2\alpha + (\dot{\kappa}_1 Y_y - \dot{\kappa}_2 Y'_x)\sin^2\alpha.$$

Equation (8.1), with $\partial\gamma/\partial\alpha = 0$, gives

$$\sin 2\alpha[\dot{\kappa}_1(Y_y - Y_x) + \dot{\kappa}_2(Y'_y - Y'_x)] = 0.$$

Assuming that the quantity in brackets does not vanish, we conclude that $\alpha = 0 + n\pi/2$, n an integer. Because no positive reinforcement is needed in direction 1, and no negative reinforcement is needed in direction 2, we verify that the bracket is essentially negative. The case $\dot{\kappa}_2 > 0 > \dot{\kappa}_1$ is treated similarly.

References

Apostol, T. M., 1974, *Mathematical Analysis*, Second Edition, Addison-Wesley, Reading, MA.

Armand, J. L., and Lodier, B., 1978, Optimal design of bending elements, *Int. J. Num. Meth. Eng.* **13**, 373–384.

Banichuk, N. V., 1976, Optimum design of elastic bars in torsion, *Int. J. Solids Struct.* **12**, 275.

Banichuk, N. V., and Karihaloo, B. L., 1976, Minimum-weight design of multipurpose cylindrical bars, *Int. J. Solids Struct.* **12**, 267–273.

Barnett, R. L., 1961, Minimum-weight design of beams for deflection, *Proc. Am. Soc. Civ. Eng.* **87**, No. EM 1, 75–109.

Batterman, S. C., and Felton, L. P., 1971, *Optimal Plastic Design of Doubly Symmetric Closed Structures*, Report No. 126, Technion (Israel Institute of Technology), Haifa.

Betti, E., 1872, Teoria dell'Elasticità, in *Opere*, Vol. 2, pp. 291–390.

Bleich, H., 1932, Uber die Bemessung statisch unbestimmter Stahltragwerke unter Berucksichtigung des elastischplastischen Verhaltens des Baustoffs, *Bauing.* **19/20**, 261–267.

Borkauskas, A., and Atkociunas, J., 1975, Optimal design for cyclic loading, in *Optimization in Structural Design* (Sawczuk, A., and Mroz, Z., Eds.), Springer, pp. 433–440.

Brach R. M., 1968, On the extremal fundamental frequencies of vibrating beams, *Int. J. Solids Struct.* **4**, 667–674.

Brach, R. M., 1973, On optimal design of vibrating structures, *J. Optimiz. Theory Appl.* **11**, 662–667.

Brotchie, J. F., 1967, Discussion, *Proc. ASCE, EM5, J. Eng. Mech. Div.* 173–175.

Cantu, E., and Cinquini, C., 1979, Iterative solutions for problems of optimal elastic design, *Comp. Meth. Appl. Mech. Eng.* **20**, 257–266.

Cea, J., 1978, *Optimization Theory and Algorithms*, Tata Institute of Fundamental Research, Bombay.

Chern, J. M., 1971, Optimal structural design for given deflection in presence of body forces, *Int. J. Solids Struct.* **7**, 363–382.

Chern, J. M., and Prager, W., 1970, Optimal design of rotating disk for given radial displacement of edge, *J. Optimiz. Theory Appl.* **6**, 161–170.

Chern, J. M., and Prager, W., 1971, Minimum-weight design of statically determinate trusses subject to multiple constraints, *Int. J. Solids Struct.* **7**, 931–940.

Chern, J. M., and Prager, W., 1972, Optimal design of trusses for alternative loads, *Ing. Arch.* **41**, 225–231.

Chern, J. M., Dafalias, J. F., and Martin, J. B., 1973, Structural design for bounds on dynamic response, *J. Eng. Mech. Div., Proc. ASCE*, **99**, 261–270.

Cinquini, C., 1979, Optimal elastic design for prescribed maximum deflection, *J. Struct. Mech.* **7**, 21-24.

Cinquini, C., 1981, Structural optimization of plates of general shape by finite elements, *J. Struct. Mech.* **9**, 465-481.

Cinquini, C., and Kouam, M., 1983, Optimal plastic design of stiffened shells, *Int. J. Solids Struct.*, to appear.

Cinquini, C., and Sacchi, G., 1980, Problems of optimal design for elastic and plastic structures, *J. Méc. Appl.* **4**, 1-29.

Cinquini, C., Lamblin, D., and Guerlement, G., 1977, Variational formulation of the optimal plastic design of circular plates, *Comput. Meth. Appl. Mech. Eng.* **11**, 19-30.

Collins, I. F., 1968, An optimum loading criterion for rigid-plastic materials, *J. Mech. Phys. Solids* **16**, 73-80.

Cox, H. L., 1958, *The Theory of Design*, Report No. 19791, Aeronautical Research Council, Great Britain.

Cyras, A., 1975, Optimization theory in the design of elastic-plastic structures, *Int. Cent. Mech. Sci. (C.I.M.S.) Courses and Lectures*, No. 237, *Structural optimization* (Brousse, Ed.), Springer, pp. 80-150.

Dems, K., 1980, Multiparameter shape optimization of elastic bars in torsion, *Int. J. Num. Meth. Eng.* **15**, 1517-1539.

Dems, K., and Mroz, Z., 1980, Optimal shape design of multicomposite strucutres, *J. Struct. Mech.* **8**, 309-329.

Dokmeci, M. C., 1966, A shell of constant strength, *Z. Angew. Math. Phys.* **17**, 545-547.

Dow, M., Nakamura, H., and Rozvany, G. I. N., 1982, Optimal shape of copulas of uniform strength, *Ing. Archiv.* **52**, 335-353.

Drucker, D. C., and Shield, R. T., 1956, Design for minimum weight, *Proceedings of the Ninth International Congress on Applied Mechanics*, Brussels, Book 5, pp. 212-222.

Drucker, D. C., and Shield, R. T., 1957, Bounds on minimum weight design, *Q. Appl. Math.* **15**, 269-281.

Drucker, D. C., Prager, W., and Greenberg, H. J., 1952, Extended limit design theorems for continuous media, *Q. Appl. Math.* **9**, 381-389.

Eason, G., 1960, The minimum-weight design of circular sandwich plates, *Z. Angew. Math. Phys.* **11**, 368-375.

Erbatur, F., and Mengi, Y., 1977a, On the optimal design of plates for a given deflection, *J. Optimiz. theory Appl.* **21**, 103-110.

Erbatur, F., and Mengi, Y., 1977b, Optimal design of plates under the influence of dead weight and surface loading, *J. Struct. Mech.* **5**, 345-356.

Foulkes, J., 1954, The minimum-weight design of structural frames, *Proc. R. Soc. London, Ser. A*, **223**, 482-494.

Flugge, W., 1960, *Stresses in Shells*, Springer, New York.

Freiberger, W., 1956, Minimum-weight design of cylindrical shells, *J. Appl. Mech.* **23**, 576-580.

Freiberger, W., 1957, On the minimum-weight design problems of cylindrical sandwich shells, *J. Aero. Sci.* **24**, 847-848.

Freiberger, W., and Tekinalp, B., 1956, Minimum-weight design of circular plates, *J. Mech. Phys. Solids* **4**, 294-299.

Garstecki, A., and Gawecki, A., 1978, Experimental study on optimal plastic rings in the range of large displacements, *Int. J. Mech. Sci.* **20**, 823-832.

Gierlinski, J., and Mroz, Z., 1981, Optimal design of elastic plates and beams taking large deflections and shear forces into account, *Acta Mech.* **39**, 77-92.

Greenberg, H. J., and Prager, W., 1949, *Limit Design of Beams and Frames*, Technical Report No. A18-1; Brown University printed with discussion in *Trans. Am. Soc. Civ. Eng.* **117**, (1952), 447-484.

Guerlement, G., 1975, *Contribution à l'analyse limite des coques cylindriques*, Thèse de doctorat, Faculté Polytechnique de Mons, Belgium.

Gunneskov, O., 1975, *Optimal Design of Rotating Disks in Creep*, D.C.A.M.M. Report No. 87, Technical University, Denmark.

Gvozdev, A. A., 1938, The determination of the value of the collapse load, of statically indeterminate systems undergoing plastic deformation (in Russian), *Proceedings of a Conference on Plastic Deformation, Akad. Nauk, USSR*, p. 19.

Haug, E. J., 1981, A unified theory of optimization of structures with displacement and compliance constraints, *J. Struct. Mech.* **9**, 415–437.

Haug, E. J., and Arora, J. S., 1979, *Applied Optimal Design*, Wiley, New York.

Haug, E. J., and Cea, J., Eds., 1981, *Optimization of Distributed Parameter Structures*, Vol. 1 and Vol. 2, Strijthoff and Noordhoff, Alpen aan den Rijn, Netherlands.

Hegemier, G. A., and Prager, W., 1969, On michell trusses, *Int. J. Mech. Sci.* **11**, 209–215.

Hemp, W. S., 1968, *Abstract of Lecture Courses: Optimum Structures*, University of Oxford.

Hemp, W. S., 1973, *Optimum Structures*, Clarendon Press, Oxford.

Hencky, H., 1923, Ueber einige statisch bestimmte Falle des Gleichgewichts in Plastischen Korpern, *Z. Angew. Math. Mech.* **3**, 241–251.

Hestenes, M. R., 1966, *Calculus of Variations and Optimal Control Theory*, Wiley, New York; reprint edition; R. E. Krieger, 1980.

Heyman, J., 1958, Rotating disk, insensitivity of design, *Proc. 3rd U.S. Nat. Cong. Appl. Mech.*, Brown University, Providence, RI.

Heyman, J., 1959, On the absolute minimum-weight design of framed structures, *Q. J. Mech. Appl. Math.* **12**, 314–324.

Hill, R., 1950, *The Mathematical Theory of Plasticity*, Clarendon Press, Oxford.

Hopkins, H. G., and Prager, W., 1955, Limits of economy of materials in plates, *J. Appl. Mech.* **22**, 372–373.

Horne, M. R., 1950, Fundamental propositions in plastic theory of structures, *J. Inst. Civ. Eng.* **34**, 174–177.

Hu, T. C., and Shield, R. T., 1961, Minimum volume of discs, *J. Appl. Math. Phys.* **12**, 414–433.

Huang, N. C., 1971, Optimal design of beams for minimum-maximum deflection, *J. Appl. Mech.* **38**, 1078–1081.

Huang, N. C., and Sheu, C. Y., 1970, Optimal design of elastic circular sandwich beams for minimum compliance, *J. Appl. Mech.* **37**, 569–577.

Huang, N. C., and Tang, H. T., 1969, Minimum-weight design of elastic sandwich beams with deflection constraints, *J. Optimiz. Theory Appl.* **4**, 277–298.

Icerman, L. J., 1969, Optimal structural design for given dynamic deflection, *Int. J. Solids. Struct.* **5**, 473–490.

Igic, T., 1980, *Doprimos Optimalnom Dimenzionisanju Konstrukcija*, Doktorska disertacija, Gradjevinski Fakultet, University of Nis, Yougoslavia.

Issler, W., 1964, Membraschalen gleicher Festigkeit, *Ing. Archiv.* **33**, 330–345.

Kaliszky, S., 1981, Optimal design of rigid-plastic solids and structures under dynamic pressure, *Z. Angew. Math. Mech.* **61**, T100–T101.

Koiter, W. T., 1956, A new general theorem on shakedown of elastic-plastic structures, *Proc. K. Ned. Akad. Wet., Ser. B* **59**, 24–34.

Konig, J. A., 1966, Theory of shakedown of elastic-plastic structures, *Arch. Mech. Stosow.* **18**, 227–238.

Korn, G. A., and Korn, T. M., 1961, *Mathematical Handbook for Scientists and Engineers*, McGraw-Hill, New York.

Kozlowski, W., and Mroz, Z., 1969, Optimal design of solid plates, *Int. J. Solids Struct.* **5**, 781–794

Kozlowski, W., and Mroz, Z., 1970, Optimal design of disk subject to geometric constraints, *Int. J. Mech. Sci.* **12**, 1007-1021.

Krzys, W., 1964, Optimum design of the box-section of a beam bent in elastic-plastic range, *Bull. Acad. Pol. Sci., Sér. Sci. Tech.* **12**, 261-271.

Krzys, W., and Zyczkowski, M., 1963, A certain method of parametrical structural optimum design, *Bull. Acad. Pol. Sci. Sér. Sci. Techn.* **10**, 335-345.

Lamblin, D., 1972, Minimum-weight plastic design of continuous beams subjected to one single moveable load, *J. Struct. Mech.* **1**, 133-157.

Lamblin, D., 1975, *Analyse et dimensionnement plastique de coùt minimum de plaques circulaires*, Thèse de Doctorat en Sciences Appliquées, Faculté Polytechnique de Mons, Belgium.

Lamblin, D., and Guerlement, G., 1976, Dimensionnement plastique de volume minimal sous contraintes de plaques sandwich circulaires soumises à des charges fixes ou mobiles, *J. Mec.* **15**, 55-84.

Lamblin, D., and Save, M. A., 1971, Minimum-volume plastic design of beams for moveable loads, *Meccanica* **6**, No. 3, 157-163.

Lamblin, D., Cinquini, C., and Guerlement, G., 1980, Finite element iterative method for optimal elastic design of circular plates, *Comput. Struct.* **12**, 85-92.

Lamblin, D., Save, M. A., and Guerlement, G., 1985, Solutions de dimensionnement plastique de volume minimal de plaques circulaires pleines et sandwich en présence de contraintes technologiques, *J. Mec. Théorique Appliquée.*

Latta, G. E., 1962, *Ordinary differential equations*, in *Handbook of Engineering Mechanics* (W. Flugge, Ed.), Mcgraw-Hill, New York, pp. 10-13ff.

Lekszycki, T., and Olhoff, N., 1980, Optimal design of viscoelastic structures under forced steady state vibration, D.C.A.M.M. Report No. 195, Technical University, Denmark.

Leipik, U., and Mroz, Z., 1977, Optimal design of plastic structures under impulsive and dynamic pressure loading, *Int. J. Solids. Struct.* **13**, 657-674.

Libove, C., 1962, Elastic stability, in *Handbook of Engineering Mechanics* (W. Flugge, Ed.), McGraw-Hill, New York, pp. 44-47ff.

Lowe, P. G., and Melchers, R. E., 1974, On the theory of optimal, edge beam supported, fiber-reinforced plates, *Int. J. Mech. Sci.* **16**, 627-641.

Marcal, P. V., 1967, Optimal plastic design of circular plates, *Int. J. Solids Struct.* **3**, 427-443.

Marcal, P. V., and Prager, W., 1964, A method of optimal plastic design, *J. Méc.* **3**, 509-530.

Martin, J. B., 1964, A displacement bound principle for elastic continua subjected to certain classes of dynamic loading, *J. Mech. Phys. Solids.* **12**, 165-175.

Martin, J. B., 1968, Displacement bounds for dynamically loaded elastic structures, *J. Mech. Eng. Sci.* **10**, 213-218.

Martin, J. B., 1971, The optimal design of beams and frames with compliance constraints, *Int. J. Solids Struct.* **7**, 63-81.

Martin, J. B., 1972, *On the Application of the Bounding Theorems of Plasticity to Impulsively Loaded Structures* (Herrman, J., and Perrone, N., Eds.), Pergamon Press, New York, pp. 73-93.

Martin, J. B., and Ponter, A. R. S., 1972, The optimal design of a class of beams structures for a nonconvex cost function, *J. Méc.* **11**, 341-360.

Masur, E. F., 1970, Optimum stiffness and strength of elastic structures, *J. Eng. Mech. Div., ASCE* **96**, No. EM5, 621-640.

Masur, E. F., 1974, Optimal structural design for a discrete set of available structural members, *Comput. Meth. Appl. Mech. Eng.* **3**, 195-207.

Masur, E. F., 1975a, Optimality in the presence of discreteness and discontinuity, in *Proc. I.U.T.A.M. Symp. on Optimization in Structural Design* (Sawczuk, A., and Mroz, Z., Eds.), Springer, pp. 441-453.

Masur, E. F., 1975b, Optimal placement of available sections in structural eigenvalue problems, *J. Optimiz. Theory Appl.* **15**, 69-84.

Masur, E. F., 1978, Optimal design of symmetric structures against postbuckling collapse, *Int. J. Solids Struct.* **14**, 319–326.

Maxwell, J. C., 1890, On reciprocal figures, frames, and diagrams of force, in *Scientific Papers*, Vol. 2, Universtiy Press, Cambridge, pp. 161–207.

Mayeda, R., and Prager, W., 1967, Minimum-weight design of beams for multiple loading, *Int. J. Solids Struct.* **3**, 1001–1011.

Megarefs, G. J., 1966, Method for minimal design of axisymmetric plates, *Proc. ASCE* **92**, No. EM6, 79–99.

Megarefs, G. J., 1967, 1968, Minimal design of sandwich axisymmetric plates, *Proc. ASCE* **93**, No. EM6, 245–269; **94**, EM1, 177–198.

Megarefs, G. J., and Hodge, P. G., Jr., 1963, Singular cases in the optimum design of frames, *Q. Appl. Math.* **21**, 91–103.

Melan, E., 1936, Theorie statisch unbestimmter Tragwerke aus ideal-plastischem Baustoff, *Sitzungsber. Akad. Wiss. Wien, Math.-Naturiviss. Kl., Abt. 2A* **145**, 195–218.

Melan, E., 1938, Zur Plastizitat des raumlichen Kontinuums *Ing.-Arch.* **9**, 116–126.

Michell, A. G. M., 1904, The limits of economy in frames-structures, *Philos. Mag.* **8**, 589–597.

Miele, A., Mangiavacchi, A., Mohanty, B. I., and Wu, A. K., 1978, Numerical determination of minimum mass structures with specified natural frequencies, *Int. J. Num. Meth. Eng.* **13**, 265–282.

Milankovic, M., 1980, Arbeiten aus der Jugoslawischen Akademie der Wissenschaften, *Agram* 175.

Morley, C. T., 1966, The minimum reinforcement of concrete slabs, *Int. J. Mech. Sci.* **8**, 305–319.

Mroz, Z., 1974, Optimal design criteria for reinforced plates and shells, in *Problems of Plasticity* (Sawczuk, ed.), Noordhoff, Leyden, pp. 425–429.

Mroz, Z., and Gawecki, A., 1975, Post-yield behaviour of optimal plastic structures, *Proc. I.U.T.A.M. Symp. on Optimization in Structural Design* (Sawczuk, A., and Mroz, Z., Eds.), Springer, pp. 518–540.

Mroz, Z., and Rozvany, G. I. N., 1975, Optimal design of structures, with variable support conditions, *J. Optimiz. Theory Appl.* **15**, 85–101.

Nagtegaal, J. C., 1972, On optimal design of prestressed elastic structures, *Int. J. Mech. Sci.* **14**, 779–781.

Nagtegaal, J. C., 1973, A superposition principle in optimal plastic design for alternative loads, *Int. J. Solids Struct.* **9**, 1465–1471.

Nagtegaal, J. C., and Prager, W., 1973, Optimal layout of a truss for alternative loads, *Int. J. Mech. Sci.* **15**, 583–592.

Nakamura, H., Dow, M., and Rozvany, G. I. N., 1981, Optimal spherical copula of uniform strength: allowances for self-weight, *Ing. Archiv.* **51**, 159–181.

Nemirovskii, Y. V., 1971, Design of optimum disks in relation to creep, in *Strength of Materials*, Vol. 3, pp. 891–894.

Niordson, F., 1981, Optimal design of elastic plates with a constraint on the slope of the thickness function, D.A.C.M.M. Report No. 225, Technical University of Denmark.

Odqvist, F. K. G., 1966, *Mathematical Theory of Creep and Creep Rupture*, Clarendon Press, Oxford.

Olhoff, N., 1970, Optimal design of vibrating circular plates, *Int. J. Solids Struct.* **6**, 139–156.

Olhoff, N., 1974, Optimal design of vibrating rectangular plates, *Int. J. Solids Struct.* **10**, 93–109.

Olhoff, N., 1975, On singularities, local optima and formation of stiffners, *Optimization in Structural Design* (Sawczuk, A., and Mroz, Z., Eds.), Springer, pp. 82–103.

Olhoff, N., 1976, A survey of the optimal design of vibrating structural elements, *Shock and Vibration Digest* **8**, No. 8, 3–10 and No. 9, 3–10.

Olhoff, N., and Ramussen, S. H., 1977, On single and biodal optimal buckling loads of clamped columns, *Int. J. Solids Struct.* **13**, 605–614.

Olhoff, N., and Taylor, J. E., 1978, Designing continuous columns for minimal total cost of material and interior supports, *J. Struct. Mech.* **4**, 367–382.

Onat, E. T., and Prager, W., 1956, Limits of economy of materials in cylindrical shells, *Ingenieur* **67**, 46–49.

Onat, E. T., Shield, R. T., and Schumann, W., 1957, Design of circular plates for minimum weight, *Zt. Angew. Math. Phys.* **8**, 485–499.

Plaut, R. H., 1970, On minimizing the response of structures to dynamic loadings, *J. Appl. Math. Phys.* **21**, 1004–1010.

Prager, W., 1957, Shakedown in elastic–plastic media subject to cycles of load and temperature, *Symposium sulla Plasticita nella Scienza della Costruzioni*, N. Zanichelli, Bologna, pp. 239–244.

Prager, W., 1958, On a problem of optimal design, *Proc. Symp. Nonhomogeneity in Elasticity and Plasticity*, Warsaw.

Prager, W., 1961, *Introduction to Mechanics of Continua*, Ginn and Co., Lexington, Massachusetts.

Prager, W., and Shield, R. T., 1967, The general theory of optimal plastic design, *J. Appl. Mech.* **34**, 184–186.

Prager, W., 1968, Optimal structural design for given stiffness in stationary creep, *J. Appl. Math. Phys.* **19**, 252–256.

Prager, W., 1969a, Optimal plastic design of rings, in *Contributions to Mechanics* (D. Abir, Ed.), Pergamon Press, New York, pp. 163–169.

Prager, W., 1969b, The deformation of rigid-workhardening optimal structures, in *L. I. Sedov Annual Volume*, Moscow, pp. 393–396; see also *Problems of Hydrodynamics and Continuum Mechanics*, S.I.A.M., pp. 563–567.

Prager, W., 1970, Optimal thermo-elastic design for given deflection, *Int. J. Mech. Sci.* **12**, 705–709.

Prager, W., 1974, Limit analysis: the development of a concept, *Problems of Plasticity* (A. Sawczuk, ed.), Noordhof, pp. 3–24.

Prager, W., 1976, Geometric discussion of the optimal design of a simple truss, *J. Struct. Mech.* **4**, 57–63.

Prager, W., 1977, Optimal layout of cantilever trusses, *J. Optimiz. Theory Appl.* **23**, 111–117.

Prager, W., 1978, Nearly optimal design of trusses, *Comput. Struct.* **8**, 451–454.

Prager, W., 1981, Unexpected results in structural optimization, *J. Struct. Mech.* **9**, 71–90.

Prager, W., and Rozvany, G. I. N., 1977, Optimal layout of grillages, *J. Struct. Mech.* **5**, 1–18.

Prager, W., and Rozvany, G. I. N., 1980, Optimal spherical copula of uniform strength, *Ing. Archiv.* **49**, 287–294.

Prager, W., and Taylor, J. E., 1968, Problems of optimal structural design, *J. Appl. Mech.* **35**, 102–106.

Prandtl, L., 1923, Anwendungsbeispiele zu einem Henckyschen Satz über plastisches Gleichgewicht, *Z. Angew. Math. Mech.* **3**, 401–407.

Rayleigh, J. W., 1878, *Theory of Sound*, Chapter 4.

Reiss, R., 1974, Minimum-weight design for conical shells, *J. Appl. Mech., Trans. ASME* **41**, 599–603.

Reiss, R., 1976, Optimal compliance criterion for axisymmetric solid plates, *Int. J. Solids Struct.* **12**, 319–329.

Reiss, R., and Megarefs, G. J., 1971, Minimal design of sandwich axisymmetric plates obeying Mises criterion, *Int. J. Solids Struct.* **7**, 603–623.

Rozvany, G. I. N., 1972a, Grillages of maximum strength and maximum stiffness, *Int. J. Mech. Sci.* **41**, 651–666.

Rozvany, G. I. N., 1972b, Optimal loads transmission by flexure, *Comput. Meth. Appl. Mech. Eng.* **1**, 253–263.

Rozvany, G. I. N., 1973a, Nonconvex structural optimization problems, *Proc. ASCE* **99**, No. EM1, 243–248.

Rozvany, G. I. N., 1973b, Optimal plastic design of partially preassigned strength distribution, *J. Optimiz. Theory Appl.* **11**, 421-436.

Rozvany, G. I. N., 1973c, Optimal force transmission by flexure-clamped boundaries, *J. Struct. Mech.* **2**, 99-124.

Rozvany, G. I. N., 1973d, Basic geometrical properties of optimal flexural force transmission fields, *J. Struct. Mech.* **2**, 259-264.

Rozvany, G. I. N., 1974, Optimal plastic design with discontinuous cost functions, *J. Appl. Mech.* **41**, 309-310.

Rozvany, G. I.N., 1975, Analytical treatment of some extended problems in structural optimization, *J. Struct. Mech.* **3**, 359-385.

Rozvany, G. I. N., 1976, *Optimal Design of Flexural Systems*, Pergamon Press, Oxford.

Rozvany, G. I. N., 1981, A general theory of optimal structural layout, *Int. Symp. Opt. Struct. Design*, University of Arizona, pp. 337-446.

Rozvany, G. I. N., and Hill, R. D., 1978, Optimal plastic design: superposition principles and bounds on the minimum cost, *Comput. Meth. Appl. Mech. Eng.* **13**, 151-173.

Rozvany, G. I. N., and Mroz, Z., 1977a, Column design: optimization of support conditions and segmentation, *J. Struct. Mech.* **5**, 279-290.

Rozvany, G. I. N., and Mroz, Z., 1977b, Analytical methods in structural optimization, *Appl. Mech. Rev.* **30**, 1461-1470.

Rozvany, G. I. N., and Prager, W., 1976, Optimal design of partially discretized grillages, *J. Mech. Phys. Solids* **24**, 125-136.

Rozvany, G. I. N., and Prager, W., 1979, A new class of structural optimization problems: optimal archgrids, *Comput. Meth. Appl. Mech. Eng.* **19**, 127-150.

Rozvany, G. I. N., and Wang, C. M., 1982, Extensions of Prager's layout theory, in *Optimization Methods in Structural Design* (Eschenauer, H., and Olhoff, N., Eds.), Mannheim, Wien, Zurich, pp. 103-110.

Rozvany, G. I. N., Hill, R., and Gangadhariah, C., 1973, Grillages of least-weight-Simply supported boundaries, *Int. J. Mech. Sci.* **15**, 665-677.

Rozvany, G. I. N., Nakamura, H., and Kuhnell, B., 1980, Optimal archgrids: allowance for selfweight, *Comput. Meth. Appl. Mech. Eng.* **24**, 287-304.

Rozvany, G. I. N., Olhoff, N., Cheng, K. T., and Taylor, J., 1982, On the solid plate paradox in structural optimization, *J. Struct. Mech.* **18**, 1-32.

Rozvany, G. I. N., Wang, C. M., and Dow, M., 1982, Prager structures: archgrids of optimal layout, *Comput. Meth. Appl. Mech. Eng.* **31**, 91-113.

Sacchi, G., and Save, M., 1969, Le problème du poids minimum d'armature des plaques en béton armé, *Mémoires de l'AIPC*, Vol. 29-II.

Save, M. A., 1968, Some aspects of minimum-weight design, in *Engineering Plasticity* (Heyman and Leckie, Eds.), Cambridge University Pres , New York, pp. 611-626.

Save, M. A., 1972, A unified formulation of the theory of optimal plastic design with convex cost function, *J. Struct. Mech.* **1**, 267-276.

Save, M. A., 1975, A general criterion of optimal structural design, *J. Optimiz. Theory Appl.* **15**, 119-129.

Save, M. A., 1983, Remarks on the minimum-volume designs of a three-bar truss, *J. Struct. Mech.* **11**, 101-110.

Save, M., and Igic, T., 1982, Exemples de poutres optimales à deux fonctions, *J. Méc. Theor. Appliquée* **1**, 311-321.

Save, M. A., and Massonnet, C. E., 1972, *Plastic Analysis and Design of Plates, Shells and Disks*, North-Holland, Amsterdam.

Save, M. A., and Prager, W., 1963, Minimum-weight design of beams subjected to fixed and moving loads, *J. Mech. Phys. Solids.* **11**, 255-267.

Save, M. A., and Shield, R. T., 1966, Minimum-weight design of sandwich shells subjected to fixed and moving loads, *Proc. Eleventh Int. Cong. Appl. Mech., Munich, 1964,* Springer, Berlin, pp. 341–349.

Shamiev, F. G., 1975, Optimal design of plates loaded by two opposite sets of loads, in *Optimization in Structural Design* (Sawczuk, A., and Mroz, Z., Eds.), Springer, pp. 579–585.

Sheu, C. Y., and Prager, W., 1969, Optimal plastic design of circular and annular sandwich plates with piecewise constant cross section, *J. Mech. Phys. Solids* **17**, 11–16.

Shield, R. T., 1960a, Plate design for minimum weight, *Q. Appl. Math.* **18**, 131–144.

Shield, R. T., 1960b, *Optimum design methods for structures,* in *Plasticity,* Pergamon Press, pp. 580–591.

Shield, R. T., 1960c, On the optimum design of shells, *J. Appl. Mech.* **27**, 316–331.

Shield, R. T., 1963, Optimum design methods for multiple loadings, *Z. Angew. Math. Phys.* **14**, 38–45.

Shield, R. T., and Prager, W., 1970, Optimal structural design for given deflection, *J. Appl. Math. Phys.* **21**, 513–523.

Smith, D. L., 1974, *Plastic Limit Analysis and Synthesis of Structures by Linear Programming,* Ph.D. Thesis, University of London, Imperial College of Science and Technology.

Stampacchia, G., and Cecconi, J., 1976, *Analisi Matematica* (Liguori, Ed.), Napoli.

Sved, G., and Ginos, Z., 1968, Structural optimization under multiple loading, *Int. J. Mech. Sci.* **10**, 803.

Synge, J. L., and Schild, A., 1949, *Tensor Calculus,* University of Toronto Press, Toronto.

Taylor, J. E., 1969, Maximum-strength elastic structural design, *J. Eng. Mech. Div., Proc. ASCE* **95**, 653–664.

Thompson, J. M. T., and Supple, W. J., 1973, Erosion of optimum design by compound branching phenomena, *J. Mech. Phys. Solids* **21**, 135–144.

Timoshenko, S., and Woinowsky-Krieger, S., 1959, *Theory of Plates and Shells,* McGraw-Hill, New York.

Ting, S., and Reiss, R., 1977, Design of axisymmetric sandwich plates for alternative loads, *J. Optimiz. Theory Appl.* **21**, 385–421.

Venkayya, V., 1978, Structural optimization: a review and some recommendations, *Int. J. Num. Meth. Eng.* **13**, 203–228; special issue on structural optimization.

Way, S., 1962, Plates, in *Handbook of Engineering Mechanics* (Flügge, Ed.), Chap. 39, McGraw-Hill, New York.

Wood, R. H., 1961, *Plastic and Elastic Design of Slabs and Plates,* Thames and Hudson, London.

Ziegler, H., 1958, Kuppeln gleicher Festigkeit, *Ing. Archiv.* **26**, 378–382.

Index